D1754790

# THE CAUSES AND PROGRESSION OF DESERTIFICATION

*Meine geliebten Mutter, Maria Geist, geborene Weigl*

# The Causes and Progression of Desertification

HELMUT GEIST
*University of Louvain, International Project Office of the
IGBP-IHDP Land Use/Cover Change (LUCC) Project*

**ASHGATE**

© Helmut Geist 2005

All rights reserved. No part of this publication may be reproduced, stored in a retrieval system or transmitted in any form or by any means, electronic, mechanical, photocopying, recording or otherwise without the prior permission of the publisher.

Helmut Geist has asserted his right under the Copyright, Designs and Patents Act, 1988, to be identified as the author of this work.

Published by
Ashgate Publishing Limited
Gower House
Croft Road
Aldershot
Hants GU11 3HR
England

Ashgate Publishing Company
Suite 420
101 Cherry Street
Burlington, VT 05401-4405
USA

Ashgate website: http://www.ashgate.com

**British Library Cataloguing in Publication Data**
Geist, Helmut, 1958-
  The causes and progression of desertification. - (Ashgate studies in environmental policy and practice)
  1. Desertification
  I. Title
  333.7'36

**Library of Congress Cataloging-in-Publication Data**
Geist, Helmut, 1958-
  The causes and progression of desertification / by Helmut Geist.
    p. cm. -- (Ashgate studies in environmental policy and practice)
  Includes bibliographical references and index.
  ISBN 0-7546-4323-9
  1. Desertification. I. Title. II. Series.

GB611.G45 2004
551.41'5--dc22

2004017163

ISBN 0 7546 4323 9

Printed and bound in Great Britain by Antony Rowe Ltd, Chippenham, Wiltshire

# Contents

| | | |
|---|---|---|
| *List of Figures* | | *vii* |
| *List of Tables* | | *ix* |
| *Acknowledgements* | | *xiii* |
| 1 | The Problem and Its Identification | 1 |
| 2 | Research Design | 29 |
| 3 | Initial Conditions | 53 |
| 4 | Causes and System Properties | 95 |
| 5 | Syndromes and Process Rates | 133 |
| 6 | Pathways | 163 |
| 7 | Indicators | 187 |
| 8 | Discussion | 201 |
| 9 | Conclusions | 221 |
| *Bibliography* | | *229* |
| *Index* | | *243* |

# List of Figures

2.1  Location of case studies                          33
2.2  Framing the causes of desertification             41

# List of Tables

| | | |
|---|---|---|
| 2.1 | Data statistics | 32 |
| 2.2 | Regional differences of land use systems – Managed ecosystems | 38 |
| 2.3 | Regional differences of land use systems – Infrastructure and land use in natural ecosystems | 39 |
| 2.4 | Typology of syndromes of land change – Slow variables | 43 |
| 2.5 | Typology of syndromes of land change – Fast variables | 44 |
| 2.6 | Selected cases measured against the world's drylands | 47 |
| 2.7 | Selected cases measured against aridity zones of the world | 48 |
| 2.8 | Dominant types of land use categories found in the cases | 49 |
| 2.9 | Frequency of occurrence of broad underlying causes of desertification measured against the author's disciplinary background | 50 |
| 2.10 | Frequency of occurrence of applied indicators measured against the author's disciplinary background | 51 |
| 4.1 | Frequency of broad clusters of proximate causes of desertification | 96 |
| 4.2 | Frequency of specific agricultural activities causing desertification | 98 |
| 4.3 | Frequency of increased aridity causing desertification | 100 |
| 4.4 | Frequency of specific infrastructure activities causing desertification | 101 |
| 4.5 | Frequency of specific extractional activities causing desertification | 102 |
| 4.6 | Frequency of broad clusters of underlying driving forces of desertification | 104 |
| 4.7 | Frequency and mode of climatic factors driving desertification | 106 |
| 4.8 | Frequency of specific technological factors driving desertification | 107 |
| 4.9 | Frequency of specific institutional and policy factors driving desertification | 109 |
| 4.10 | Frequency of specific economic factors driving desertification | 111 |
| 4.11 | Frequency of specific demographic factors driving desertification | 112 |
| 4.12 | Frequency of specific cultural or socio-political factors driving desertification | 115 |
| 4.13 | Chain-logical connection of broad proximate causes | 118 |
| 4.14 | Chain-logical connection of broad underlying causes | 120 |
| 4.15 | Chain-logical connection of broad underlying forces driving broad proximate factors leading to desertification | 122 |
| 4.16 | Driving forces of desertification by scale of influence | 123 |
| 4.17 | Mediating factors in desertification | 125 |
| 4.18 | Thresholds of desertification | 129 |
| 5.1 | Slow/fast rates of desertification causes in Central Asia | 134 |

| | | |
|---|---|---|
| 5.2 | Slow/fast rates of desertification causes in Tibet, the former Soviet Union, India, Jordan and the Arabian Peninsula | 135 |
| 5.3 | Frequency of syndromes associated with slow land change – Resource scarcity | 136 |
| 5.4 | Frequency of syndromes associated with slow land change – Changing markets and policy interventions | 137 |
| 5.5 | Frequency of syndromes associated with slow land change – Adaptive capacities and social organization | 139 |
| 5.6 | Frequency of syndromes associated with fast land change – Resource scarcity, changing markets and policies | 141 |
| 5.7 | Frequency of syndromes associated with fast land change – Adaptive capacities and social organization | 142 |
| 5.8 | Annual decrease in rainfall (%) | 144 |
| 5.9 | Annual decrease in the areal extent of undegraded vegetation cover (%) – Slow change (<3.4%) | 146 |
| 5.10 | Annual decrease in the areal extent of undegraded vegetation cover (%) – Fast change (>3.4%) | 147 |
| 5.11 | Annual degradation of *Populus euphtatica* (*Pe*) vegetation in the Tarim River Basin of Taklimakan Desert in N-China, 1958-78 | 148 |
| 5.12 | Annual increase in the areal extent of degraded vegetation cover (%) | 149 |
| 5.13 | Annual increase in the areal extent of sand cover (%) – Fast change (> 4.7%) | 150 |
| 5.14 | Annual increase in the areal extent of sand cover (%) – Slow change (< 4.7%) | 151 |
| 5.15 | Annual increase in the areal extent of eroded, bare, rocky ground cover (%) | 153 |
| 5.16 | Annual changes (%) related to water degradation – Salinity increase in groundwater | 154 |
| 5.17 | Annual changes (%) related to water degradation – Areal spread of salinity in irrigation schemes | 155 |
| 5.18 | Annual changes (%) related to water degradation – Decrease of water surface | 156 |
| 5.19 | Annual changes (%) related to water degradation – Decrease of river discharge | 157 |
| 5.20 | Annual decline (%) in pastoral suitability – Areal decrease of residual rangeland | 158 |
| 5.21 | Annual decline (%) in pastoral suitability – Areal invasion of undesirable (impalatable) species | 159 |
| 5.22 | Annual decline (%) in agricultural suitability – Decline in crop yields | 160 |
| 5.23 | Annual decline (in %) in agricultural suitability – Loss of cultivated land | 161 |

## List of Tables

| | | |
|---|---|---|
| 6.1 | Basic and distinct features of dryland regions – Ecological factors shaping the environmental history (climate, topography, and vegetation) | 176 |
| 6.2 | Basic and distinct features of dryland regions – Ecological factors shaping the environmental history (soil, water and changes in species habitat) | 177 |
| 6.3 | Basic and distinct features of dryland regions – Changing institutional and socio-economic contexts (land use history) | 178 |
| 6.4 | Basic and distinct features of dryland regions – Changing institutional and socio-economic contexts (current land tenure) | 179 |
| 6.5 | Basic and distinct features of dryland regions – Pre-historic, ancient and historical dryland formation and degradation | 180 |
| 6.6 | Basic and distinct features of dryland regions – Contemporary dryland formation and degradation | 181 |
| 6.7 | Basic and distinct features of dryland regions – Proximate causes of desertification | 182 |
| 6.8 | Basic and distinct features of dryland regions – Underlying driving forces of desertification | 183 |
| 6.9 | Basic and distinct features of dryland regions – Major syndromes and mediating factors | 184 |
| 6.10 | Basic and distinct features of dryland regions – System properties of desertification | 185 |
| 7.1 | Frequency of desertification indicators | 188 |
| 7.2 | Vegetation change – Ecological degrees of dryland degradation | 190 |
| 7.3 | Erosion – Ecological degrees of dryland degradation | 191 |
| 7.4 | Water degradation – Ecological degrees of dryland degradation | 192 |
| 7.5 | Other deterioration of terrestrial ecosystem functions – Ecological degrees of dryland degradation | 193 |
| 7.6 | Pastoral suitability – Socio-economic degrees of dryland degradation | 194 |
| 7.7 | Agricultural suitability – Socio-economic degrees of dryland degradation | 195 |
| 7.8 | Other human welfare degradation – Socio-economic degrees of dryland degradation | 196 |
| 7.9 | Meteorological degrees of dryland degradation | 197 |
| 7.10 | Indicated types of regional dryland degradation | 199 |
| 8.1 | Livestock transition towards increasing intensity in Asia and Africa | 204 |
| 8.2 | Livestock transition towards increasing intensity in Europe, Australia, North and Latin America | 205 |
| 8.3 | Cropping transition towards increasing intensity in Asia | 207 |
| 8.4 | Cropping transition towards increasing intensity in Africa | 208 |
| 8.5 | Cropping transition towards increasing intensity in Europe, Australia, North and Latin America | 209 |

# Acknowledgements

The author is grateful for the support from the Services of the Prime Minister of Belgium, Office for Scientific, Technical, and Cultural Affairs. The book has greatly benefited from ideas developed within the LUCC (Land-Use/Cover Change) project (www.geo.ucl.ac.be/LUCC) of the IGBP and IHDP programmes (International Geosphere-Biosphere Programme, International Human Dimensions Programme on Global Environmental Change). Comments on earlier versions of the manuscript by past and present members of the LUCC Scientific Steering Committee (SSC) are particularly acknowledged. The author is grateful for an endorsement of the book by Coleen Vogel, chair of IHDP, and Guy Brasseur, chair of IGBP. Special thanks go to Eric Lambin, chair of the LUCC project, for carefully reading various earlier drafts of the study and for reviewing the final version. Nonetheless, any omissions, misunderstandings or typographical errors fall under the sole responsibility of the author.

Chapter 1

# The Problem and Its Identification

**Introduction**

Studies located at the local- to national-scale demonstrate the importance and socio-ecological significance of desertification. Nonetheless, land-use/cover change in drylands remains poorly documented at the global scale, and causes are hardly understood (Puigdefábregas, 1995; Warren, 1996; Thomas, 1997; Reynolds and Stafford Smith, 2002; Lambin, Geist and Lepers, 2003). Consequently, the word 'myth' has been used (Forse, 1989; Helldén, 1991; Thomas and Middleton, 1994; Swift, 1996; Lambin, Turner, Geist, Agbola, Angelsen, Bruce, Coomes, Dirzo, Fischer, Folke, George, Homewood, Imbernon, Leemans, Li, Moran, Mortimore, Ramakrishnan, Richards, Skånes, Steffen, Stone, Svedin, Veldkamp, Vogel and Xu, 2001). The chapter presents what desertification is, why it is important, what studies have been carried out already, why there is still an information gap, and how a new approach like the one used in this book can contribute to fill the gap.

**What is Desertification?**

The process of desertification is thought to be widespread, such as tropical deforestation, but considerably less well documented and defined. More than one hundred formal definitions have been proposed so far, each emphasizing either an enormous breadth of the topic or focussing on unique issues, and most of them (often) displaying many particular spatial and temporal scales of interest, thus representing various disparate viewpoints in total. Identifying the differences of and distilling the commonalities from these definitions, Soulé (1991) and Reynolds (2001) state that desertification principally consists of three major components, i.e., the meteorological, ecological and human dimensions of desertification. The meteorological dimensions relate, for example, to drought, atmospheric dust, air temperature, elevated atmospheric $CO_2$, and intra- and interannual variability in precipitation. The ecological dimensions include, for example, nutrient cycling, plant growth, regeneration and mortality, microbial dynamics, plant cover, herbivory life cycles, and evapotranspiration. And, the socio-economic dimensions are often related to the loss of habitat, the fragmentation of crucial habitat, issues of overexploitation (such as overgrazing by domestic animals), the spread of exotic organisms (pests and weeds), air, soil and water pollution, and issues linked to climate change. Rather than proposing yet another definition, Reynolds and

Stafford Smith (2002) select examples of definitions and classify them in terms of the emphasis given on the ecological, meteorological, and/or human or socio-economic dimensions of the problem. For example, Graetz (1991) emphasizes mainly meteorological dimensions by stating that desertification means 'the expansion of desert-like conditions and landscapes to areas where they should not occur climatically'. Others explore and define mainly the human dimensions of desertification, for example, in terms of 'lower useful productivity' for humans (Johnson, 1977), or by focussing on 'land degradation ... resulting from adverse human impact' (Middleton and Thomas, 1997). Following Reynolds (2001), each of the components – be they ecological, meteorological, or human – is complex, difficult to predict, and highly interdependent, but a failure to recognize the simultaneous role of (and feedbacks between) these different components has led to many of the controversies and misconceptions alluded to previously.

Until today, the most authoritative definition of desertification remains that of the United Nations Convention to Combat Desertification (UNCCD): 'land degradation in arid, semi-arid and dry sub-humid areas resulting from various factors, including climatic variations and human activities' (UNEP, 1994). Thus, desertification refers to the degradation of land in dry zones. Conventionally, following the United Nations Conference on Desertification (UNCOD), held in Nairobi in 1977, desertification embraces 'the diminution or destruction of the biological potential of the land [which] can lead ultimately to desert-like conditions' (UNEP, 1977). According to the United Nations Environment Programme (UNEP), drylands climatically denote an area with rainfall up to 600 mm per year, and have a ratio of average annual precipitation to potential evapotranspiration (P/Etp; also referred to as the index of aridity) between 0.05 and 0.65. In other words, desertification is presumed to result in a reduction in the biological and, hence, economic potential of the land to support human populations, livestock, and wild herbivores in drylands of the world (Toulmin, 1998). While, by definition, desertification relates to drylands only, some scientists have expanded this definition to subsume the impoverishment of any terrestrial ecosystem which can be measured in reduced productivity of desirable plants, undesirable alterations in biomass and plant and fauna diversity, and accelerated erosion (Dregne, 1983; Dregne, Kassa and Rozanov, 1991; Watts, 2001; Dregne, 2002).

Desertification is about biophysical and socio-economic linkages, and how they affect human welfare (Blaikie and Brookfield, 1987; Puigdefábregas, 1995; Vogel and Smith, 2002; Mooney, Cropper and Reid, 2003; Turner, Kasperson, Matson, McCarthy, Corell, Christensen, Eckley, Kasperson, Luers, Martello, Polsky, Pulsipher and Schiller, 2003). It is a multidimensional issue, and several other concepts are important to this definition such as capability (or the quality of land to satisfy a particular use), sustainability (or the ability of the land to continue to produce over two generations, at least), vulnerability (or the exposure to hazards such as perturbations and stresses, and the sensitivity of the coupled human-environmental system experiencing hazards such as drought), resilience (or that quality of a resource that makes it sustainable or resistant to land degradation), and carrying capacity (or the number of people and animals the land can normally support without being significantly stressed). For example, responses and adaptation to land degradation are part of what desertification constitutes, or, in other words, the determinate political, economic and institutional capabilities of people in specific places at specific times (Watts and Bohle, 1993; Vogel, 1995). This implies that desertification is also about the social resilience of groups in dryland areas that are most vulnerable (Bohle, Downing, Field and Ibrahim, 1993; Vogel and Smith, 2002; Downing and Lüdeke, 2002).

The point has to be made here that different segments of society, or 'stakeholders', will view the issue with differing degrees of concern and interest (Stafford Smith and Reynolds, 2002). This poses a particular problem of how to research desertification as an outcome of dryland degradation. Reynolds, Stafford Smith and Lambin (2003) make the point that 'what superficially appears straightforward may in fact be multifaceted, eschewing overly simplistic answers', and Shi and Shao (2000) provide some exemplification of this in terms of soil and water losses from the Loess Plateau in China. Nonetheless, in terms of values implied, desertification is about endangerment in both social and environmental terms, and the term degradation needs to be clearly defined (Lohnert and Geist, 1999). A still very helpful example is the definition as put forward by Blaikie and Brookfield in their seminal work on land degradation and society (1987, p. 6).

[D]egradation is defined as a reduction in the capability of land to satisfy a particular use. If land is transferred from one system of production or use to another, say, from hunter-gathering to agriculture, or from agricultural to urban use, a different set of its intrinsic qualities become relevant and provide the physical basis of capability. Land may be more or less capable in the new context [...] Socially, degradation must relate to capability.

In other words, desertification is not only about biogeophysical processes such as soil crusting and compaction, loss of soil structure and cohesion, gullying, sheet erosion, soil erosion by ablation, dune formation, local deposition in outwash fans, addition of sediment to water bodies, loss of productivity of croplands, pastures and woodlands, dust storms, increased atmospheric aerosol loadings, loss of surface roughness, increased albedo, decreased convection, reduced rainfall, and changed atmospheric conditions. By definition, desertification is about dryland ecology and capability. At the heart of desertification lies ecology, including the meteorological dimensions, grounded in the web of social relations that ties, for example, pastoral and/or farming households together, and links them to larger economic and political entities, namely, the market, the access to assets, land tenure, and the state (Swift, 1996; Turner, 2003). It has long been known that this 'demands a careful study of local-level processes and demonstrates that environmental change needs to be carried out not in vague terms but at specific locations and among specific segments of ... society' (Watts, 1985).

## Why is it Important?

Desertification is often seen as one of the most serious environmental problems confronting the world from at least the 1920s onwards. The process, however, is not confined to this period of time. Processes of desertification are of great antiquity, and the study of ancient desertification can be important *per se*, for example, to potentially derive generic principles for the co-evolution of human societies with their natural environments. Ancient or pre-historic desertification led to the collapse of civilizations and empires, and historic desertification led to the localized collapse of farming communities and population displacements. Clearly, the importance of these processes relate to 'land degradation ... resulting from various factors, including climatic variations and human activities' (UNEP, 1994), and they do not include the formation of deserts *per se*, i.e., decoupled from human impacts such as the 'browning of the Sahara'. The latter example of a purely biophysical process means the rapid conversion about 5,500 years ago of the Sahara, with a vegetation cover resembling that of a modern African savanna and carrying significant populations of large animals and humans, into its present desert condition. The ultimate cause was a small, subtle change in the Earth's orbit, leading to a small change in the distribution of solar radiation on Earth's surface, but triggering deleterious vegetation cover changes (Steffen, Sanderson, Tyson, Jäger, Matson, Moore, Oldfield, Richardson, Schellnhuber, Turner and Wasson, 2004).

*Collapse of Ancient Societies*

There are abundant examples of ancient societies which collapsed due to factors related to land use and dryland degradation (Runnels, 1995; Redman, 1999). For example, Bunney (1990) offers evidence of land degradation from early human history in the area surrounding Lake Patzcuaro in Mexico, and McAuliffe, Sundt, Valiente-Banuet, Casas and Viveros (2001) demonstrate how a very severe episode of erosion of cultivated uplands further south had turned, approximately 900 years ago, drainage basins of the Tehuacán Valley, the 'cradle of maize', into today's degraded and impoverished areas. Olson (1981) examines clues from the collapse of ancient civilizations such as the Sardis in Turkey which suggest that overexploitation of land resources played a significant role. In the following, two detailed examples are given of the pre-historic importance of desertification for human development and civilizations.

The 'Fertile Crescent' in Southwestern Asia, or today's Near East, is probably the most intensively studied and best understood part of the globe as regards the rise and collapse of agricultural land use in what is now a fragile and widely degraded ecological dryland setting. The region appears to have been the earliest site for a whole string of developments, including animal as well as plant domestication and innovative food production, cities, states, writing, metallurgy, wheels, and what is termed civilization. However, the Sumerian civilization which exploited the Tigris and Euphrates river basin, for example, suffered the consequences of poor irrigation and salinity-induced desertification 6,000 years ago, and deforestation went hand in hand with the proliferation of desert-like conditions in Attica, which was noted by Plato some 2,500 years ago. Diamond (1999) provides some arguments on why domesticated plants and animals in Southwestern Asia and the eastern Mediterranean region gave the regions such a 'potent head start' with view upon modern world development, including the favourable Mediterranean climate, large stands of already highly productive wild ancestors of crops, etc. Turning to the ultimate causes why the region lost its enormous lead of thousands of years to late-starting Europe, proximate factors need to be addressed such as the development of a merchant class, capitalism, patent protection for inventions, failure to develop crushing taxation, tradition of critical empirical enquiry, etc.: why did they arise in Europe rather than Mesopotamia? Diamond (1999, pp. 410-11) identifies a westward shift of 'civilization', at the heart of which lies desertification. Actually, Europe received crops, livestock, technology, and writing systems from the Fertile Crescent, which then gradually eliminated itself as a major center of power and innovation.

> After the rise of Fertile Crescent states in the fourth millennium B.C., the center of power initially remained in the Fertile Crescent (...). With the Greek conquest of all advanced societies from Greece east to India ... in the late fourth century B.C., power finally made its first shift westward. It shifted farther west with Rome's conquest of Greece in the second century B.C., and after the fall of the Roman Empire it eventually moved again, to western and northern Europe. (...) Today, the expressions 'Fertile Crescent' and 'world leader in food production' are absurd. Large areas of the former Fertile Crescent are now desert, semi-desert, steppe, or heavily eroded or salinized terrain unsuited for agriculture. (...) In ancient times, however, much of the Fertile Crescent and eastern Mediterranean region, including Greece, was covered with forest. (...) Its woodlands were cleared for agriculture, or cut to obtain construction timber, or burned as firewood or for manufacturing plaster. Because of low rainfall and hence low primary productivity (...), regrowth of vegetation could not keep pace with its destruction, especially in the presence of overgrazing by abundant goats. With the tree and grass cover removed, erosion proceeded and valleys silted up, while irrigation agriculture in the low-rainfall environment led to salt accumulation. These processes, which began in the Neolithic era, continued into modern times. (...) Thus, Fertile Crescent and eastern Mediterranean societies had the misfortune to arise in an ecologically fragile environment. They committed ecological suicide by destroying their own resource base.

Another prominent example of ancient (to historical) desertification outcomes linked to coupled human activities and climatic variations stem from the upland steppe plateaus of northern China (Sheehy, 1992; Zhou, Dodson, Head, Li, Hou, Lu, Donahue and Jull, 2002; Lin and Tang, 2002; Jiang, 2002). The zone has supported century, if not millennia-old nomadic or semi-nomadic grazing, and parts of it have now turned into the most desertified areas of China. During the Quaternary, the climate had changed several times, and the last episode of these climatic changes, the dry interval dating back to ca. 7,000 years before present, has been the current arid climate. Reinforcing climate predisposement was a change in the Quaternary geological environment, sometimes even called decisive for desertification in northern China. This had been the uplifting of the Tibet (Qinghai-

Xizang) Plateau from 1,000 m (in the Pliocene) to the current 5,000 m, mainly during the past 10,000 years. The event strenghtened the Mongolian higher pressure zone and makes the climate drier and less rainy in northwest and north China. The uplifting impacted directly upon vegetation and surface soil cover, and thus became an immediate cause leading to present desertification. Mainly at locations in northwestern China such as the Ordos Plateau, pre-historic predisposement is reinforced by oscillating desert margins due to the coupled human-environmental impacts in ancient times. These had been both drastic climatic variations in the $11^{th}/13^{th}$ centuries, and again in the $17^{th}$ century, and destructive land uses under several dynasties, but especially in the Tang period (618-907 AD). What had previously been the Mu Us grasslands or meadows were located at the juncture of the Inner Mongolia, Shanxi and Ningxia provinces in northern China, which delineate the transitional zone between the sub-humid core areas of Chinese civilization, based on irrigated and sedentary agriculture, and the northern (semi)arid grasslands, best suited for nomadic and extensive pastoralism. Inappropriate land use in concomitance with biophysical changes has influenced the position of the desert margin. Desert ecosystems shifted southward by about three degrees of latitude to the present. Archaeological record proves that the natural vegetation of the semi-arid transitional zone was severely affected by the combined effects of human population increases, dry farming activities, forest clearance, and frequent warfares at around 3000-2000 $^{14}C$ yr BP. Wind and water erosion became intense, and grassland ecology in some places changed to semi-desert or desert conditions, with mobile sand dunes starting to develop. The Xia Imperial Dynasty, for example, established in Shanxi Province in AD 413, was exterminated by wind-drift sand, and chaos was caused by subsequent war for 400 years. Nonetheless, at the beginning of the $5^{th}$ century AD, the natural environment at most locations had still fertile land with plenty of available freshwater and vast meadows with well-developed animal husbandry. Under the Tang Dynasty (established in AD 618), natural undergrowth was cleared, and grain was planted. Following frequent warfare during which forests and forage got burned (to destroy resources required by nomadic pastoralists) and natural grasslands trodden by warhorses, agricultural land was abandoned. Farmland, including irrigation ditches, became covered with sand transported by the winter monsoons. The western part of the Mu Us grasslands were reclaimed and cultivated during the Tang and the following Song dynasties (AD 960-1126) in succession, which again triggered desertification. By AD 822, high sand dunes had been created, with the Great Wall buried by sand in places at the middle of the $16^{th}$ century. Over this time period, alternations between livestock and grain farming caused the opening of more land for farming, more wood required for fuel, and more livestock for grazing. Further land reclamation was carried out during the Ming (AD 1368-1644) and Qing (AD 1661-1911) dynasties, exploiting what had been called 'Mongolian uncultivated land'. The goals were to build a strong military and to boost the economy, which is why the Qing government strongly exploited the grasslands. It resulted in the complete desertification of an area previously called Mu Us meadow. It caused deserts to expand southeast towards the remaining grasslands, and made the active sand dunes move southwards. Within a period of 1,000 years, the desert had annexed land from the northwest to the southeast by as much as 150 km. The Mu Us meadow is now called Mu Us desert, the most seriously desertified area in China at present.

## Dooming Prospects of Contemporary Desertification

Not the collapse of a whole society, but localized economic collapse, farm foreclosures and population displacements characterized a prolonged drought in the 1930s in the US-American Midwest. The historical case is typical for some features of contemporary desertification. First, it shows that land degradation relates to market systems which had not been characteristic for ancient societies (Polanyi, 1944). Second, it indicates that desertification occurs in some areas undergoing rapid land cover change (at changing times). Third, 'hot spots' appear to be spread over more than one hundred countries around the globe, thus constituting a truly global phenomenon (Lepers, Lambin, Janetos, DeFries, Achard, Ramankutty and Scholes, 2003). And, fourth, dryland degradation potentially triggers 'crisis' conditions. Henceforth, the interest in the historical 'dust bowl' case in the Great Plains of the United States of America, as described by Puigdefábregas (1995, p. 311).

> [F]armers, encouraged by the high prices of cereals, after the shortage caused by World War I, plowed up large surfaces of marginal land and grasslands. By 1920, the cropping area on the southern plains was double that of 1910 (...), and during the widespread drought that arrived in the 1930s, the wind blew up enormous quantities of soil. This 'dust bowl' led to a mass exodus of farmers and brought the erosion risks to public knowledge.

Desertification reached the world stage in the 1970s, in the aftermath of the great Sahelian drought/famine (1968-73) and the prospect of a global food crisis. The Sahelian drought led to the death of untold numbers of livestock and, indirectly, of pastoralists and villagers who sought to survive the ravages of the countryside by fleeing southwards to cities and refugee camps (Watts, 1987; Bohle, Downing, Field and Ibrahim, 1993; Oyowe, 1998; Dregne, 2002). The Sahelian drought offers one of the best case histories showing that the process of desertification is triggered by anthropogenic disturbances, with similar sequences of events reported from other parts of the world (Puigdefábregas, 1995, p. 311).

> During the period 1945-68, rainfall was higher than normal in the region. (...) People moved to the north, plowing land and raising livestock in areas where only seminomadic husbandry had been practiced before. The drought that started in the 1970s trapped those people in a dead end, between the desert to the north and the already settled agricultural land to the south. (...) No other choice was left to them than to go forward to the full exhaustion of resources.

Both research- and policy-driven interests in dryland environments were mobilized then, such as the 1972 Stockholm Conference on the Human Environment and the United Nations Resolution 3337 in 1974, calling for a global action on desertification and recommending UNCOD. It took another fifteen years, however, for an internationally sanctioned plan of action to be developed. At the global level, more impetus to examine land degradation has been triggered only by the widespread adoption of the term 'sustainable development' and the various proposed actions outlined at the Rio Conference in 1992. Thus, UNCCD became one of the three 'Rio Conventions'. Consequently, efforts such as Agenda 21 were set in motion to improve land management, culminating in the signing of CCD in the mid-1990s. Leadership in efforts to stop and reverse the environmental deterioration in drylands has come from UN agencies since UNCOD was convened in 1977. UNEP originally has the primary role, and, at present, the CCD secretariat in Bonn is expected to carry out that function (UNEP, 1977; UNEP, 1994; UN, 1994; Briceño, 1998; Chasek and Corell, 2002).

Desertification in today's drylands is considered to be a global problem, characterized, for example, by the deterioration and degradation of soil and vegetative cover. It is not confined to deserts *per se*, but can occur in any dryland region (Glantz, 1977; Watts, 2001). Some argue that contemporary desertification is one of the most critically important global environmental change issues, potentially affecting more than one third of the earth's surface and about one fifth of the world's human population (Kassas, 1995; Dregne, 1996; Darkoh, 1998). Some 25 years ago, Peter Shaw Thacher, UNEP's deputy executive officer, even used to claim that 'desertification ... is probably the greatest single environmental threat to the future well-being of the Earth' (Shaw Thacher, 1979).

Drylands cover more than one third of the earth's surface, large parts of which are claimed to be heavily and perhaps irreversibly degraded (Mainguet, 1999). Still, the most comprehensive study to date is that of the UN Environmental Programme (Middleton and Thomas, 1997). It is estimated that global drylands cover about 5,160 million ha, and it claimed that 70% of the drylands suffer from some form of land degradation. By land use category, these would be irrigated croplands at 146 million ha (30% degraded), rainfed cropland at 458 million ha (47% degraded), and rangelands at 4,556 million ha (73% degraded). Historical claims, persisting until now, state that these drylands include large numbers of rural poors. For example, it has been estimated in 1996 by the UNCCD secretariat that 250 million people are directly affected by desertification, and that one billion are at risk. Some figures rank even higher. Adams and Eswaran (2000), for example, use to claim that up to 2.6 billion people are potentially affected by desertification in more than 110 countries around the world. In response, the UN established CCD, whose aim is to 'target poverty, drought and food security in dryland countries experiencing desertification' (UN, 1994). A recent example is the semiarid zone of south-central Mexico, which had actually been the 'cradle of maize' during historic and ancient times. Some of the land in drainage basins of the Tehuacán Valley had already been abandoned 45 years ago, and the recurrent failure of maize for the last several years caused hunger in parts of the local population. The currently cultivated lands are estimated to supply only about one fifth and one third of the community's subsistence needs for maize and beans, respectively. The shortage of arable land is acute, and as the population continues to grow, poverty is extreme and increasing (McAuliffe, Sundt, Valiente-Banuet,

Casas and Viveros, 2001). In more general terms, Vogel and Smith (2002, p. 150) describe the conditions of endangered livelihoods in dry environments as follows.

> Those making a living from the land in semi-arid or arid environments are often constrained by a number of risks, threats, and difficult daily realities. Some of these include poverty, the risk of drought or an extreme number of dry years, civil strife, and the reality of living with diseases and illness, including HIV/AIDS. To live and derive a livelihood in such environments has usually resulted in a variety of coping and/or adaptive strategies that often stretch the very fabric of survival.

In Subsaharan Africa, desertification is believed to be especially widespread, affecting some 200 million people (Adams and Eswaran, 2000). Cleaver and Schreiber (1994) claim that half of the farmland in Subsaharan Africa is affected by soil degradation and erosion, 'and up to 80% of its pasture and range areas show signs of degradation'. Dryland degradation is also believed to be widespread and severe in Asia and Latin America, as well as in other parts of the globe (Adams and Eswaran, 2000). Rather than being spread homogenously across all drylands, contemporary desertification appears to be largely confined to a few areas undergoing rapid land cover change, such as the vast plain and basin sites in the Central Asian desert and steppe region (Lepers, Lambin, Janetos, DeFries, Achard, Ramankutty and Scholes, 2003).

Finally, the importance of contemporary desertification can be taken from some degrees of dryland degradation which have been identified in the 'dust bowl' and 'Sahel' cases, but also in a wide array of local- to subnational-scale case studies further explored in this book. The spectrum ranges from slight to extreme reductions in the biological and socio-economic potential or 'capability' of a given piece of land. For example, vegetation change can slightly affect the canopy of undegraded perennial plant cover, extending over up to 70% of the surface (or more) in the beginning, but can be reduced to less than 5% of the surface over time. Similarly, tree density or timber stock may decrease by more than three fourths of the initial density or volume over time. Increases of bare, rocky ground cover or eroded land may extend from less than one fourth of a hitherto slightly affected area to more than three fourths of the area, leaving behind land which is unreclaimable and impossible to restore. Also, there may be hardly any sand deposits in the beginning and no activation of (semi)fixed sand dunes, while dune activation extends to more and more sites and frequent, periodic sand deposits could occur on pastoral and cultivated land, with related siltation of low-lying zones and/or burial of agricultural terrain and human settlements even. Translated into terms of socio-economic potential, these biophysical changes could generate terrain with somewhat reduced pastoral and/or agricultural suitability in the beginning (still suitable for use in some local grazing and/or cropping systems), but may quickly turn into terrain with greatly reduced suitability or may be no more reclaimable at the household level. Also, increasing economic losses can occur as well as wide-scale out-migration from rural to urban areas. Degraded rangelands may suffer from insect or rodent damage and/or be increasingly abandoned.

Increased conflicts may arise between herders and cultivators about shrinking natural resources, and major improvements and/or engineering works could be required to restore the productivity of the land. All these changes, including the abandonment of human settlements even, may reflect moderate to strong degrees of desertification already, while restoration to full productivity might still be possible through modifications of the currently practiced management system. As an extreme, the terrain can be pushed beyond a threshold where any suitability for cropping, pastoral and even wild animal uses might be lost. It might turn out in the end, that the terrain is unreclaimable and beyond restoration for any human use.

**What Studies have been Carried out Already?**

*Multiple Impacts*

Concerns about changes of land use and cover in drylands emerged in the research agenda on global environmental change several decades ago with the realization that land surface processes influence climate (Lambin, Geist and Lepers, 2003; Steffen, Sanderson, Tyson, Jäger, Matson, Moore, Oldfield, Richardson, Schellnhuber, Turner and Wasson, 2004). For example, in the mid 1970s, it was recognized that land cover change modifies surface albedo and thus surface-atmosphere energy exchanges, which have an impact on regional climate (Otterman, 1974; Charney and Stone, 1975; Sagan, Toon and Pollack, 1979). Later, the important contribution of local evapotranspiration to the water cycle – that is precipitation recycling – as a function of land cover highlighted yet another considerable impact of land use and land cover change on climate, at a local to regional scale in this case (Eltahir and Bras, 1996). Increasingly, atmosphere-vegetation interactions and greenhouse-gas induced climate changes were explored and discussed as a function of land cover such as in the case of North Africa (Claussen, Brovkin and Ganopolski, 2002).

Later, a much broader range of impacts of land use and land cover change in drylands on ecosystem goods and services were further identified (Lambin, Geist and Lepers, 2003; Steffen, Sanderson, Tyson, Jäger, Matson, Moore, Oldfield, Richardson, Schellnhuber, Turner and Wasson, 2004). Of primary importance are impacts on carbon sequestration (McGuire, Sitch, Clein, Dargaville, Esser, Foley, Heimann, Joos, Kaplan, Kicklighter, Meier, Melillo, Moore, Prentice, Ramankutty, Reichenau, Schloss, Tian, Williams and Wittenberg, 2001; Lal, 2002), on biotic diversity (Briceño, 1998; Sala, Chapin, Armesto, Berlow, Bloomfield, Dirzo, Huber-Sanwald, Huenneke, Jackson, Kinzig, Leemans, Lodge, Mooney, Oesterheld, Poff, Sykes, Walker, Walker and Wall, 2000; Dirzo and Raven, 2003), soil degradation (Lal, 1988; Trimble and Crosson, 2000; Niemeijer and Mazzucato, 2002), and the ability of biological systems to support human needs (Vitousek, Mooney, Lubchenco and Melillo, 1997; Mooney, Cropper and Reid, 2003). It was found, for example, that the expansion of woody shrubs in the western United States grasslands, following fire suppression and overgrazing, contributed to a large carbon sink (Houghton, Hackler and Lawrence, 1999; Pacala, Hurtt, Baker,

Peylin, Houghton, Birdsey, Heath, Sundquist, Stallard, Ciais, Moorcroft, Caspersen, Shevliakova, Moore, Kohlmaier, Holland, Gloor, Harmon, Fan, Sarmiento, Goodale, Schimel and Field, 2001). Further, time series of remote sensing data revealed that land cover changes do not always occur in a progressive and gradual way, but they may show periods of rapid and abrupt change followed either by quick recovery of ecosystems or by a non-equilibrium trajectory. Such short-term changes, often caused by the interaction of climatic and land use factors, were identified to have an important impact on ecosystem processes (Taylor, Lambin, Stephenne, Harding and Essery, 2002). For example, droughts in the African Sahel and their effects on vegetation are reinforced at the decadal timescale through a feedback mechanism that involved surface changes caused by the initial decrease in rainfall (Zeng, Neelin, Lau and Tucker, 1999). Grazing and conversion of semiarid grasslands to row-crop agriculture are the source of another positive desertification feedback by increasing heterogeneity of soil resources in space and time (Schlesinger, Reynolds, Cunningham, Huenneke, Jarrell, Virginia and Whitford, 1990). And, the reduction of precipitation from clouds affected by desert dust can cause drier soil, which in turn raises more dust, thus providing a feedback loop to further decrease precipitation, with land use change exposing the topsoil initiating such a desertification feedback (Rosenfeld, Rudich and Lahav, 2001).

*Multiple Causes*

Recent research on the causative factors of dryland change has largely dispelled simplifications such as that land change consisted mostly in the destruction of natural vegetation by overgrazing which leads to desert conditions or desertification. These conversions were assumed to be irreversible and spatially homogenous and to progress linearly, and only the growth of the local population was thought to drive changes in land conditions. However, work on pastoralism in the Sahel, for example, replaced these simplifications by a representation of much more complex and sometimes intricate, processes of land use and land cover change (Watts, 1987; Mortimore, 1998; Niamir-Fuller, 1999; Mortimore and Adams, 2001; Bassett and Zuéli, 2000; Hiernaux and Turner, 2002; Robbins, Abel, Jiang, Mortimore, Mulligan, Okin, Stafford Smith and Turner, 2002; Turner, 1999, 2003). New research challenges commonly accepted 'wisdoms' such as that land degradation happens everywhere and all the time in drylands (Mortimore, 1993, 1998; Benjaminsen, 1993, 2001; Tiffen, Mortimore and Gichuki, 1994; Leach and Mearns, 1996; Raynaut, 1997; Gray, 1999; Howorth and O'Keefe, 1999; Mortimore and Adams, 2001).

A careful study, adopting the framework of analysis developed under the auspices of the Land-Use/Cover Change (LUCC) project of the International Geosphere-Biosphere Programme (IGBP) and the International Human Dimensions Programme (IHDP) on Global Environmental Change, for example, used 'multiplicity' as the most appropriate approach or theme, when exploring the causes of major land use and land cover changes in an agropastoral environment of northwestern Syria, framing findings along pathway thinking (Nielsen and

Zöbisch, 2001). It was found that, in the recent past, multi-factorial driving forces were in operation, including 'climatic conditions, population growth, government laws, market relations and forces, technology changes and increased crop-livestock integration', and that 'both the degradation and intensification pathways of land-use change can be observed'.

A group of experts distilled a finite set of pathways from the complex processes of land change leading to rangeland modification. Taking the case of sub-Saharan Africa, state policies are considered to be framed under the assumption that pastoralists overstock rangelands, leading to degradation, with resulting management strategies tending to control, modify, and even obliterate traditional patterns of pastoralism, including the development of watering points or long-term exclusion. Two common pathways that follow are described (Lambin, Turner, Geist, Agbola, Angelsen, Bruce, Coomes, Dirzo, Fischer, Folke, George, Homewood, Imbernon, Leemans, Li, Moran, Mortimore, Ramakrishnan, Richards, Skånes, Steffen, Stone, Svedin, Veldkamp, Vogel and Xu, 2001, p. 264).

> Weakened indigenous pastoral systems undermine local economies and resource institutions or precipitate urban migration with rural remittances, either of which may lead to land alienation an conversion, with concentration in the remaining areas, local overstocking and degradation. Alternatively, exclusion and reduced grazing lead to changes in species diversity, vegetation cover and plant production, with implications for biodiversity conservation and/or animal production. In wetter rangelands, reduced burning leads to increasing woodland. Evidence indicates that grazing, rather than being inherently destructive, is necessary for the maintenance of tropical rangelands.

Similarly, Kates and Haarmann (1992) found some generic principles, or a set of common interactive processes linking environmental degradation and poverty, through a careful comparison of diverse case studies. These case studies told common tales of poor people's displacement from their lands, the division of their resources, and the degradation of their environments, which culminated in three major spirals of household impoverishment and land degradation driven by combinations of development and commercialization, population, poverty, and natural hazards. Barbier (2000a), by examining case study evidence on the effects on land degradation of economic liberalization and globalization – trade liberalization reforms, in particular, to 'open up' the agroindustrial sector – identifies modification as well as conversion of land cover as the immediate and principal impact on rural resources. Direct and indirect effects can be held apart: directy, increased agricultural productivity triggers land degradation from unsustainable production methods, while, indirectly, agroindustrial development displaces (near) landless and rural poors, who are then pushed to marginal agricultural lands. Further, Barbier (2000b) shows how 'good' and 'bad' policies can affect the economic incentives determining poor rural household's decisions to conserve or degrade their land.

Related to research on causative mechanisms are general insights into the role of natural variability and natural environmental change as a causative factor of human-environmental co-evolution in drylands. Obvious interactions with human causes of land use change were found. In particular under dry to subhumid climatic conditions, highly variable ecosystem conditions driven by climatic variations amplify the pressures arising from high demands on land resources. Natural and socio-economic changes may operate as synchronous but independent events (Jiang, 2002; Lambin, Geist and Lepers, 2003). For example, in the Iberian Peninsula during the sixteenth and seventeenth centuries the peak of the Little Ice Age occurred almost simultaneously with large-scale clearing for cultivated land following the consolidation of Christian rule over the region. This cultivation triggered changes in surface hydrology and significant soil erosion (Puigdefábregas, 1998). Natural variability may also lead to socio-economic unsustainability, for example when usually wet conditions alter the perception of drought risks and generate overstocking on rangelands. When drier conditions return, the livestock management practices are ill adapted and cause land degradation. This overstocking happened several times in Australia, and, in the 1970s, in the African Sahel (Puigdefábregas, 1998). Similarly, agricultural expansion into marginal rangeland areas during wet periods leaves farmers more seriously exposed to hazard when drought or drier conditions return than pastoralists would have been (Glantz, 1994), given the higher degrees of flexibility and mobility of transhumant land use systems to respond to droughts (Niamir-Fuller, 1999).

*Vulnerability, Criticality, Syndromes*

It has further been identified that land use and land cover changes, such as cropland extension into drylands, also determine the vulnerability of places and people to climatic, economic, or socio-political perturbations (Vogel and Smith, 2002; Downing and Lüdeke, 2002). The study of regions at risk and environmental criticality by Kasperson, Kasperson and Turner (1995, 1999) was an approach that provided a classification of the situations in which environmental degradation occurs. Several case studies of dryland regions under environmental degradation – such as the Ukambani Region of Kenya, the Ordos Plateau of China, the Aral Sea Basin, and the Llano Estacado in the United States – were described qualitatively by their histories (Rocheleau, Benjamin and Diang'a, 1995; Jiang, Zhang, Zheng and Wang, 1995; Glazovsky, 1995; Brooks and Emel, 1995). These qualitative trajectories were represented in terms of development of the wealth of the inhabitants and the state of the environment. A 'critical environment' (Kasperson, 1993) was defined as one in which the extent or the rate of degradation precludes the maintenance of current resource-use systems or levels of human well-being, given feasible adaptations and the community's ability to mount a response. Different typical time courses of these variables were identified and interpreted with respect to more or less problematic future development of the regions. The Aral Sea, for example, was unquestionably a critical region after a few decades of Soviet-sponsored, ill-conceived large-scale irrigation schemes (Glazovsky, 1995).

Related to the research on causative factors of desertification, but also to the analysis of the vulnerability of people and places, are insights from studies which show that not all causes of land change and all levels of organization are equally important. For any given human-environment system, a limited number of causes are essential to predict the general trend in land use (Stafford Smith and Reynolds, 2002). This is the basis, for example, of the syndrome approach which describes archetypical, dynamic, co-evolutionary patterns of human-environment interactions, taken from the analysis of case studies (Petschel-Held, Lüdeke and Reusswig, 1999; Lüdeke, Moldenhauer and Petschel-Held, 1999). A taxonomy of syndromes – such as the 'overexploitation syndrome', the 'dust bowl' syndrome, or the 'Sahel syndrome' – links processes of degradation to both changes over time and the status of state variables. By definition, patterns or syndromes do not solely refer to, for example, the Sahel region in West Africa as in the case of the Sahel syndrome. This syndrome is based on a particular pattern of smallholder agriculture in developing countries, strongly seasonal climatic patterns, poor soils and an arid to semiarid vegetation (Petschel-Held, 2004). The approach aims at a high level of generality in the description of mechanisms of environmental degradation and is applied at the intermediate functional scales that reflect processes taking place from the household level up to the international level (Lambin, Chasek, Downing, Kerven, Kleidon, Leemans, Lüdeke, Prince and Xue, 2002).

*Rates, Extent, Severity*

A consensus has been progressively reached on the rate and location of some of the main land changes worldwide, but desertification is still unmeasured or poorly documented at the global scale (Lambin, Geist and Lepers, 2003). Various attempts have been made to assess the magnitude of the problem and to provide a baseline for monitoring. In an early attempt, Mabbut and Flores (1980) collected case studies from national governments, focussing on randomly selected sites – such as Luni Development Block in Rajasthan or the Greater Mussayeb project in Iraq – rather than applying a well-conceived framework for analysis. The latter has been improved for pilot and case studies related to the Land Degradation Assessment in Drylands (LADA) project, with national case studies stemming from Argentina, Egypt, Mexico, Uzbekistan, Senegal, Kenya, South Africa, and Malaysia, mainly aimed at programmatic efforts (Nachtergaele, 2002). Following an analysis of all available global information sources, Dregne (2002) concludes that there is a pressing need for more reliable data. Only very few global assessments of the extent and severity of desertification have been published until now. To document the extent of desertification, UNCOD commissioned several studies such as those of Lamprey (1975) and Mabbutt (1984). Some vulnerability or hazard maps, based on preliminary data such as the one by Dregne (1977), were included in the documents for UNCOD in 1977. A few years later, the first country-by-country assessment was made of land degradation in 100 countries. Four degradation classes for three major land uses were distinguished, again on the basis of qualitatively derived data which included informed opinion, anecdotal evidence, observations by travellers and (un)published reports (Dregne, 1983). The inclusion of field experiments and improved literature searches led to a much better assessment in 1992, but its accuracy was still low and there was no way to check with other databases.

The Global Assessment of Human-induced Soil Degradation (GLASOD) was the hitherto last global survey conducted under UNEP sponsorship. It has actually been an analysis of human-induced soil degradation for both arid and humid regions, including the type, degree, extent, rate and even cause of soil degradation. The product was a map derived from the informed opinion of hundreds of soil scientists and others (Oldeman, Hakkeling and Sombroek, 1990). GLASOD triggered the first edition of an UNEP word atlas of desertification, and a greatly revised second edition attempts to rectify the overemphasis on soils by including vegetation degradation. Nearly all of the information in this last review comes from published reports of varying quality (Middleton and Thomas, 1997). The GLASOD database is still 'far and away the best representation of world soil degradation' (Dregne, 2002), and even today's model-based desertification assessments – such as the IMAGE Land Degradation Model (Hootsmans, Bouwman, Leemans and Kreileman, 2001) – still depend on the existence of it (Leemans and Kleidon, 2002). Following GLASOD, there have only been national assessments, such as for China (Zhu and Wang, 1993), and descriptive region-by-region assessments, such as for Africa, Asia, Australia, Europe, North and South America, provided, for example, by Dregne (2002) on the basis of published information sources.

An effort to overcome the limitations of the GLASOD database relates to the quantification of areas of most rapid and most recent land cover changes including desertification. The work was done under the auspices of the Millenium Ecosystem Assessment (MEA) which is about 'strengthening capacity to manage ecosystems sustainably for human well-being' (Mooney, Cropper and Reid, 2003). It was a global effort, with any information on drylands compiled from regional, remote sensing-driven databases and other sources. As a result, quantification of contemporary 'hot spots' of desertification still suffers from large uncertainties (Lepers, Lambin, Janetos, DeFries, Achard, Ramankutty and Scholes, 2003).

For the study of desertification at the local to national – rather than global – scales, there has been growing recognition over the past decade of the importance of the processes of modification of land attributes rather than conversion (i.e., the complete replacement of once cover type by another). Land cover modifications are more subtle changes that affect the character of the land cover without changing its overall classification (Lambin, Geist and Lepers, 2003). For example, in northwestern Senegal, situated in the West African Sahel, declines in tree density and species richness were measured and provided evidence of desertification in that region (Gonzalez, 2001). Another study in western Sudan, a region that was allegedly affected by desertification (Ibrahim, 1984; Lamprey, 1975, 1988), did not find any decline in the abundance of trees despite several decades of droughts (Helldén, 1991; Schlesinger and Gramenopoulos, 1996).

Similarly, analyses of multiyear time series of land surface attributes, their fine-scale spatial pattern, and their seasonal evolution have led to a broader view of land-cover change (Lambin, Geist and Lepers, 2003). As for remote sensing of vegetation productivity, in general there are three major approaches, using reduced net primary production (NPP): rain use efficiency, comprehensive biogeochemical models, and locally scaled NPP, all of them acting 'as a means to focus attention on the areas most likely to be desertified' (Prince, 2002). Remote sensing data highlight high temporal frequency land-cover modifications of great importance for earth system processes. In particular, data from wide-field-of-view satellite sensors reveal patterns of seasonal and interannual variations in land surface attributes that are driven not just by land-use change but also by climatic variability. These variations include the impact of the El Niño Southern Oscillation (ENSO) phenomena (Plisnier, Serneels and Lambin, 2000; Behrenfeld, Randerson, McClain, Feldman, Los, Tucker, Falkowski, Field, Frouin, Esaias, Kolber and Pollack, 2001), natural disasters such as droughts (Lambin and Ehrlich, 1997; Lupo, Reginster and Lambin, 2001; Pickup, 1998), and changes in vegetation productivity due to erratic rainfall fluctuations in the African Sahel, which lead to an expansion and contraction of the Sahara (Tucker, Dregne and Newcomb, 1991). An examination of AVHRR satellite data covering the southern fringe of the Sahara shows that, while the fringe of the desert (as marked by the 200 mm rainfall isoline) has fluctuated markedly, there has been virtually no net increase in the desert's area from 1980 to 1997. In fact, the trend line suggests that the Sahel, the region just to the south of the Sahara, has actually become slightly greener (moister) during this time period (Tucker and Nicholson, 1999). Another study linked coarse resolution remote sensing data with rainfall data and tested whether there was a decadal trend in the rain-use efficiency of the African Sahel region. It revealed the absence of widespread subcontinental-scale dryland degradation, although some areas did show signs of degradation (Prince, De Colstoun and Kravitz, 1998; Prince, 2002). These results prove a highly variable, event-driven non-equilibrium process, or, in other words, suggest that the resilience of the Sahel in primary production per unit rainfall has not changed despite serious droughts in the 1970s and 1980s. The studies focussing on rates and magnitudes of change conform with insights from new research questioning the commonly accepted wisdom of causes of land degradation in areas where it has been presumed to be widespread (Schulz, 1994; Rasmussen, Fog and Madsen, 2001).

Finally, the impact of mostly anthropogenic fires on land cover in tropical drylands – resulting from a combination of climatic factors, which determine fuel availability, fuel flammability, and ignition by lightning, and factors related to land use and land cover change that control fire propagation in the landscape and human ignition – has also been well documented with remote sensing data (Pereira, Pereira, Barbosa, Stroppiana, Vasconcelos and Grégoire, 1999; Dwyer, Pereira, Grégoire and De Camara, 2000).

## Why is there an Information Gap?

*Simplifications and Contradictions*

It can be argued that much of the controversy still surrounding desertification can be attributed to a veritable 'shopping list approach to questions of desertification dynamics', because only few careful studies of local-level processes were carried out, and because environmental change was too often studied in vague terms and not at specific locations and among specific segments of rural societies or ecosystems (Watts, 1985, 2001).

Since the 1920s at least, concerns about dryland degradation and desertification have often persisted in the form of references to causes and events that were not well grounded in social and ecosystems theory and, thus, not supported by theory-guided empirical research (Vogel and Smith, 2002). For example, manyfold reference has been made to simplifications such as 'marching deserts' (Lamprey, 1975, 1988; Suliman, 1988; Keita, 1998), the 'irreversible spread of desert-like conditions' (Eckholm and Brown, 1977; Mainguet and Chemin, 1991), to the 'ruination of the land by ignorant peasants' (Mensching and Ibrahim, 1976; Ibrahim, 1978; Cleaver and Schreiber, 1994), 'ill-advised human actions' or 'mismanagement' (Mainguet, 1999) such as the 'irrational' behaviour of pastoral groups overstocking their land (Lavauden, 1927; Eckholm and Brown, 1977; Cleaver and Schreiber, 1994), farmers deforesting their land (Aubreville, 1949; Eckholm and Brown, 1977; Ringrose and Matheson, 1992; Mwalyosi, 1992; Venema, Schiller, Adamowski and Thizy, 1997; Gonzalez, 2001), or unwise, wide-scale technological applications such as irrigation schemes (Szabolcs, 1990; Glazovsky, 1995; Pickup, 1998; Genxu and Guodong, 1999; Saiko and Zonn, 2000), replacement of wild herbivores by domestic livestock (Le Houérou, 1993; Pickup, 1998; Fredrickson, Havstad, Estell and Hyder, 1998; Manzano, Návar, Pando-Moreno and Martinez, 2000), human misuse through excessive population growth (Ibrahim, 1984; Le Houérou, 1991; Cleaver and Schreiber, 1994; World Bank, 1996), increased desiccation in isolation from human factors, linked to decreases in rainfall and the drying up of water tables (Thébaud and Toulmin, 1994), and a purely man-made deterioration of dryland conditions (Eckholm and Brown, 1977; World Bank, 1996; Le Houérou, 2002; Breckle, Veste and Wucherer, 2002).

Adding to this theoretically untidy 'shopping list' of causal elements has been the culmination of dryland degradation as a global interest, with a particular focus on Africa, though. First, there was the notion of desert encroachment in West Africa due to local over-exploitation as a threat to the colonies during European colonization of Africa (Stebbing, 1935; Stamp, 1940). In the aftermath of the dust bowl event in the US Midwest, the stereotypes of 'marching deserts' and 'ignorant peasants' got revitalized and triggered several campaigns and policies to halt the negative impact on natural resources in the United States as well as in West and southeastern Africa, with examples being the Rhodesian soil conservation efforts, the land husbandry program of Malawi, and the establishment of the Soil Conservation Service in the United States (Stocking, 1985, 1996; Swift, 1996; Fairhead and Leach, 1996). Second, in the aftermath of the Sahelian drought, there had been the arguments brought forward by policy makers who combined the issues of poverty and economic growth with notions of desiccation (Warren and Agnew, 1988; Fairhead and Leach, 1996; Vogel and Smith, 2002). Not denying the fact that many people are failing to thrive and, indeed, survive in drylands, many national governments of countries with large dryland zones and international development agencies have been using the coupled notion of desiccation and impoverishment from the 1970s onwards to rescue an ideology of authoritarian intervention in rural land use and to claim rights to stewardship over resources previously outside their control. For example, Bassett and Zuéli (2000) explain the presence of the desertification narrative in the Côte d'Ivoire National Environmental Action Plan, despite of the fact that the landscape there is becoming more wooded and not desertified. Still there is an explicit focus of intervention measures on Africa, which is directly tied to the political and managerial outcomes of the Rio Conference rather than based on empirical evidence (Adger, Benjaminsen, Brown and Svarstad, 2001). Accordingly, the full text of the CCD reads as the 'United Nations Convention to Combat Desertification in those Countries Experiencing Serious Drought and/or Desertification, Particularly in Africa' (UNEP, 1994).

In many instances, various elements of causal simplifications were mixed arbitrarily and repeated over and over again. Some thirty years ago, for example, Lamprey (1975) claimed that current, man-made desertification in Sudan had been linked to past human events, stating that 'sand encroachment ... is the result of several thousand years of abuse of the fragile ecosystems which formerly existed in the Sahara and Nubian areas'. Eckholm and Brown (1977, p. 9) easily took up the notion then of desert encroachment as a mainly man-made phenomenon caused by the misuse of land.

> [T]hat the desert's edge is gradually shifting southward there is little doubt. The spread of the Sahara has probably been measured most precisely in Sudan. There, as elsewhere, vegetational zones are shifting southward as a result of overgrazing, woodcutting and accelerated soil erosion ... [D]esert creeps into steppe, and while steppe loses ground to the desert it creeps into the neighbouring savanna which, in turn, creeps into the forest.

However, there are studies today which show that this environmental narrative was misconceived, at least for some locations in the savanna regions of West Africa. In contrast to the view that wooded savannas are becoming desertified, these studies argue that the landscape is becoming more wooded (Leach and Mearns, 1996; Fairhead and Leach, 1996; Bassett, 1993; Bassett and Zuéli, 2000). Still, it is not clear to which degree these new insights can be generalized.

Another desertification narrative depicts overpopulation in drylands as the main problem, merging it with the notion of carrying capacity (World Bank, 1996, p. 24).

> [The Sudano-Sahelian] belt features one of the most rapid annual growth rates [of population] of the continent, despite the fact that in many areas the mainly rural population ... is already beyond the carrying capacity at current technological levels. This growth has resulted in a downward spiral of extensive land degradation and fuelwood shortage.

Focussing on political factors rather than on population growth, another narrative runs directly counter to the carrying capacity notion (Lappé, Collins and Rosset, 1998, pp. 42-3).

In West Africa, colonial administrators imposed on local farmers monocultures of annual crops for export, notably peanuts for cooking oil and livestock feed and cotton for French and British textile mills. But growing the same crops year after year on the same land, without any mixing of or rotation of crops, trees and livestock, rapidly ruined the soils. Just two successive years of peanuts robbed the soil in Senegal of almost a third of its organic matter. Rapidly depleting soils drove farmers to push export crops onto even more vulnerable lands ... Furthermore, the spread of export crops, by crowding livestock herders into even smaller areas, has contributed to overgrazing.

Although root causes of desertification have often been assumed to be poverty and food insecurity interlinked with harsh climatic conditions, the very specific nature of interrelationships and thresholds beyond the climate-poverty nexus remains poorly understood. As a result of the prevailing 'shopping list' approach in desertification research, there has been no consistent information at the international level until now, i.e., land degradation information is not comparable, and it is of insufficient resolution for setting investment priorities or defining and implementing regional and national actions plans for international conventions such as UNCCD. Relating to the example above, scholarly work on the Republic of Sudan, as reported by Helldén (1991, p. 379), could not verify the magnitude and even the character of the process of desertification.

None of these studies verified the creation of long lasting desert-like conditions in the Sudan during the 1962-84 period, which corresponded to the magnitude described my many authors. There was no trend in the creation or possible growth of desert patches around 103 examined villages and water holes over the period 1961-83. No major shifts in the northern cultivation limit was identified, no major sand-dune transformations or Sahara desert encroachment was identified. No major changes in vegetation cover and crop productivity was identified, which could not be explained by varying rainfall characteristics.

## Causes

Another recent example of a theoretically untidy shopping list approach to the causative mechanisms of desertification is the summarization of man-made causes by Kadomura, Miyazaki, Fujimori, Sauer and Kawaguchi (1997, p. 21), as if these causes apply to all places at any time and are equally important at all levels of hierarchical organization.

Degradation mainly results from over-exploitation of land resources through over-cultivation, over-grazing, poor irrigation practices, mismanagement and deforestation and the destruction of woody vegetation. Protectionism and structural imbalances in the international economic system have also contributed to the over-exploitation of land and water resources. Furthermore socio-economic problems such as population growth and migration, low income and educational level as well as socio-political problems such as land tenure, property rights and legislation also play an important role in the desertification process.

Historically, one of the major contentious issues surrounding the debate on desertification has been the question of what is the particular mix of major proximate and underlying causes of land degradation in drylands, e.g., what is the weighting of processes such as agricultural production in marginal areas (over cultivation), overgrazing, poor irrigation practice, deforestation, population growth (by migration or natural increment), state policies, and surplus extraction, what are the key socio-economic key drivers of land use that lead to desertification, and what is the relative importance of natural, climate-driven *versus* human-made processes. A major drawback of dealing with the issue has been that the socio-economic and biophysical dimensions of dryland change have been studied in isolation from each other (Glantz, 1987; Watts, 2001; Reynolds and Stafford Smith, 2002). Until now, the two major, mutually exclusive – and still unsatisfactory – explanations for desertification are single factor causation and irreducible complexity.

*Single factor causation* On the one hand, proponents of single-factor causation suggest various primary causes, such as irrational or unwise land management by nomadic pastoralists and growing populations in semi-arid ecosystems. Central to this understanding is the overworking of land by ever increasing numbers of rural poors, creating in its extreme the notion of 'man-made deserts', i.e., the human-driven, irreversible extension of desert landforms and landscapes. A variant of – an assumed, rather than proven – key factor causation has persisted until today in the form of bad land management exerted by pastoral groups. This is true even despite of much new work on desertification, especially among Sahelian pastoral communities and on integrated livestock-cropping systems in the Sudan-Sahel, or elsewhere. For example, dealing with land use and land cover changes of global relevance, nomadic pastoralism still is reported as the assumed single causative factor related to desertification in the work of Mannion (2002, p. 77).

> Nomadic pastoralism in the developing world occurs mostly in the arid and semi-arid regions of Asia and Africa; it has a long history, extending back at least $7 \times 10^3$ years BP … and probably grew out of hunting strategies. In essence, this is an agricultural system that manipulates a naturally occurring land cover rather than one that alters it substantially. With injudicious management, however, especially in relation to carrying capacity, major alterations of land cover may ensue; apart from changes in the species composition of rangelands, desertification may occur, which not only impairs productivity but may also lead to land abandonment.

Overgrazing and livestock management, often put into the context of an imbalance between growing human populations and land carrying capacity, have underpinned many studies as the key causative mechanism, sometimes identified as the 'population-environment nexus'. Similarly, the exploration of associated impacts such as vegetation degradation and declining biological productivity of the land have often been coupled to an acknowledgement of increasing population pressure as the dominant underlying driving force (Mortimore, 1993). The work on Sahelian pastoralists, integrated land use systems in the Sudan-Sahel and other studies, however, suggest that some narratives of desertification which were encapsulated in the UNCOD report became dominant and need to be revised. These narratives 'were alarmist in their predictions, rooted in limited research, and rested on simple-minded presumptions of drought causality, poor local resource management by farmers and pastoralists, and unregulated population growth' (Watts, 2001). Rural people of arid ecosystems have often been portrayed as both 'victims' and 'villains' of various events and processes such as drought. Only recently, a more informed and realistic understanding of households and communities in drylands has been gained, by focusing on adaptive capacity and capabilities (Watts and Bohle, 1993; Vogel, 1995; Downing and Lüdeke, 2002; Vogel and Smith, 2002).

Furthermore, the notion of relatively fixed carrying capacities got dismissed by insights from range ecologists according to which rangelands in arid or semi-arid zones are increasingly seen as non-equilibrium ecosystems (Oba, Weladji, Lusigi and Stenseth, 2003). Modification in the biological productivity of these rangelands at the annual to decadal time scales is mainly governed by biophysical drivers such as interannual rainfall variability and ENSO events, with stocking rates having less long-term effect on productive potential (Behnke, Scoones and Kerven, 1993). This implies that the intrinsic variability of rangeland ecology makes it difficult to distinguish directional change such as soil or vegetation degradation from readily reversible fluctuations (Ellis and Galvin, 1994; Puigdefàbregas, 1998). Rather than considering human impact on rangelands, such as grazing, to be inherently destructive, more and more evidence indicates that grazing is necessary for the maintenance of tropical rangelands, in particular (Oba, Stenseth and Lusigi, 2000). On the other hand, less arid ecosystems in tropical and subtropical areas are increasingly seen as governed by a combination of human and biophysical drivers (Solbrig, 1993; Sneath, 1998), and may be more prone to being developed through intensification and conversion (Bassett and Zuéli, 2000).

*Irreducible complexity* On the other hand, desertification has been attributed to multiple causative factors that are specific to each locality, revealing no distinct pattern (Dregne, 2002; Warren, 2002). For example, Pickup (1998) claims that desertification is a global problem, but 'its causes are complex, frequently local, and vary from one part of the world to another'. Also, there is a great deal of debate, not only on whether the causes of desertification lie in the socio-economic or biophysical spheres (human-induced land degradation *versus* climate-driven desiccation), but also over the degree to which causes are local or remote, and how variables interact across organizational levels in different regions of the world and at different time periods. One can reasonably assume, though it has not yet been quantified sufficiently, that interaction across hierarchical scales varies in different regions of the world and at different times (Stafford Smith and Reynolds, 2002). Clearly, the exploration of the causes of desertification has not yet received the level of attention that, for example, tropical deforestation has (Geist and Lambin, 2002). In particular, case studies of desertification do not always make a difference between proximate, underlying and intermediate factors, which helps unravelling the complexity of land change processes (Turner, Skole, Sanderson, Fischer,

Fresco and Leemans, 1995; Lambin, Geist and Lepers, 2003; Geist, Lambin, Palm and Tomich, 2005).

*Rates and Extent*

Rates of desertification indicate the speed at which dryland degradation occurs. Reynolds (2001) made the point that 'current estimates of the rates ... of global desertification are based on limited data and thus are generally rough at best or (at worse) inaccurate'. For example, the severity of soil erosion and its impact on associated resources in the United States has been debated because of discrepancies between estimates based on models and observed sediment budgets (Trimble and Crosson, 2000). In some instances, gross exaggerations were produced, for example, on areas that had been arid for centuries but have been claimed to have become 'desertified', such as large parts of the Horn of Africa (Biswas, Masakhalia, Odego-Ogwal and Palnagyo, 1987). In other cases, observations were made on short-term ecosystem dynamics, especially vegetation changes (rather than on longer term responses such as soil degradation), which have been cited as evidence of desertification (Reynolds and Stafford Smith, 2002). Examples of alarmist rates of advancing deserts stem, for example, from the claimed advance of the Sahara southwards 'by an average of about 90-100 km in the last 17 years' (Lamprey, 1975, 1988), repeated in several variants (Smith, 1986; Suliman, 1988), including official government documents, which state that the desert is 'currently advancing at a rate of 5 to 6 km per year' (Desert Encroachment Control and Rehabilitation Programme, 1976).

Therefore, it should not come as a surprise that the extent of desertification evokes much disagreement and controversy. For example, irreversible degradation or 'desert-like conditions' – such as ruinous water erosion at subhumid agricultural sites or the formation of mobile sand dunes in pastoral drylands – can be the end point of dryland degradation processes, if they are not stopped (UNEP, 1977). However, some scientists continue to claim that most, if not all desertification, especially in an advanced state, 'is often essentially irreversible' (Phillips, 1993). This runs contrary to evidence that 'for the vast majority of the drylands that wasteland end point never occurs' (Dregne, 2002). It has been estimated that the very severe or 'irreversible' desertification class included about 780,000 $km^2$, which means about 1.5% of the global drylands only (Dregne, 1983; Dregne and Chou, 1992). Some sources, as reported earlier in this chapter, routinely report that up to 70% of all drylands are 'desertified', while others suggest that the figure is no more than 17% (Dregne and Chou, 1992). Recent studies about the scale of desertification differ by 50% (Dregne, 2002). It is clear, for example, that the Sahelian zone has experienced a decline in annual precipitation from the 1950s until the mid-1990s, but this downturn may still fall within the bounds of statistical expectation, i.e., precipitation is increasing again and back to the long-term average (Grainger, 1990; Watts, 2001; Reynolds and Stafford Smith, 2002). One of the latest assessments of desertification by the European Commission (2000, p. 4) reports this wide divergence in the percentage of land affected by dryland degradation, whether or not vegetation changes are included.

[T]he GLASOD (global assessment of soil degradation) commissioned by the FAO ... showed that 19.5% of drylands worldwide were suffering from desertification. A second survey carried out by the ICASALS group in Texas [The International Center for Arid and Semiarid Land Studies at Texas Tech University; HG] came up with a much higher figure of 69.5% of drylands worldwide suffering from desertification, due to their inclusion not only of areas affected by soil erosion, but also where a change in vegetation had occurred.

Consequently, it should not come as a surprise again that the numbers of people potentially affected by desertification differ considerably. Although some UN sources report that up to 2.6 billion people are affected by desertification (Adams and Eswaran, 2000; Nachtergaele, 2002), other UN sources downscale the figure by more than 50%, with Kofi Annan, for example, being quoted in the CCD that 'drought and desertification threaten the livelihood of over 1 billion people ...' (Reynolds, Stafford Smith and Lambin, 2003).

*Indicators*

An environmental indicator is commonly defined as a phenomenon or statistic so strictly associated with a particular environmental condition that its presence can be taken as indicative of that condition. Indicators are used for the purpose of monitoring, for example, land-cover changes or assessing the status and damage costs of degradation. Various attempts have been made to inventory desertification in order to assess the magnitude of the problem and to provide a baseline for monitoring (Reining, 1978; Mabbut and Floret, 1980; Mabbutt, 1986; Sharma, 1998). However, only few of the proposed desertification indicators seem to be specific to dryland degradation. They appear to be difficult to quantify, and none of the indicators is yet able to capture the character of a coupled human-environmental system, or, in reverse, to separate the effect of human activities from biophysical or geochemical effects such as climatic oscillations (Reynolds, 2001; Diouf and Lambin, 2001).

Moreover, prevailing land quality indicators hardly capture the most common type of land change taking place in semi-arid regions, which is land-cover modification rather than conversion. Therefore, desertification indicators need to be improved and further developed. Clearly, 'the lack of any readily measured, objective indicators applicable at a regional scale has inhibited progress' (Prince, 2002). There is a tendency now to search for a limited number of objective, quantitative key indicators (comparable to, for example, the annual rate of deforestation), which should better capture globally valid interactions between core variables involved in desertification processes, and be comparable to those in use in the economic and social sciences. There is also a need to adopt a different approach than that of the GLASOD database, which means the hitherto focus on the physical, chemical and biological processes of soil degradation. In this study, indicators are presented which originate from a wide array of case studies of desertification. They are used to hold apart slow *versus* fast rates of change in indicators, which describe the process of desertification.

*Pathways*

Despite views to the contrary, desertification is not a homogenous process throughout all drylands. The succession of causes and events leading to desertification vary substantially between regions. In fact, what could be called pathways of desertification are nearly as diverse as the histories, cultures, and ecosystems of the major dryland areas in the world. Understanding these pathways is crucial for designing appropriate policy interventions. Three sets of factors are seen here to lead to these regional variations – see Figure 1.1: first, the environmental and land use history of each region, which defines the initial conditions for each subsequent round of land use change; second, the particular combination of causes, triggering and driving land use change; and third, the feedback structure, i.e., the social and ecological responses to changes of dryland cover affecting the rates of change. Rather than treating dryland degradation as a globally homogenous process, the baseline of this study is as follows. Within each region, there is a great diversity of situations and several variants to dominant pathways, but in general, the pathways represent frequent successions of events that have led to desertification. Processes of land use change in drylands have multiple causes, originate from a variety of organizational levels, and stem from different time periods. In addition, they include amplifying mechanisms, which lead to an acceleration of change, and attenuating mechanisms, which dampen the human impact on the natural environment (Lambin, Geist and Lepers, 2003).

Opinions are still divided though whether desertification constitutes a phenomenon of irreducible complexity, or whether some generalizations can be gained. As for the latter, there has been much debate on whether desertification necessarily implies a uni-directional or irreversible trajectory of land change. The issue of pathways also relates to insights from the theory of non-equilibrium range ecology: it is hypothesized that in arid regions with high rainfall variability, the ecology is mainly determined by climatic and not biotic factors such as animal grazing (Behnke, Scoones and Kerven, 1993; Walker, 1993). The argument carried even further could also imply that event-driven changes operating at different spatial and temporal scales are distinct from long-term and potentially irreversible trends of (uniform, unidirectional) desertification. In this study, a framework for pathway analysis is adopted from deforestation research that combines both quantitative and qualitative (or narrative) approaches (Lambin and Geist, 2003). Namely, these are causative mechanisms driving desertification, initial conditions which set a pattern that contemporary desertification has followed, built upon, or broken, and built-in system properties such as feedbacks or feedback loops (that amplify or attenuate the impact of the causes of desertification), control points or switch and choke points (at which sudden, abrupt and irreversible shifts from one land use into another occur), and thresholds (steering fundamental changes in system behaviour) (Steffen, Sanderson, Tyson, Jäger, Matson, Moore, Oldfield, Richardson, Schellnhuber, Turner, and Wasson, 2004).

## How a New Approach can Fill the Gap

In addition to chronicling the struggle to arrive at an agreed upon understanding of the causes and progression of desertification, the literature is rich in local-scale case studies investigating the causes and progression of dryland change in specific localities. The aim with this book is to generate from these local-scale case studies a general understanding of the proximate causes and underlying driving forces as well as of the progression of desertification, including cross-scalar interactions of causative factors and addressing system properties of the process such as feedbacks, while preserving the descriptive richness of these studies. Some attempts had already been carried out to produce a synthesis of cases studies of desertification, but they were not done in a systematic manner. Only few standardized data protocols were used to run a limited number of original case studies, but no standardized framework has so far been applied for a meta-analytical comparison of a large array of local-scale case studies. The standardized framework used here combines the analysis of causes with an exploration of the social and ecological responses (or feedbacks, affecting the progression of desertification) as well as the initial conditions that are assumed to fundamentally shape the typical pathways of dryland change.

The new approach adopted in this book is that 'multiplicity' is the most common theme as reported in empirically supported narratives of the impact on drylands of actors and activities: multiple agents, multiple uses of land, multiple responses to social, climatic and ecological changes, multiple spatial and temporal scales in the causes of and responses to desertification, multiple connections in social and geographic space, and multiple ties between people and land in drylands (Rindfuss, Walsh, Mishra, Fox and Dolcemascolo, 2003). The theoretical framework that best accounts for this complexity is seen to be system dynamics, with special emphasis on the history of the system (initial conditions), heterogeneity between actors, hierarchical levels of organization, non-linear dynamics caused by feedback mechanisms, and system adaptation (Lambin, Geist and Lepers, 2003). The most important observation is that complexity due to multiplicity is not irreducible, but associated with a limited number of recurrent pathways of desertification, which makes the problem tractable (Stafford Smith and Reynolds, 2002).

The approach of multiplicity is best exemplified in terms of some general insights into the multiple causes of land change (Ojima, Galvin and Turner, 1994; Turner, Skole, Sanderson, Fischer, Fresco and Leemans, 1995 ; Lambin, Baulies, Bockstael, Fischer, Krug, Leemans, Moran, Rindfuss, Sato, Skole, Turner and Vogel, 2001; Lambin, Turner, Geist, Agbola, Angelsen, Bruce, Coomes, Dirzo, Fischer, Folke, George, Homewood, Imbernon, Leemans, Li, Moran, Mortimore, Ramakrishnan, Richards, Skånes, Steffen, Stone, Svedin, Veldkamp, Vogel and Xu, 2001; Lambin, Geist and Lepers, 2003). Land use change in drylands and elsewhere is always caused by multiple interacting factors originating from different levels of organization of the coupled human-environment systems. The mix of driving forces of land use change varies in time and space, according to specific human-environment conditions. Driving forces underpin the proximate causes of land change, which are the immediate land use activities or human actions impacting directly on land cover. Driving forces can be slow or fast variables, and biophysical drivers may be as important as human drivers. The latter define the natural capacity or predisposing conditions for land use change. The set of abiotic and biotic factors that determine this natural capacity varies among localities and regions. Trigger events, whether these are biophysical (such as a drought) or socio-economic (such as an economic crisis or war), also drive land use changes. Changes are generally driven by a combination of factors that work gradually and factors that happen intermittently.

In the following, proximate causes are seen to be mainly human activities or immediate actions at the local level, such as grazing or irrigated cropland expansion, that originate from intended land use and directly impact dryland cover. Climate-driven increased aridity can also be considered as a proximate factor or direct driver of ecosystem change (Mooney, Cropper and Reid, 2003), because of the reduced climatic potential to sustain natural vegetation and crops beyond some critical threshold (Nicholson, 2002). Underlying driving forces are mainly fundamental social processes, such as human population dynamics or agricultural policies, that underpin the proximate causes and either operate at the local level or have an indirect impact from the national or global level. Various meteorological factors or aspects of the climate leading to increased aridity are to be considered, too. For qualification as an underlying driving force or indirect driver of ecosystem change, it has to include any aspect which reduces the amount or regularity in moisture supply, temperature and rainfall regime (Nicholson, 2002). Mediating factors may shape or modify the interplay between the underlying and proximate causes. They include factors such as gender, access to resources, wealth status, or ethnic affiliation. In addition, causal factors exert impact at differing organization levels and need to be studied across interacting scales (Lambin, Geist, and Lepers, 2003). Feedbacks relate to the social and ecological responses to land-cover changes affecting the rates or progression of change. They can amplify or dampen land use change. Typical pathways of dryland change were identified, i.e., particular chains of events and sequences of causes and effects resulting in specific outcomes of desertification such as decreases in vegetation cover, exposure of bare, rocky ground cover, increases in sand cover, or salinization. These pathways are seen to be made up of initial conditions, causes and feedbacks, with the environmental and land use history of each region defining the initial conditions for each subsequent round of land use and ecosystem change (Lambin, Geist, and Lepers, 2003).

The approach used in this study was already applied in deforestation studies (Lambin and Geist, 2003) and has been more or less developed in parallel to the so-called Dahlem Desertification Paradigm (DDP) (Stafford Smith and Reynolds, 2002). The framework of analysis here does not necessarily follow all of the DDP assertions, but fits into the DDP overall thinking. The main points of the paradigm are summarized by Reynolds, Stafford Smith and Lambin (2003, p. 7) as follows.

> [First] socio-ecological systems in drylands of the world are not static; [second] while change is inevitable, there does exist a constrained set of ways in which these socio-

ecological systems function at different, interlinked scales, thereby allowing us to understand and manage them; [third] an integrated approach, which simultaneously considers both biophysical and socio-economic attributes in these systems, is absolutely essential to understand land degradation; [fourth] the biophysical and socio-economic attributes that govern or cause land degradation in any particular region are invariably 'slow' (e.g., soil nutrients) relative to those that are of immediate concern to human welfare (e.g., crop yields, the 'fast' variables). It is necessary to distinguish these in order to identify the causes of land degradation from its effects; and [fifth] restoring degraded socio-ecological systems to a sustainable state requires outside intervention.

While DDP is conceived to drive several networks of new, original case studies of desertification (www.biology.duke.edu/aridnet), the meta-analytical approach used here can inform the design of future case studies. In doing so, it can also contribute to, or better inform programmatic efforts such as the Land Degradation Assessment in Drylands (LADA) project which responds to the need to strengthen support to combat land degradation. Nachtergaele (2002, p. 15) describes the goals of LADA (www.fao.org/ag/agl/agll/lada) as follows.

LADA aims to generate up-to-date ecological, social, and economic and technical information, including a combination of traditional knowledge and modern science, to guide integrated and cross-sectoral management planning in drylands. The principle objective ... is foremost to develop and implement strategies, tools and methods to assess and quantify the nature, extent, severity and impacts of dryland degradation on ecosystems, watersheds and river basins, and carbon storage in drylands at a range of spatial and temporal scales. The project will also build national, regional and global assessment capacities to enable the design and planning of interventions to mitigate land degradation and establish sustainable land use and management practices.

With an approach as adopted in this book – i.e., analyzing the frequency of proximate causes and underlying driving forces of desertification, including their interactions and system properties (such as feedbacks upon land use), as well as the frequency of occurrence of any information on the progression of desertification over time, as reported in 132 subnational case studies – an existing gap of desertification research is about to be filled. It is shown that desertification is driven by a limited suit of recurrent core variables, of which the most prominent are, at the underlying level, climatic factors, economic factors, institutions, national policies, population growth, and remote influences, driving, at the proximate level, cropland expansion – at the expense of grazing land and natural grassland, thus leading to overstocking – and infrastructure extension. There are identifiable regional patterns of causal factor synergies, which, in combination with feedback mechanisms and regional land use and environmental histories, make up specific pathways of land change per region and time period. The progression of desertification is linked to these pathways and found to be dominated by clusters of slow and, to a lesser degree, fast variables of land change. Rather than describing an 'alarmist' process at the global scale, slow or gradual progression in contemporary hot spots of change is much more typical for desertification. Understanding these pathways of dryland change is crucial for designing appropriate policy interventions. To achieve a sustainable management of dryland ecosystems, interventions have to be fine-tuned to the region-specific dynamical patterns associated with desertification (Geist and Lambin, 2004), and trade-offs need to be considered between what is to be developed and what needs to be sustained (Lambin, Geist and Lepers, 2003; Geist, Lambin, Tomich and Palm, 2005).

Chapter 2

# Research Design

**Introduction**

Cases studies of dryland degradation for 132 sites under human uses were analysed to get a better understanding of desertification indicators used, to determine whether there are general patterns of proximate causes and underlying driving forces of desertification, to identify regional pathways of dryland change, and to estimate rates of dryland change. An overview is provided on the research design used, explaining the general methodology, data selection procedure, and data analysis, giving data statistics, and discussing data bias.

**General Methodology**

Meta-analyses of case studies on global environmental change issues have been increasingly recognized as an important analytical framework for comparative research that aims to draw inferences on common issues with different but allied empirical backgrounds (Matarazzo and Nijkamp, 1997). For example, Parmesan and Yohe (2003) as well as Root, Price, Hall, Schneider, Rosenzweig and Pounds (2003) evaluated results from thousands of cases linking climate change and population ecology to better understand the assumed relevance of an ecological footprint of climate change. In an earlier study, causes of tropical deforestation were systematized using 152 case studies (Geist and Lambin, 2002), and, further, to identify trajectories of forest cover change (Lambin and Geist, 2003).

The purpose of meta-analytical research is to combine findings from separate but largely similar studies, here done on the subject of desertification. Studies were selected that allowed for the application of a variety of analytical techniques such as structured literature review on the initial conditions of desertification and formal statistical approaches to the study of causative mechanisms behind desertification. The intent of the study is to use a large body of local-scale case studies investigating dryland degradation in specific localities, generating a general understanding of the causes, rates, and pathways of global desertification, while preserving the descriptive richness of these studies. From a broad set of more than 100 site-specific cases, it was attempted to combine, compare, select and seek out common as well as distinct elements, relevant results and cumulative properties of desertification. Strictly selected case studies of desertification were used that were published in international, peer-reviewed journals, that allow for both a structured literature review (e.g., to gain narratives) and the application of formal statistical procedures (e.g., to estimate the frequency of occurrence of causative factors). Thus, the methodological approach moves beyond several other attempts which had been done in the past, either through a non-systematic compilation of national-scale case studies required from governments of states that had been affected by desertification (e.g., Mabbutt and Floret, 1980), or through compiling all available information of varying quality from various published sources, mainly in the form of narrative literature reviews (e.g., Le Houérou, 2002; Dregne, 2002). In doing so, I adhere to the notion as put forward by Stern, Young and Druckman (1992, p. 92) on the human causes of global environmental change in general.

> The task is relatively simple in the sense that the initial accounts need not have great precision. For social scientific work to begin, it will be sufficient to know whether a particular human activity contributes on the order of 20 percent, 2 percent, or 0.2 percent of humanity's total contribution to a global change. Such knowledge will allow ... to set worthwhile research priorities until more precision is available.

Methodologically, a middle way is taken between (qualitative, narrative, descriptive) case study and (quantitative) variable-oriented research (Ragin, 1989). The vision of such a middle path is to combine the strengths of within-case analysis (such as coupled environmental and land use narratives) and cross-case analysis (such as the identification of causes and pathways of desertification). According to Ragin (2003), the approach could be called 'configurational comparative research', which has the following characteristics:

- proximate goal: comparative research uses substantively defined categories of cases (usually five to 50, or more; here: 132 cases), with the goal of making sense of individual cases as well as clusters of similar cases in the light of knowledge of cross-case patterns, and vice versa;
- N of cases: to make strategic comparisons, comparative researchers use diverse cases (here: from various dryland regions of the world having differing time horizons); at the same time, they establish case homogeneity because the cases should all be instances of, or candidates for the same outcome (i.e., desertification); when delimiting the set of relevant cases, one thus has to balance conflicting pressures (to be discussed later under data bias);
- role of theory: existing theory (of desertification as well as of land change in total) is rarely well-formulated enough to provide explicit hypotheses to be tested; therefore, the primary theoretical objective in comparative research is not theory testing, but elaboration, refinement, concept formation, and thus contributing to theory development;
- conception of outcomes: comparative research starts by intentionally selecting cases that do not differ greatly from each other with respect to the outcome that is being investigated, i.e., 'positive cases' only; the constitution and analysis of these positive cases is usually a prerequisite for the specification of relevant negative cases;

- understanding of causation: comparative research looks at causation in terms of multiple pathways, with positive cases often easily classified according to the general path which each travelled to reach the outcome; each path, in turn, involved a different combination of relevant causal conditions and other factors such as system properties and pre-disposing initial conditions; and
- within versus cross-case analysis: comparative research focuses on configurations of causally relevant characteristics of cases, with the goal of determining how relevant aspects fit together; cross-case analysis is used to strengthen and deepen within-case analysis, and vice-versa; to the extent possible, comparative research tries to balance both analyses.

From six theoretically possible configuration research strategies (Ragin, 2003), the research design used here could be characterised as 'multiple-conjunctural causation', with positive cases only. This means that the approach aims at identifying the different combination of causal conditions linked to a single outcome (of dryland degradation), which is desertification.

**Data Selection**

In total, 132 cases of desertification were selected (representing 132 sites), which were taken from 54 articles published in 28 journals covered by the citation index of the Institute for Scientific Information in the period 1992-2002 – see Table 2.1, and 'case studies' in the bibliography section. Entering 'desertification' as search word, 620 articles were identified in March 2002. A strict selection of the studies was operated. Each journal abstract was searched for indication of the following criteria:

- sites under human uses or subject to human impacts (no undisturbed 'wild places' or studies of desert formation without reference to land use);
- studies including in-depth field investigations;
- consideration of clearly named factors as (potential) causes of desertification, including basic features of the socio-economic setting and the natural resource endowment;
- investigation method based on both quantitative data and consideration of clearly named biophysical (or geochemical) and/or socio-economic indicators, processes, or narratives;
- quantification of the rates of land change; and
- absence of obvious disciplinary bias.

It was assumed that each study revealed information on actual causes, rates, and other factors making up the pathways of desertification in the study area. Therefore, the comparative analysis of cases evaluates which causal patterns, pathways, rates, and indicators associated with desertification are most often found in different regions of the world, and in the course of time.

**Table 2.1    Data Statistics (N = 132 sites)**

| Journal[a] | [b] | [c] | Volume: Pages[c] |
|---|---|---|---|
| Agr Ecosyst Environ | 1 | 4 | 69(1):55-67[4] |
| Am Midl Nat | 1 | 1 | 144(2):273-85[1] |
| Ambio | 5 | 15 | 20(8):372-83[1]; 21(4):303-7[3]; 22(6):395-403[2]; 29(8):468-76[8]; 30(6):376-80[1] |
| Ann Arid Zone | 1 | 6 | 39(3):285-304[6] |
| Appl Geogr | 2 | 2 | 16(3):225-42[1]; 20(4):349-67[1] |
| Arid Soil Res Rehab | 1 | 5 | 12:95-122[5] |
| Biodivers Conserv | 1 | 3 | 8:1479-98[3] |
| Bodenkultur | 1 | 2 | 52(1):37-44[2] |
| Catena | 3 | 7 | 39(3):147-67[1]; 40(1):19-35[3], 51-68[3] |
| Clim Res | 3 | 8 | 11(1):51-63[5]; 17(2):217-28[1], 195-208[2] |
| Earth Surf Process Landf | 1 | 2 | 25(11):1201-20[2] |
| Environ Conserv | 2 | 2 | 19(2):145-52[1]; 27(2):208-15[1] |
| Environ Geol | 4 | 9 | 40(7):884-90[1]; 41(1/2):229-38[4]; 1(3/4):314-20[1]; 41(7):806-15[3] |
| Environ Monit Assess | 1 | 1 | 48(2):139-56[1] |
| Geoforum | 1 | 5 | 24(4):397-409[5] |
| Global Ecol Biogeogr | 2 | 5 | 2(1):16-25[3]; 8(3/4):243-56[2] |
| Global Environ Change | 1 | 6 | 11(4):271-82[6] |
| Holocene | 1 | 1 | 12(1):107-12[1] |
| Hum Ecol | 1 | 1 | 27(2):267-96[1] |
| Hydrogeol J | 1 | 1 | 9(2):202-7[1] |
| Int J Remote Sensing | 3 | 6 | 16(4):651-72[1]; 21(13/14):2645-63[1]; 22(6):1005-27[4] |
| J Arid Environ | 11 | 29 | 36(2):367-84 & 41(4):463-77[3]; 38(3):397-409[5]; 39(2):143-53[2], 165-78[2], 191-207[1]; 39(4):623-29[6]; 43(2):121-31[6]; 47(1): 47-75[2]; 47(2):123-44[1]; 49(2):413-27[1] |
| J Develop Stud | 1 | 3 | 37(3):99-133[3] |
| J Environ Manage | 1 | 2 | 49(1):125-55[2] |
| Proc Natl Acad Sci USA | 1 | 1 | 94(18):9729-33[1] |
| Remote Sens Environ | 1 | 1 | 74 (1):26-44[1] |
| Soc Natur Resour | 1 | 2 | 12(7):643-57[2] |
| Sov Soil Sci | 1 | 2 | 23(7):22-31[2] |

[a] Abbreviations according to the Institute for Scientific Information (ISI) (www.isinet.com).
[b] Number of articles.
[c] Number of sites.

**Figure 2.1  Location of Case Studies***

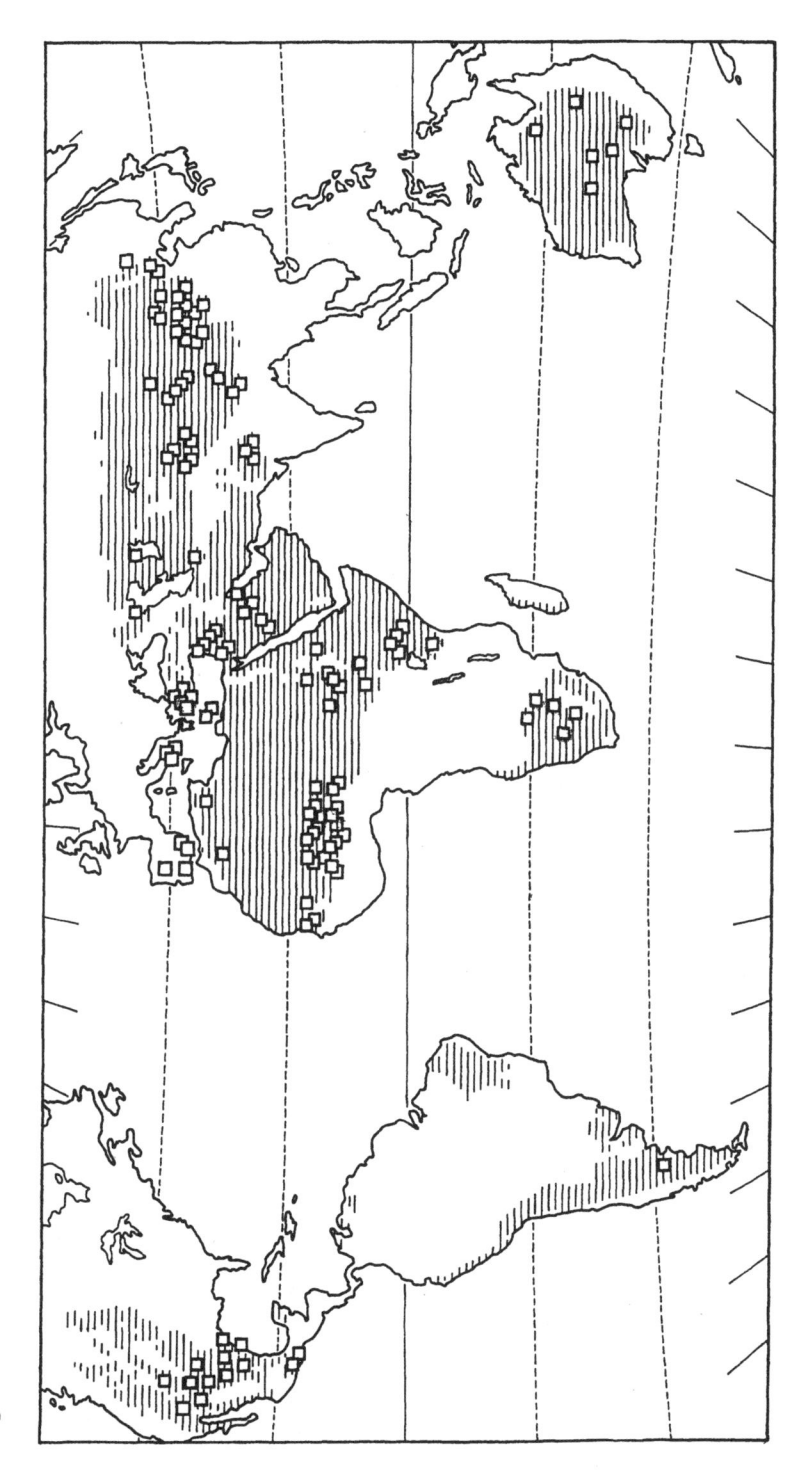

* Hatched areas indicate dryland ecosystems such as dry grassland, shrubland, savanna, steppe and (sub- or semi-) desert.

## Data Analysis and Statistics

Most of the information compiled from the 132 cases was coded using the Statistical Package for the Social Sciences (SPSS), version 10 (Norusis, 2000). Mainly descriptive statistics was used to estimate the frequency, and thus relative importance, of various factors in the cases for purposes of cross-case analysis. Within-case analysis was done qualitatively, mainly in the form of narratives, or following a structured guideline used for a semi-quantitative literature review. For details of the data analysis such as formation and clustering of variables, see the subsections following the discussion of data bias (i.e., initial conditions, causes, etc.).

Only cases from the subnational scale were considered. Thus, no countrywide analysis was used. Rather, the comparative analysis was limited to desertification processes reported in case study areas which range from a site as small as one hectar – i.e., a transect of 25 to 30 meters length situated on Otero Mesa of the Chihuahan Desert in the US-American Southwest (Ludwig, Muldavin and Blanche, 2000) – to multi-province areas as large as 1,065 million hectars – i.e., the Songnen Plain of Songliao Basin situated on the Inner Mongolian Plateau of Northeast China (Lin and Tang, 2002). From cases in which the exact size of the study area was given (n=70, or 53%), the mean study area was calculated to be 28,000,000 hectars (median = 570,400 ha, modus = 57,600 ha).

The cases span time periods from 1700 to 2000 (n= 132, or 100%). The mean period studied stretches from 1915-94 over almost eighty years (median and mode = 35), and the most frequently covered time period spans 1700-1995.

Results were broken down by broad geographic dryland regions, the categorisation of which follows Reynolds and Stafford Smith (2002), based upon UNEP (1997) – see also Tables 2.6 and 2.7. The regional breakdown was used mainly to give a preliminary structure to initial conditions and causative factors (though, in the end, it turned out that the pathways of desertification do not necessarily follow the regional breakdown as adopted). The regional categorisation also helps to avoid duplication of narratives or specific examples of factors or factor combinations. Examples – such as specific agricultural policies underlying the expansion of cropland at the expense of rangelands, or specific mediating factors such as poverty *versus* wealth conditions – are compiled in the section on initial conditions, and can easily traced back from other parts of the analysis if referred to, for instance, as 'cases from the Sudano-Sahelian zone in West Africa'.

*Location of case studies* As can be taken from Figure 2.1, dryland cases from Asia (n=51) stem mainly from the Central Asian desert and steppe region, with northern China, namely the Ordos Plateau, holding the major part of them, and including cases from Russia, Turkmenistan, Kazakstan, and Usbekistan. Further considered are cases from the East Mediterranean steppe zone, the Arabian Peninsula, and the Thar Desert in India. Most of the Asian cases originate from the central Asian desert steppe zone, including cases from various, mainly topographically defined ecosystems such as plateaus, basins, plains, uplands, and mountains. The major part of these cases stems from northern and western China, i.e., the Ordos Plateau – including the Mu Us Region (desert or 'meadow'), parts of the Yellow River Valley, and parts of adjacent landscapes such as the Loess Region, and the Helan and Liupan mountains (Sheehy, 1992; Runnström, 2000; Ho, 2001; Zhou, Dodson, Head, Li, Hou, Lu, Donahue and Jull, 2002; Lin and Tang, 2002) – the Inner Mongolian Plateau – including the Keerqin (Korqin) Desert (or 'meadow') and the Songnen Plain of the Songliao Basin (Sheehy, 1992; Lin and Tang; 2002) – the Qinghai-Xizang (Tibet) Plateau (Holzner and Kriechbaum, 2001; Liu and Zhao, 2001; Wang, Qian, Cheng and Lai, 2001), the Kashi Plain and various reaches of the Keriya and Tarim Rivers in the Tarim River Basin of the Taklimakan Desert (Yang, 2001; Feng, Endo and Cheng, 2001; Lin, Tang and Han, 2001), and the so-called Hexi Corridor with various reaches of the Hei River in the Hei River Basin Region (Genxu and Guodong, 1999). Cases outside of China are sites in plains such as the Circum Aral Sea Region (Saiko and Zonn, 2000), the Caspian Sea Region and Plain (Rozanov, 1991), the Turkmenia Plain around the Zone of the Karakum canal (Rozanov, 1991), northwestern Jordan in the East Mediterranean steppe zone (Khresat, Rawajfih and Mohammad, 1998), the Saudi Desert of Kuwait and Saudi Arabia (Brown and Schoknecht, 2001; Weiss, Marsh and Pfirman, 2001), and the Thar Desert of NW-India (Tsunekawa, Kar, Yanai, Tanaka and Miyazaki, 1997; Ram, Tsunekawa, Sahad and Miyazaki, 1999).

Dryland cases from Africa (n =42) stem mainly from the Sahelian and Sudano-Sahelian zones of West Africa (Mauritania, Senegal, Mali, Burkina Faso, Niger, and Nigeria) and the Republic of Sudan. They do further originate from the western Mediterranean Basin in North Africa (Algeria, Morocco), the East African grassland zone (Kenya, Tanzania), and the Kalahari sandveld in southern Africa (Botswana, South Africa). Most of the West African cases stem from three areas, namely, the alluvial inland plain (or delta) of the Niger River in Central Mali (Benjaminsen, 1993; Turner, 1999a; Ringrose and Matheson, 1992), the central and southwestern parts of Niger (Ringrose and Matheson, 1992; Turner, 1999b), and the Oudalan zone of northern Burkina Faso (Rasmussen, Fog and Madsen, 2001). The other cases are from the western Sahel in southern Mauritania and northwestern Senegal (Venema, Schiller, Adamowski and Thizy, 1997; Gonzalez, 2001), and the Kano Close-Settled Zone in northern Nigeria (Mortimore, Harris and Turner, 1999). Similar to the zoning of landscapes in West Africa are cases from the sub-humid to hyperarid parts of the Republic of Sudan (Helldén, 1991; Olsson, 1993; Ayoub, 1998). The North African cases cover the Middle and High Atlas Region of Morocco, and the Aures area of northern Algeria (Gauquelin, Bertaudière, Montes, Badri and Asmode, 1999). The East African cases stem from the Turkana Steppe and Masailand in northern Kenya and Tanzania, respectively (Mwalyosi, 1992; Keya, 1998). Cases from southern Africa are all settled in various parts of the Kalahari such as the Okavango Delta, central and southern Kalahari sites (Ringrose, Vanderpost and Matheson, 1996; Palmer and Rooyen, 1998; Dube and Pickup, 2001).

Dryland cases from Europe (n=13) stem exclusively from countries which are situated or hold territorial shares in the Mediterranean Basin. From the west to east, these are Portugal – Bragança Province in the north (Vandekerckhove, Poesen,

Wijdenes, Nachtergaele, Kosmas, Roxo and Figueiredo, 2000) and the Alentejo Region further in the south (Seixas, 2000); Spain – Ebro Basin and Andalucia in its central part (Gauquelin, Bertaudière, Montes, Badri and Asmode, 1999) and Almeria Province (Vanderkerckhove, Poesen, Wijdenes, Nachtergaele, Kosmas, Roxo and Figueiredo, 2000) as well as Zarcilla de Ramos (Wijdenes, Poesen, Vandekerckhove and Ghesquiere, 2000) in the southeast; southern Italy – various parts of the Agri Valley in the Basilicata Region (Basso, Bove, Dumontet, Ferrara, Pisante, Quaranta and Taberner, 2000); and Greece, with the Asteroussia and Psiloriti Mountains on the island of Crete (Hill, Hostert, Tsiourlis, Kasapidis, Udelhoven and Diemer, 1998) and various sites on the island of Lesvos in the northeast of the Aegean Sea (Kosmas, Gerontidis and Marathianou, 2000).

Dryland cases from Australia (n=6) stem from almost all major areas on the continent, namely the Northern Territory (in Central and North Australia), New South Wales in the southeastern part, Queensland in the Northeast, South and Western Australia (Bastin, Pickup and Pearce, 1995; Pickup, 1998).

Dryland cases from North America (n=6) stem from various locations in the Great Basin Region of the US-American Southwest, i.e., the Chihuahua Desert (Rango, Chopping, Ritchie, Havstad, Kustas and Schmugge, 2000; Ludwig, Muldavin and Blanche, 2000; Brown, Valone and Curtin, 1997), the Sonora Desert (Fredrickson, Havstad, Estell and Hyder, 1998), the Mojave Desert (Okin, Murray and Schlesinger, 2001), and the Colorado Plateau (Mouat, Lancaster, Wade, Wickham, Fox, Kepner and Ball, 1997), situated in the states of southeastern California, Arizona, New Mexico, and Utah.

Dryland cases from Latin America (n=14) stem from various sites in two countries, i.e., Mexico and Argentina. In Mexico, these are small drainage basins of the Rio Zapotitlán, associated with the Tehuacán Valley, in the south-central part of the country (McAuliffe, Sundt, Valiente-Banuet, Casas and Viveros, 2001), and the Chihuahuan and Sonoran Deserts, mountains and high uplands in Durango State, Nuevo Leon State, and the coastal lowlands of the Gulf of Mexico (Coahuila State), called the Tamaulipan matorral, all of them situated in northern Mexico (Manzano, Návar, Pando-Moreno and Martinez, 2000). All cases in Argentina stem from the Patagonian rangelands located in the southern part of the country, namely the Chubut, Neuquén, Rio Negro, Santa Cruz and Tierra del Fuego Provinces (Valle, Elissalde, Gagliardini and Milovich, 1998; Aagesen, 2000).

*Frequency of occurrence of land uses* Due to a strategic decision for case study selection not to consider 'undisturbed' sites, 'wild places' or studies of desert formation without any reference to human uses, all land uses analysed (N=132, or 100%) fall under the broad category of 'productional activities' carried out in 'managed ecosystems', with some land uses also settled in 'unmanaged' or natural ecosystems, or being settlement, infrastructure and related uses (Pieri, Dumanksi, Hamblin and Young, 1995) – see Tables 2.2 and 2.3.

The cases report that livestock production was slightly more widespread than crop production, but with considerable overlaps between the two modes of land uses – see Table 2.2. Grazing is a 'robust' dryland use, and regional distinctions emerge only when it comes to specific modes of pastoralism. Extensive grazing activities outweigh nomadic pastoralism by factor two. They encompass both sedentary livestock raising and variants of transhumance. The first means that farmers either keep their livestock on grazingland situated on or next to their farm, or drive their herds to farther remote rangelands in the morning, and return to the village, when the sun sets. The latter means that herds are moved, for example, from valley locations to mountain meadows all over the hot and dry season, while they are returned to the farmsteads in the valley grounds for the wetter and colder season. Nomadic grazing was found to be an important land use in about two fifths of the cases only, featuring half of the African and Asian cases each (it was not reported from other regions). However, pure nomadic land use was found to be rare, and the combination of nomadic and extensive grazing had been much more common. In two fifths of the cases, extensive grazing constitutes the sole livestock-based land use. Consequently, combinations with other modes of grazing were important in the remaining cases, with the nomadic-extensive grazing combination being the most prevalent combination. In whatever combination, expensive grazing was reportedly a robust land use. It had been characteristic for it that traditional nomadic grazing animals, well adapted to dryland ecosystem conditions (such as yak or camel), were replaced by less well adapted species, in particular cattle, sheep, and goats. It was further reported in the cases that intensive livestock production as well as feral grazing (i.e., by wild animals) matter far less, and, if so, that they were clustered on the American continent.

Cropping activities were found to be slightly less widespread than livestock-related activities – see Table 2.2. Crop production was reportedly important in the cases associated with desertification – with the exception, though, of Australia (where virtually no cropping existed except on an intermittent basis in favoured locations on the semi-arid fringe), the arid and semi-arid lands of the US American Southwest (where few suitable farming land is available), and Patagonia (where it was mentioned, but not described in further detail due to the overriding importance of sheep ranching). Permanent cultivation of annual crops, both for subsistence and commercial uses, reportedly dominated the cropland pattern, and could thus be labelled robust (n=83, or 63%). In contrast, not robust were the cultivation of perennials and the various modes of irrigation agriculture.

Among other land uses are forestry production (Table 2.2.), mainly from dry natural forests, which reportedly occurred in all regions (n=36, or 27%), with the exception of Australia, and the collection of plant and/or animal products from natural, mainly shrub and woodland ecosystems (Table 2.3) (n=47, or 36%).

Settlement and related infrastructure activities were found to be important land uses in three fifths of all cases (n=75, or 57%) – see Table 2.3. These non-agrarian land uses, which were mainly associated with human settlements and related agricultural irrigation infrastructures, are important but vary by regions. They feature more of the Australian cases, while mineral extraction appears to be a feature of American cases only. Residential uses of land (n=50, or 38%) and related infrastructures such as roads, water installation, electricity lines and sanitary facilities (n=46, or 35%) rank especially high.

**Table 2.2   Regional Differences of Land Use Systems – Managed Ecosystems[a]**

| | All Cases (N=132) | | Asia (n=51) | | Africa (n=42) | | Europe (n=13) | | Australia (n=6) | | N-America (n=6) | | L-America (n=14) | |
|---|---|---|---|---|---|---|---|---|---|---|---|---|---|---|
| | abs | rel | abs | rel | abs | rel | abs | rel | abs | rel | abs | rel | abs | rel |
| Forestry production | 36 | 27 | 18 | 35 | 8 | 19 | 2 | 15 | 0 | | 2 | 33 | 6 | 43 |
| … from natural forests | 35 | 27 | 18 | 35 | 8 | 19 | 2 | 15 | 0 | | 2 | 33 | 5 | 36 |
| … from planted forests | 8 | 6 | 4 | 8 | 2 | 5 | 0 | | 0 | | 1 | 17 | 1 | 7 |
| Livestock, grazing | 117 | 87 | 44 | 86 | 41 | 98 | 9 | 69 | 6 | 100 | 5 | 83 | 12 | 86 |
| … nomadic grazing | 51 | 39 | 28 | 55 | 23 | 55 | 0 | | 0 | | 0 | | 0 | |
| … extensive grazing | 92 | 70 | 31 | 61 | 34 | 81 | 9 | 69 | 6 | 100 | 5 | 83 | 7 | 50 |
| … intensive production | 15 | 11 | 0 | | 3 | 7 | 0 | | 0 | | 1 | 17 | 11 | 79 |
| … wild animals' (feral) grazing | 8 | 6 | 0 | | 3 | 7 | 0 | | 5 | 83 | 0 | | 0 | |
| Cropping and fisheries production | 100 | 76 | 41 | 80 | 33 | 79 | 10 | 77 | 5 | 83 | 2 | 33 | 9 | 64 |
| … shifting cultivation | 1 | 1 | 0 | | 1 | 2 | 0 | | 0 | | 0 | | 0 | |
| … annual cropping[b] | 83 | 63 | 29 | 57 | 28 | 67 | 10 | 77 | 5 | 83 | 2 | 33 | 9 | 64 |
| … perennial cropping[b] | 14 | 11 | 2 | 4 | 3 | 7 | 8 | 62 | 0 | | 1 | 17 | 0 | |
| … wetland cropping | 10 | 8 | 5 | 10 | 4 | 10 | 0 | | 0 | | 0 | | 1 | 7 |
| … irrigated cropping (e.g., oasis) | 10 | 8 | 9 | 18 | 1 | 2 | 0 | | 0 | | 0 | | 0 | |
| … fishing | 7 | 5 | 1 | 2 | 6 | 14 | 0 | | 0 | | 0 | | 0 | |

[a]   Multiple counts possible; abs = absolute number; rel = relative percentages; cum = cumulative percentages; percentages relate to the total of all cases for each category (column), relative percentages may not total 100 because of rounding.

[b]   If not specified otherwise, annual and perennial crop production is meant to be carried out under the mode of permanent, sedentary production.

Table 2.3  Regional Differences of Land Use Systems – Infrastructure and Land Use in Natural Ecosystems[a]

| | All Cases (N=132) | | Asia (n=51) | | Africa (n=42) | | Europe (n=13) | | Australia (n=6) | | N-America (n=6) | | L-America (n=14) | |
|---|---|---|---|---|---|---|---|---|---|---|---|---|---|---|
| | abs | rel | abs | rel | abs | rel | abs | rel | abs | rel | abs | rel | abs | rel |
| *Natural ecosystems* | | | | | | | | | | | | | | |
| Wilderness areas, not used | 0 | | 0 | | 0 | | 0 | | 0 | | 0 | | 0 | |
| Conservation areas, partial | 13 | 10 | 1 | 2 | 10 | 24 | 0 | | 0 | | 2 | 33 | 0 | |
| Collection of plant/animal produce | 41 | 31 | 15 | 29 | 25 | 60 | 1 | 8 | 0 | | 0 | | 0 | |
| Total | 47 | 36 | 16 | 31 | 28 | 67 | 1 | 8 | 0 | | 2 | 33 | 0 | |
| *Settlement, infrastructure and related uses* | | | | | | | | | | | | | | |
| Settlement-related infrastructure | 72 | 55 | 28 | 55 | 24 | 57 | 3 | 23 | 6 | 100 | 3 | 50 | 8 | 57 |
| ... unspecified | 1 | 1 | 0 | | 1 | 2 | 0 | | 0 | | 0 | | 0 | |
| ... residential | 50 | 38 | 16 | 31 | 21 | 50 | 3 | 23 | 5 | 83 | 2 | 33 | 3 | 21 |
| ... commercial | 3 | 2 | 3 | 6 | 0 | | 0 | | 0 | | 0 | | 0 | |
| ... industrial, including oil/gas | 19 | 14 | 9 | 18 | 0 | | 0 | | 5 | 83 | 0 | | 5 | 36 |
| ... road, water, electricity, sanitary | 46 | 35 | 21 | 41 | 10 | 24 | 2 | 15 | 6 | 100 | 2 | 33 | 5 | 36 |
| Recreational uses | 10 | 8 | 0 | | 0 | | 0 | | 0 | | 3 | 50 | 7 | 50 |
| Military uses | 4 | 3 | 1 | 2 | 1 | 2 | 0 | | 0 | | 2 | 33 | 0 | |
| Mineral extraction[b] | 12 | 9 | 2 | 4 | 0 | | 0 | | 0 | | 2 | 33 | 8 | 57 |
| Total | 75 | 57 | 29 | 57 | 24 | 57 | 3 | 23 | 6 | 100 | 4 | 67 | 9 | 64 |

[a] Multiple counts possible; abs = absolute number; rel = relative percentages; cum = cumulative percentages; percentages relate to the total of all cases for each category (column); relative percentages may not total 100 because of rounding.
[b] Mining and quarrying, including the cutting of turf/peat.

*Initial Conditions*

Initial conditions of dryland sites were defined by the land use and environmental history of each case. They were seen to set a pattern that desertification has followed, built upon, or broken. Ideally, the environmental history of each case addresses climate, topography, vegetation, soil, water, and species habitat. The land use history of each case addresses key proximate and underlying factors of land use change in both managed and natural ecosystems. It provides a narrative of change for both the livestock and cropping sector. Since a strategic decision was made not to focus on undisturbed 'wild places', or studies of desert formation without any reference to land use, but studying cases of dryland sites under human uses instead, the chapter on initial conditions is concluded with an overview on the frequency of occurrence of major land use categories as described in the cases.

As much as possible, a categorisation was avoided by separate entities such as climate, vegetation, population, economy, etc. Rather, it was decided to write up a coupled human-environmental history which allowed within-case insights as well as cross-case comparisons within and across broad geographic regions such as Asia, Africa, Europe, Australia, North America, and Latin America. Striking examples of region-specific land change are pointed out in boxes. In a further step of data aggregation, differences in the regional histories of both ecological factors shaping the environment and the changing context of institutional and socio-economic factors were brought into a tabulated format – see Tables 6.1 to 6.4. Finally, the regional histories were used as important building blocks – together with causes, rates, and system properties of dryland change – to identify typical pathways of desertification.

*Causes*

A distinction was made between proximate causes and underlying driving forces of desertification, including mediating factors that may shape or modify the interplay between these two broad groups. Proximate causes are analysed according to four broad groups: agricultural activities, infrastructure extension, extractional activities, and increased aridity. Classification and coding of the first three factor groups – which actually constitute land use activities associated with land degradation – follows the categories for Land Quality Indicators (LQI) of the Food and Agriculture Organisation of the United Nations (FAO) (Dumanski and Pieri, 2000) – see Tables 2.2 and 2.3. As for 'increased aridity', the variable was subdivided into two broad categories, i.e., the impact of climatic variability (and accompanying land surface feedbacks to the atmosphere), and those aspects of climate which directly influence surface vegetation such as climate-induced changes in fire regime or the occurrence of prolonged droughts (Nicholson, 2002). Underlying driving forces fall into six broad categories, i.e., demographic, economic, technological, institutional/policy, socio-cultural, and meteorological factors. The broad categories of both proximate and underlying factors were further subdivided into specific activities such as nomadic grazing, annual cropping, agricultural modernization policies, migration, or public attitudes, etc., with third order subdivisions. To allow for cross-case analysis, all subdivisions were identical for all the cases – see Figure 2.2.

**Figure 2.2  Framing the Causes of Desertification**

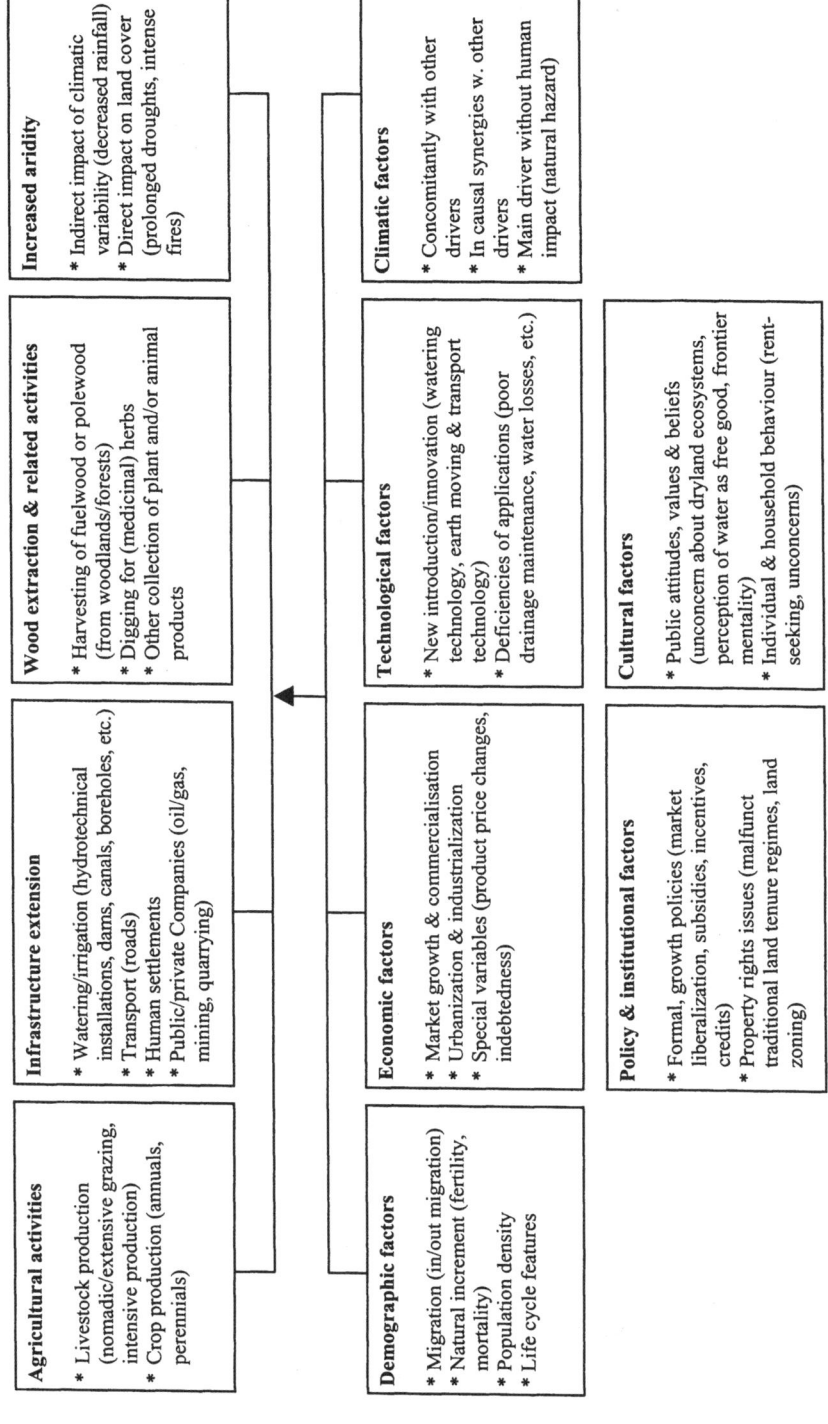

**Agricultural activities**
* Livestock production (nomadic/extensive grazing, intensive production)
* Crop production (annuals, perennials)

**Infrastructure extension**
* Watering/irrigation (hydrotechnical installations, dams, canals, boreholes, etc.)
* Transport (roads)
* Human settlements
* Public/private Companies (oil/gas, mining, quarrying)

**Wood extraction & related activities**
* Harvesting of fuelwood or polewood (from woodlands/forests)
* Digging for (medicinal) herbs
* Other collection of plant and/or animal products

**Increased aridity**
* Indirect impact of climatic variability (decreased rainfall)
* Direct impact on land cover (prolonged droughts, intense fires)

**Demographic factors**
* Migration (in/out migration)
* Natural increment (fertility, mortality)
* Population density
* Life cycle features

**Economic factors**
* Market growth & commercialisation
* Urbanization & industrialization
* Special variables (product price changes, indebtedness)

**Technological factors**
* New introduction/innovation (watering technology, earth moving & transport technology)
* Deficiencies of applications (poor drainage maintenance, water losses, etc.)

**Climatic factors**
* Concomitantly with other drivers
* In causal synergies w. other drivers
* Main driver without human impact (natural hazard)

**Policy & institutional factors**
* Formal, growth policies (market liberalisation, subsidies, incentives, credits)
* Property rights issues (malfunct traditional land tenure regimes, land zoning)

**Cultural factors**
* Public attitudes, values & beliefs (unconcern about dryland ecosystems, perception of water as free good, frontier mentality)
* Individual & household behaviour (rent-seeking, unconcerns)

Causal factors were quantified by determining the most frequent proximate and underlying factors in each case. The major interactions (and feedbacks) between the causative factors were also identified to reveal the systems dynamics that is commonly associated with desertification (explored in more detail in chapter 6 on pathways). As with initial conditions, a breakdown of causes by broad geographical regions was done to show the absolute number as well as the relative percentages of the frequency of causative variables, their modes of causation and interactions, as reported in the case studies. A causative factor was called 'robust' if the factor showed little regional variation across the regions (Asia, Africa, Latin America, North America, Australia, and Europe). Different from the initial condition section, cause analysis was much more grounded in the application of formal, mainly descriptive statistical procedures.

*Syndromes*

For any given human-environment system in drylands, a limited number of causes are essential to predict the general trend in land use (Stafford Smith and Reynolds, 2002). This is the basis, for example, for the syndrome approach, which describes archetypical, dynamic, coevolutionary patterns of human-environment interactions (Petschel-Held, Lüdeke, and Reusswig, 1999). A taxonomy of syndromes links processes of degradation to both changes over time and status of state variables. The approach is applied at the intermediate functional scales that reflect processes taking place from the household level up to the international level. The syndrome approach aims at high levels of generality in the description of mechanisms of environmental degradation, and the typology of syndromes obviously reflects expert opinion based on local case examples. In the following, a syndrome of land change constitutes the particular combination of specific causal conditions, involving both proximate and underlying factors, and rates of change – see chapter 5. On the latter, a difference has been made between slow and fast causative variables of desertification, to be held apart from process rates of change. In hardly any of the cases, however, quantitative data were available, and case study authors only rarely labelled causative factors to operate at a high or low speed. Therefore, we applied a matrix or qualitative categorization of rates of change which is based upon a typology of factors and processes driving land-use change in tropical regions at various speed (Lambin, Geist and Lepers, 2003). It has six broad categories of slow *versus* fast variables linked to specific causal combinations of proximate and underlying factors – see Tables 2.4 and 2.5. The subdivisions of these broad categories or syndromes of land change actually constitute path sections of land change, which are important cornerstones of multiple-conjunctural causation when it comes to identifying the pathways of dryland change or desertification.

**Table 2.4**   **Typology of Syndromes of Land Change – Slow Variables**

| 1<br>Resource scarcity causing a pressure of production on resources | 2<br>Changing opportunities created by markets | 3<br>Outside policy intervention | 4<br>Loss of adaptive capacity, increased vulnerability | 5<br>Changes in social organisation, in resource access, and in attitudes |
|---|---|---|---|---|
| Natural population growth, division of land parcels; | Increase in commercialisation and agroindustrialisation; | Economic development programmes; | Impoverishment (e.g., creeping household debts, no access to credit, lack of alternative income sources; weak buffering capacity); | Changes in institutions governing access to resources (e.g., shift from communal to private rights, tenure, holdings and titles); |
| Domestic life cycles leading to changes in labour availability; | Improvement in accessibility through road construction; | Perverse subsidies, price distortions, fiscal incentives; | | |
| Loss of land productivity on sensitive areas following excessive or inappropriate use; | Changes in market prices for inputs or outputs (e.g., erosion of prices of primary production, unfavourable global or urban-rural terms of trade); | Frontier development (e.g., for geopolitical reasons or to promote interest groups); | Breakdown of informal social security networks; | Growth of urban aspirations; |
| | | | Dependence on external resources or on assistance; | Breakdown of extended family; |
| Failure to restore or to maintain protective works; | | Poor governance and corruption; | | Growth of individualism and materialism; |
| | | Insecurity in land tenure. | Social discrimination (ethnic minorities, women, lower class people). | |
| Heavy surplus extraction away from the land manager. | Off-farm wages and employment opportunities. | | | Lack of public education and poor information flow on the environment. |

*Source*: Lambin, E.F., Geist, H.J., Lepers, E. (2003), 'Dynamics of land-use and land-cover change in tropical regions', *Annual Review of Environment and Resources*, Vol. 28, p. 224.

**Table 2.5    Typology of Syndromes of Land Change – Fast Variables**

| 1<br>Resource scarcity causing a pressure of production on resources | 2<br>Changing opportunities created by markets | 3<br>Outside policy intervention | 4<br>Loss of adaptive capacity, increased vulnerability | 5<br>Changes in social organisation, in resource access, and in attitudes |
|---|---|---|---|---|
| Spontaneous migration, forced population displacement, refugee movements;<br><br>Decrease in land availability due to encroachment by other land uses – e.g., natural reserves ('tragedy of enclosure'). | Capital investments;<br><br>Changes in macroeconomic and trade conditions leading to changes in prices (e.g., energy prices, global financial crisis);<br><br>New technologies for intensification of resource use. | Rapid policy changes (e.g., devaluation);<br><br>Government instability;<br><br>War. | Internal conflicts;<br><br>Illness (e.g., HIV/AIDS);<br><br>Risks associated with natural hazards (e.g., leading to a crop failure, loss of resource or loss of productive capacity). | Loss of entitlements to environmental resources (e.g., expropriation for large-scale agriculture, large dams, forestry projects, tourism and wildlife conservation), leading to an ecological marginalisation of poors. |

*Source*: Lambin, E.F., Geist, H.J., Lepers, E. (2003), 'Dynamics of land-use and land-cover change in tropical regions', *Annual Review of Environment and Resources*, Vol. 28, p. 224.

*Indicators*

Following Reynolds (2001), the indicators used in the case studies – see chapter 7 – were grouped into three broad clusters, showing the ecological, meteorological, and human or socio-economic dimensions of desertification. Each broad cluster was further subdivided into more specific categories such as vegetation change, wind/water erosion, water degradation, and other degradation of terrestrial ecosystem functions, as in the case of ecological indicators. Only direct indicators were considered here, and consequently coded. This means that, for example, gradients in cover and/or rainfall response that may be indirect – or assumed – indicators of bush encroachment and change in pasture species composition were not considered. Indicators may overlap (e.g., soil erosion and increased bare, rocky, eroded ground cover), but since most of them were given different quantitative measures (e.g., tons per hectare eroded, or percentage of ground cover affected), they were coded individually.

Any qualitative as well as quantitative information available from the cases was taken to cluster indicators according to four degrees of dryland degradation, i.e., from slight to extreme. This remained a purely academic exercise meant for the improvement and further development of desertification indicators for monitoring purposes. If cases would have provided more quantified data on indicators and their change, including detailed information on the extent of the case study areas, and the areas within affected by desertification, the proposed matrix could be used to better assess extent and severity of desertification. Since environmental indicators are a phenomenon or statistic so strictly associated with a particular environmental condition that its presence can be taken as indicative of that condition, typically indicated degradation pattern per major geographic region were identified. The major purpose for dealing with indicators in this study, though, was a first preparatory step to hold apart slow *versus* fast rates of change in indicators, i.e., identify process rates which describe the speed at which desertification occurs (different from the pace of change inherent to syndromes of desertification as described above).

*Rates*

A difference has been made between slow and fast rates of change in indicators of desertification to describe the speed at which desertification happens – see chapter 5. Cases were explored in terms of whether they included quantified information on the amount of land change per time period. The classes of land change were taken from the matrix of indicators, with which the section starts, i.e., meteorological changes (mainly, annual rainfall decrease), ecological changes (vegetation change, wind and/or water erosion, water degradation), and changes in socio-economic conditions (such as pastoral and agricultural suitability). Rates of change were calculated for further subdivisions of these still broad categories. Among ecological indicators, for example, subcategories for wind and water erosion were taken to search in the cases for a quantification of the increase in sand cover and the increase in bare, rocky, eroded ground cover over time.

To allow for cross-case comparisons, a standardisation of change rates in terms of annual relative percentage values was attained, comparable to the annual rate of deforestation (expressed in % of forest cover change on a yearly basis). Only in about one fifth of the cases which had quantified data of change, these data were expressed in terms of relative change on a yearly basis as desired, though. For example, Feng, Endo and Chen (2001) measured that the annual increase in the areal extent of sand cover ('sandification') in the Tarim River Basin of the Taklimakan Desert in NW-China had been 0.7% in the upper reaches, 1.0% in the middle reaches, and 1.2% in the lower reaches of Tarim River during 1958-87. Most of the cases showed the amount of land cover change per various land change classes between two (or more) time points only. Therefore, the absolute amount of land change was transformed into relative change on a yearly basis. To gain an idea of what constitutes a high or low pace of land change, the mean annual percentage value of all rates per class of land change was taken to separate fast and slow rates. For example, 'sandification' was measured in 19 different cases (or at 19 sites). The sum of the annual increases in sand cover (expressed in %) was 0.1 + 0.1 + 0.3 + 0.5 + 0.5 + 0.7 + 1.0 + 1.2 + 1.2 + 1.4 + 1.5 + 1.6 + 2.1 + 2.9 + 3.3 + 7.8 + 10.0 + 18.1 + 34.7 = 89. Divided by the number of cases, a mean value of 89 / 19 = 4.7% was calculated. Thus, sandification of the various reaches of Tarim River over a 30-year time period occurred at a comparatively low rate at most sites. Analysis and presentation of data in a tabulated format would also have allowed for temporal within-case comparisons. However, due to the limited amount of case study data it was not further explored in the book. Given the need for uniformity and standardization, such estimations or calculations of the data ('data manipulation') are sometimes a necessary part of comparative, meta-analytical research which has to be made explicit, though (Matarazzo and Nijkamp 1997).

*Pathways*

By pathways of dryland change – see chapter 6 – particular chains of events and sequences of causes and effects were meant, or transitions, that lead to specific land cover changes, resulting here in outcomes of desertification such as decreases in vegetation cover, exposure of bare, rocky ground, increases in sand cover, or salinization. To cope with the diversity of histories, cultures, and ecosystems reflected in various pathways (or trajectories), the analysis was limited to three sets of factors. From an earlier study (Lambin and Geist, 2003), the notion was adopted that these broad sets lead to regional variations of patterns that make up the pathways of desertification: first, the environmental and land-use history of each case and major region, which defines the initial conditions for each subsequent round of land use and ecosystem change; second, the particular combination of causes triggering and driving land-use change; and, third, the feedback structure, that is, the social and ecological responses to land-cover changes affecting rates of change, in combination with thresholds or control points inherent to the desertification process. At the very beginning, data were compiled on initial conditions and related to the different combinations of causal conditions. Then, rates were added as well as system properties of land change (feedbacks, thresholds, control points) to identify typical pathways or trajectories of dryland degradation. The notion of syndromes included indicative rates of typical causal combinations, and process rates were derived from quantitatively given indicators of desertification.

**Data Bias**

The most striking bias to check is whether the cases might represent a weighting bias in the selection of case studies per broad geographic regions. It is obvious from the information above that Asian cases (n=51, or 39%) and African cases (n=42, or 32%) are more frequently represented than other cases, i.e., America (n=20, or 15%), Europe (n=13, or 10%), and Australia (n=6, or 5%). Does this constitute, for example, a bias at the expense of Australia? Ideally, the proportion of cases from a region should be (more or less) equal to the proportion of the total area of desertified land which is located in the region. Given the uncertainties surrounding the extent of desertified land, the total area of hyper-arid to dry subhumid drylands is used here as a proxy indicator – see Tables 2.6 and 2.7.

**Table 2.6   Selected Cases Measured Against the World's Drylands**

|  | All drylands | | Selected cases | |
| --- | --- | --- | --- | --- |
|  | abs.* | rel. | abs. | rel. |
| Asia | 1,949 | 32% | 51 | 39% |
| Africa | 1,959 | 32% | 42 | 32% |
| Latin America | 543 | 9% | 14 | 11% |
| North America | 736 | 12% | 6 | 5% |
| Europe | 300 | 5% | 13 | 10% |
| Australia | 663 | 11% | 6 | 5% |

\*   Millions of hectares. Percentages relate to the total of the world's drylands (6,150 mill. ha), and the total of selected cases, respectively (N=132).

*Source*:   Own data statistics; according to Reynolds and Stafford Smith (2002, p. 2), based on UNEP (1997).

The figures for total dryland areas measured against selected cases suggest that, first, no weighting bias exists for Asia and Africa (holding the bulk of global drylands) as well as Latin America, and that, second, the bias of European over North American and Australian cases is limited. Thus, if selected cases are measured against aridity zones, the bias appears to be minimal, with a slight underrepresentation of subhumid cases from North America and semiarid cases from Australia.

Second, the selected cases might include a bias related to the dominant types of land-use categories in the drylands, given that 'the overwhelming majority of drylands are rangelands (88%) and only 3% are irrigated croplands and 9% rainfed cropland' (Reynolds and Stafford Smith, 2002). If pure categories are to be considered only, the share of rangelands in our study is 28%, irrigated cropland is 1%, and rainfed cropland is 9%. – see Table 2.6. This would imply an underrepresentation of rangelands. If all combinations are to be considered, the share of rangelands amounts to 89%, irrigated croplands to 22%, and rainfed croplands to 55%. This would imply an overrepresentation of both irrigated and rainfed croplands.

**Table 2.7**  **Selected Cases Measured Against Aridity Zones of the World***

| | Asia | | Africa | | Latin America | | North America | | Europe | | Australia | |
|---|---|---|---|---|---|---|---|---|---|---|---|---|
| | abs. | % | abs. | % | abs. | % | abs. | % | abs. | % | abs. | % |
| *Aridity zones of the world's drylands* | | | | | | | | | | | | |
| Hyper-arid | 277 | 5 | 672 | 11 | 26 | <1 | 3 | <1 | 0 | | 0 | |
| Arid | 626 | 10 | 504 | 8 | 45 | 1 | 82 | 1 | 11 | <1 | 303 | 5 |
| Semi-arid | 693 | 11 | 514 | 8 | 265 | 4 | 419 | 7 | 105 | 2 | 309 | 5 |
| Dry subhumid | 353 | 6 | 269 | 4 | 207 | 3 | 232 | 4 | 184 | 3 | 51 | 1 |
| *Selected cases* | | | | | | | | | | | | |
| Hyper-arid to arid | 7 | 5 | 1 | 1 | 0 | | 0 | | 0 | | 0 | |
| Arid | 13 | 10 | 5 | 4 | 0 | | 2 | 2 | 0 | | 0 | |
| Arid to semiarid | 15 | 11 | 12 | 9 | 7 | 5 | 2 | 2 | 0 | | 5 | 4 |
| Semiarid | 9 | 7 | 20 | 15 | 6 | 5 | 2 | 2 | 8 | 6 | 1 | 1 |
| Semiarid to subhumid | 1 | 1 | 2 | 2 | 1 | 1 | 0 | | 2 | 2 | 0 | |
| Subhumid | 6 | 5 | 1 | 1 | 0 | | 0 | | 3 | 2 | 0 | |
| Subhumid to humid | 0 | | 1 | 1 | 0 | | 0 | | 0 | | 0 | |

\* Millions of hectares. Percentages relate to the total of the world's drylands (6,150 mill. ha) and the total of selected cases, respectively (N=132).

*Source*: Own data statistics; according to Reynolds and Stafford Smith (2002, p. 2), based on UNEP (1997).

This is no serious bias for the following reasons. First, global statistics of rangelands are notoriously flawed with deficiencies, the most serious one being that no difference is made between natural grasslands and actual rangeland areas. In contrast, the study here focuses on sites under human uses, so that cases of natural grasslands are not included. This might explain the underrepresentation of rangelands if only pure categories were considered. Second, as a matter of fact, it has been taken from the cases that historically rangeland expansion was dramatic, but that ranges are contemporarily shrinking at the expense of drastically expanding cropping land. If all combinations were considered, the share of rangeland fits perfectly the global statistics, and higher shares of both irrigated and rainfed croplands might indicate their growing relative importance.

Third, the selection of cases might also represent biased sampling in terms of author or academic bias. For example, is there a shared understanding of the various factors driving desertification among various disciplines, and in the use of indicators? Or, is there a bias inherent to the identification of driving forces and usage of indicators due to the disciplinary background or institutional affiliation of the respective authors? Namely, do ecologists predominantly report biophysical factors driving cases of desertification, using predominantly ecological indicators? To explore the assumption that an author's disciplinary background may have an impact on causes perceived and, thus, reported in the cases, only underlying causes (i.e., the frequency of occurrence of broad causes) is considered in the following. Also, only the application of (ecological, meteorological, and socio-economic) indicators is explored, but a potential bias in the reporting of rates and pathways is not carried out.

**Table 2.8    Dominant Types of Land Use Categories Found in the Cases***

|  | Absolute | Relative |
|---|---|---|
| Rangelands, all combinations | 117 | 89% |
| ... only rangelands | 37 | 28% |
| ... range & irrigated cropland | 22 | 17% |
| ... range & rainfed cropland | 55 | 42% |
| ... range, rainfed & irrigated cropland | 3 | 2% |
| Irrigated croplands, all combinations | 28 | 22% |
| ... only irrigated cropland | 1 | 1% |
| ... irrigated cropland & range | 22 | 17% |
| ... irrigated & rainfed croplands | 2 | 2% |
| ... range, rainfed & irrigated cropland | 3 | 2% |
| Rainfed croplands, all combinations | 72 | 55% |
| ... only rainfed cropland | 12 | 9% |
| ... rainfed cropland & range | 55 | 42% |
| ... rainfed & irrigated croplands | 2 | 2% |
| ... range, rainfed & irrigated cropland | 3 | 2% |

\* N=132; multiple counts possible; no croplands counted for Australia (since virtually no cropping, except on an intermittent basis in favoured locations on the semi-arid fringe).

As for coding the author's disciplinary background, often done in combination with the author's institutional affiliation, the following rules were applied. First, multi-author teams made up of individual contributors coming from various disciplines were coded as 'mixed teams' (found in 29 cases, or 22% of all cases). This was meant in order not to blur the difference between disciplines such as soil science, remote sensing, and agricultural economy. Second, the background of authors attached to, for example, non-governmental organizations or international research consortiums without information on their actual academic background was coded as 'unknown' (n=19, or 14%). The other easy-to-code groups were geography or geoscience (n=41, or 31%), ecology (n=21, or 16%), engineering and remote sensing (n=9, or 7%), hydrology (n=5, or 4%), agricultural science (n=3, or 2%), political science (n=3, or 2%), and soil science (n=2, or 2%) – see Tables 2.9 and 2.10.

Table 2.9  Frequency of Occurrence of Broad Underlying Causes of Desertification Measured Against the Author's Disciplinary Background[a]

|       | [a][b] | [b][b] | [c][b] | [d][b] | [e][b] | [f][b] | [g][b] | [h][b] | [i][b] |
|-------|------|------|------|------|------|------|------|------|------|
| [A][c] | 24   | 6    | 9    | 5    | 3    | 0    | 1    | 24   | 11   |
|       | 59%  | 29%  | 100% | 100% | 100% |      | 50%  | 83%  | 58%  |
| [B][c] | 23   | 13   | 3    | 5    | 3    | 3    | 2    | 13   | 14   |
|       | 56%  | 62%  | 33%  | 100% | 100% | 100% | 100% | 45%  | 74%  |
| [C][c] | 25   | 13   | 3    | 5    | 3    | 3    | 2    | 21   | 16   |
|       | 61%  | 62%  | 33%  | 100% | 100% | 100% | 100% | 72%  | 84%  |
| [D][c] | 29   | 9    | 7    | 4    | 3    | 3    | 1    | 13   | 17   |
|       | 71%  | 43%  | 78%  | 80%  | 100% | 100% | 50%  | 45%  | 90%  |
| [E][c] | 20   | 9    | 2    | 4    | 3    | 0    | 0    | 11   | 7    |
|       | 49%  | 43%  | 22%  | 80%  | 100% |      |      | 38%  | 37%  |
| [F][c] | 41   | 19   | 4    | 1    | 1    | 3    | 0    | 21   | 18   |
|       | 100% | 91%  | 100% | 100% | 100% | 100% |      | 72%  | 95%  |

[a] Percentages in [a] to [i] are column percentages; absolute values specify the frequency of occurrence of underlying causes in the cases, with multiple counts possible.

[b] [a] Geography, geoscience, [b] ecology, [c] engineering and remote sensing, [d] hydrology, [e] agricultural science, [f] political science, [g] soil science, [h] mixed teams, [i] unknown.

[c] [A] Demographic factors (n=74, or 56%), [B] economic factors (n=79, or 60%), [C] technological factors (n=91, or 60%), [D] policy or institutional factors (n=86, or 65%), [E] cultural or socio-political factors (n=56, or 42%), [F] climatic factors (n=114, or 86%).

As far as the frequency of occurrence of broad underlying drivers is concerned, the broad picture gained from all cases seems largely set by results produced in mixed teams, by geographers and geoscientists, and by the group titled 'unknown', accounting for two thirds of all author affiliations – see Table 2.7. Agricultural scientists, and, similarly, hydrologists see a greater impact of all factors in desertification than in any other disciplinary group, while soil and, similarly, political scientists tend to be more selective in reporting causative factors: the first group tends to completely disregard the impact of demographic and cultural factors, while the latter group does the same for climatic and cultural factors. Ecologists seem to report a lower impact of demographic and policy/institutional factors than any other group, and engineering and remote sensing specialists tend to do the same for economic and technological factors. With view on indicators applied, political and soil scientists also use very limited combinations of indicators to describe the process – see Table 2.8. However, these academic groups are involved in 4% of the cases only. In summary, author bias in the meta-analysis, as measured in terms of underlying causes and indicators applied, is minimal and does not contaminate the results of the study, and the conclusions to be drawn.

**Table 2.10**     **Frequency of Occurrence of Applied Indicators Measured Against the Author's Disciplinary Background**[a]

|       | [a][b] | [b][b] | [c][b] | [d][b] | [e][b] | [f][b] | [g][b] | [h][b] | [i][b] |
|-------|--------|--------|--------|--------|--------|--------|--------|--------|--------|
| [A][c] | 19     | 3      | 0      | 0      | 0      | 3      | 1      | 10     | 5      |
|       | 46%    | 14%    |        |        |        | 100%   | 50%    | 35%    | 26%    |
| [B][c] | 3      | 1      | 0      | 0      | 0      | 0      | 0      | 0      | 0      |
|       | 7%     | 5%     |        |        |        |        |        |        |        |
| [C][c] | 1      | 4      | 1      | 3      | 3      | 0      | 0      | 1      | 1      |
|       | 2%     | 19%    | 11%    | 60%    | 100%   |        |        | 3%     | 5%     |
| [D][c] | 5      | 11     | 5      | 1      | 0      | 0      | 1      | 4      | 13     |
|       | 12%    | 52%    | 56%    | 20%    |        |        | 50%    | 14%    | 68%    |
| [E][c] | 7      | 0      | 0      | 0      | 0      | 0      | 0      | 0      | 0      |
|       | 17%    |        |        |        |        |        |        |        |        |
| [F][c] | 6      | 2      | 3      | 1      | 0      | 0      | 0      | 14     | 0      |
|       | 15%    | 10%    | 33%    | 20%    |        |        |        | 48%    |        |

[a] Percentages in [a] to [i] are column percentages; absolute values specify the frequency of occurrence of underlying causes in the cases, with multiple counts possible.
[b] [a] Geography, geoscience, [b] ecology, [c] engineering and remote sensing, [d] hydrology, [e] agricultural science, [f] political science, [g] soil science, [h] mixed teams, [i] unknown.
[c] [A] Ecological indicators, only (n=41, or 31%), [B] meteorological indicators, only (n=4, or 3%), [C] ecological and meteorological indicators (n=14, or 11%), [D] ecological and socio-economic indicators (n=40, or 30%), [E] meteorological and socio-economic indicators (n=7, or 5%), [F] ecological, meteorological and socio-economic indicators (n=26, or 20%).

Fourth, for a thorough comparative analysis of cases, strict selection of cases by means of a chosen and uniformally applied formal technique is critical (Matarazzo and Nijkamp, 1997). Further, the adopted research design, following the strategy of 'multiple-conjunctural causation' (Ragin, 2003), requires 'positive cases', i.e., cases which show desertification outcomes only. Here, a potential bias cannot be excluded which relates to the strategic decision to use 'desertification' as a key search word when identifying and selecting cases. Other search words were also checked and, as a result, it was found that an abundant, not-easy-to-handle body of studies is revealed when entering a combination of terms such as 'drylands', '(land) degradation', 'land-use dynamics', or 'rangeland modification'. Therefore, it was decided to uniformly apply the term 'desertification' in the selection of cases. Agreedly, authors using the term may have a more deterministic, or even pessimistic view of the issue, as indicated by Reynolds and Stafford Smith (2002). This implies that the authors of case studies compiled here may tend to consider the complex and non-equilibrium dynamic of coupled human-environment systems in drylands to a lesser degree than authors avoiding the term 'desertification', and using, for example, 'land degradation' or 'land-use and land-cover change in drylands', instead (e.g., Nielsen and Zöbisch, 2001).

# Chapter 3

# Initial Conditions

## Introduction

The initial conditions of dryland sites, as reported in the case studies, comprise the coupled land use and environmental history of each case. They provide narratives of land change, for both the livestock and cropping sector. Within-case insights are provided as well as cross-case comparisons within and across broad geographic regions such as Asia, Africa, Europe, Australia, North America, and Latin America. The chapter concludes with an overview on the frequency of occurrence of major land use categories per broad geographic regions, as described in the cases.

## Asian Drylands

Rather than generating one coherent land use and environmental history for all the Asian drylands, five major narratives have been identified from the Asian case studies of desertification for various ecosystems:

- a century-old tradition of large-scale alternations of grain farming and livestock raising in the grassland (steppe) ecosystems, with a strong biophysical predisposition for desert formation;
- an ancient to recent transformation of traditional pastoralism towards intensified uses;
- a policy-driven extension of croplands, increasing the pressure on remaining steppe grazing land;
- a century-old tradition of irrigation or oasis farming in now degraded river ecosystems of (sub)desert ecosystems; and
- the most recent transformation of oasis and small-scale irrigation agriculture into rapidly expanding, large-scale irrigation schemes aside with booming non-agrarian uses.

### Alternations of Farming and Livestock in Steppe Ecosystems

The upland steppe plateaus or high plains make up one of the most important features of dryland ecosystems in Asia, carrying mainly dry grasslands, but also shrub vegetation (both partly scattered with mobile and semi-fixed sand dunes) and sandy soils in association with loess or loessial soils, mostly surrounded by high mountains. Originally, the steppe ecosystem has been suitable for (semi)nomadic grazing only. The arid to semi-arid steppe zones with high forage capacity are located at the fringe or transition zone of permanently settled agricultural areas. They have to be held apart from cold mountain steppe zones (including alpine meadows), with a short vegetation period and lower productivity, due to the occurrence of frozen (permafrost) soils, but characterized by easy waterlogging and high erosivity. All of the steppe areas had reportedly supported century, if not millennia-old nomadic or semi-nomadic grazing.

Typically, most of the plateaus are under the influence of both the eastern Asian summer monsoon front (governing rainfall), and the winter monsoon regime (with northerly winds generating frequent dust storms and aeolian deposits). Differently, the climate of Tibet is that of an alpine and dry, i.e., high-cold fragile ecosystem. Clearly, these cold, dry, continental climates predispose the arid and semi-arid steppe grazing lands to natural processes of desert formation (an example of how human and biophysical activities were interlinked, leading to the collapse of ancient societies, is given in Chapter 1).

As per today, increases of extreme events have been reported such as osciallations in terms of both droughts and more moisture and rainfall. The least to be said is that they make the process of desertification very dynamic, probably upscaling localised but increasing human impact in terms of areal extent and intensity of degradation. For example, the climate of the Ordos, Inner Mongolian and Tibet Plateaus grew drier and warmer since the 19$^{th}$ century, despite of more extreme rainfall events (Lin and Tang, 2002). Droughts are evident such as in the Aral Sea Region in 1974-77, 1985, 1986, and 1989 (Saiko and Zonn, 2000). Snow cover has been reduced and glaciers were retreating, on mountains of both the Tibet Plateau and on those surrounding the Tarim River Basin (Holzner and Kriechbaum, 2001; Yang, 2001). Long-term average rainfall has decreased mainly since about the 1950s (Sheehy, 1992; Wang, Qian, Cheng and Lai, 2001), and more sustained winds trigger frequent and increasing sand and salt storms, such as during the 1960-95 period in the Aral Sea Region (Saiko and Zonn, 2000), and in 1992-93, 1995 and 2000 in northern China (Zhou, Dodson, Head, Li, Hou, Lu, Donahue and Jull, 2002; Lin and Tang, 2002; Holzner and Kriechbaum, 2001; Liu and Zhao, 2001; Yang, 2001; Feng, Endo and Cheng, 2001). Typically, the Yellow River source region and upper reaches of the Tibet Plateau are characterized by rising ground temperature in the upper layer of frozen soil (0.2-0.3°C per year over the last 30 years), triggering the alteration of freeze-thaw soil processes, which concomitantly with overgrazing leads to aggravated desertification (Wang, Qian, Cheng and Lai, 2001; Holzner and Kriechbaum, 2001).

The plateaus had been cleared from forests during ancient times, and the contemporary dry grassland cover seems highly dynamic, with hardly any large-scale reduction in vegetation cover (Runnström, 2000). The cold high mountain Tibet ecosystem, however, shows overall vegetation degradation. Namely, increases were cited of desertified alpine steppe and steppified high-cold meadows, and a widespread disappearance of swamp species in the 1970/80s if compared to the 1990s. Different from geological sand formation, there are recent and frequent increases in bare (rocky) or desert-like sand ground cover (sands, coppice dunes, etc.), including increases in the volume of transported sand. They were reported from the Ordos, Inner Mongolian and Tibet Plateaus, where they exposed soils to both water and wind erosion (Sheehy, 1992; Ho, 2001; Wang, Qian, Cheng and Lai, 2001; Holzner and Kriechbaum, 2001; Liu and Zhao, 2001; Zhou, Dodson, Head, Li, Hou, Lu, Donahue and Jull, 2002). In particular, areas which are covered by blown, shifting sand expanded considerably. For example, in the $2^{nd}$ half of the $20^{th}$ century, 'sandification' led to desert expansion on the Ordos Plateau both northwards and southwards, causing the Kubuqi Desert (in the north) to connect with the northern boundary of the Loess Plateau (in the south) (Lin and Tang, 2002). Concerning water degradation, only the Tibet Plateau had been mentioned with less (potable) water in rivers and boreholes, siltation or sedimentation of low-lying zones or reservoirs, decreases of lakes and swamps as well as of stream discharges (Liu and Zhao, 2001; Wang, Qian, Cheng and Lai, 2001; Holzner and Kriechbaum, 2001).

*Transformation of Traditional Pastoralism*

Traditional nomadic grazing has been practised on most of the Arabian Peninsula (Brown and Schoknecht, 2001; Weiss, Marsh and Pfirman, 2001) and in the steppe and desert regions of northern and western China (Sheehy, 1992; Runnström, 2000; Holzner and Kriechbaum, 2001; Wang, Qian, Cheng and Lai, 2001). The animal production systems are reportedly century-old, but have hardly persisted in pure form until present. Traditional nomadic land use was characterized by well-adapted species (such as the yak in the highlands of Tibet) and seasonal movements to transitional grazing land pastures. Although large numbers of livestock died, the system was adapted to erratic environmental conditions and low human population densities. However, in most of the cases, traditional pastoralism was reported to have undergone changes in traditions and rights which reduced the possibility of controlling natural rangelands. For example, in the Inner Mongolia Autonomous Region of northern China, situated at the interface of the Mu Us Region and Tenger Desert of the Ordos Plateau, Mongol nomads used pastures for millenia in an area which had been once the famous Eerduosi Grasslands (because of the Ordos fine wool breed of sheep). The area was a vast expanse of fertile land with plenty of water and fresh grass in the first centuries AD. As by now, the number of animals has grown 10-fold, alone in the 1947-96 period, and the area is now claimed to be among the most degraded pastoral areas in semi-arid China (Runnström, 2000).

Most of the changes, however, were reportedly recent. On the Arabian peninsula, changes led to now larger nomad herd sizes due to the availability of water trucks and government subsidies for summer feed in the form of grain, for instance (Weiss, Marsh and Pfirman, 2001; Brown and Schoknecht, 2001). Increased herd sizes were also due to the introduction and addition of new and less well adapted species such as European cattle (e.g., Friesian or Simmental), horses, and sheep (e.g., Merino breed or derivatives). They were added to indigenous herbivore species and well adapted domesticated grazing animals such as traditional breeds of sheep on the Ordos Plateau, or yaks on the Tibet plateau (Sheehy, 1992; Holzner and Kriechbaum, 2001). Livestock increased drastically. In nomadic areas of the Inner Mongolian Plateau, it grew by 330% in the 1949-64 period, and by 400% in the 1960-90 period (Sheehy, 1992). What had been, in ancient and historical times, reclaiming and garrisoning the frontier, stationing troops to open up 'wasteland', and chaos caused by war, leading to desertification, was repeated through the impact of contemporary policy factors. In China, the 'Great Leap Forward' (end of 1950s) and 'The Great Cultural Revolution' (in the 1960/70s) triggered a shift from traditional nomadic to more intensive land use, which meant an inappropriate use of resources. The latter included the further destruction of meadows, for example, through overstocking sandy textured soils, thus increasing their vulnerability to soil erosion and transforming stabilized dunes into a shifting sand desert (Sheehy, 1992; Zhou, Dodson, Head, Li, Hou, Lu, Donahue and Jull, 2002; Lin and Tang 2002).

The changes implied a transition from nomadic to extensive grazing (transhumance and/or sedentary), with most steppe grazing land, however, best suited for extensively managed, but not sedentary grazing animals production (Rozanov, 1991; Runnström, 2000; Lin and Tang, 2002; Feng, Endo and Cheng, 2001). There is abundant evidence from northern China, that these changes were associated with overgrazing and impaired carrying capacity of rangelands: trampling effects by sheep, goats, camels and donkeys, causing soil compaction or loosening, finally leading to soil erosion. In several northern provinces, grasslands were turned to desert or sandy grounds several decades to a century ago, and it is claimed that an assumed one third of all desertification probably happened in the 1950-90 period due to the impact of growing livestock numbers (Runnström 2000; Lin and Tang, 2002). Changes were also recent in the Karakum Canal Zone of the Turkmenia Plain which has a naturally low range productivity. Overgrazing was reported to have started in the 1980s due to inappropriate infrastructure investments not suited well to the fragile dryland ecosystem, such as livestock buildings, mechanization of water delivery, and the built up of watering structures, in particular, shaft and pipe wells. In diameters of about 1.5 to 2 km around the ca. 3,500 wells, land got desertified (Rozanov, 1991).

*Extension of Croplands, Pressure on Grazing Land*

Following the transition from nomadic to extensive grazing, cropland uses became increasingly important and rangelands were increasingly limited to sites with marginal productivity. Examples stem from the Chinese and Russian parts of the steppe and desert region (Rozanov, 1991; Yang, 2001; Ho, 2001; Zhou, Dodson, Head, Li, Hou, Lu, Donahue and Jull, 2002). Increasing conflicts about land are reported between farmers and herders, and the overall trend has been that cropland expanded at the expense of rangeland. Thus, in addition to adding new livestock, overstocking was also caused by limited and marginal land available for pastoralists. Typically, it has been reported as a feature of former Soviet Union and Chinese territories that unidirectional land change was organized by state authorities (such as the forced immigration of Han Chinese agriculturalists).

Sedentary rainfed, less so dry farming agriculture for mainly large-scale grain production (to feed increasing human populations) was found to be located at the ecotone between steppe grazing land and sedentary agricultural areas. Examples are China with a transition (tension or contact) zone along the northern boundary of the Loess zone, or the southern boundary of desert regions, respectively, (Sheehy, 1992; Runnström, 2000; Lin and Tang, 2002), and the Kalmyk area in southern Russia with the ecotone roughly running between the Yergenin uplands and the Caspian Sea Basin (Rozanov, 1991).

Different from Russia, the land use histories of transitional zones in China date back to ancient times. Sedentary agriculture was often overtly used as a means to consolidate political, social and military goals over several thousand years. The transition zone had actually been a tension zone through much of the recorded history of northern China. The pattern was, first, the cultivation of grazing land, second, the subsequent abandonment following misuse of water resources, overcultivation, overgrazing, or warfare, and, third, translocation of people to other grazing lands where this process was then repeated (Sheehy, 1992; Zhou, Dodson, Head, Li, Hou, Lu, Donahue and Jull, 2002). Since 1949, rural peasants in China have been convinced or coerced to open up new agricultural land in the steppe zone which had low human population densities, especially through the (forced) northward movements of Han Chinese agriculturalists. However, the long, cold, windy and dry season of winter and spring, the short growing season, and the low and variable precipitation were not conducive to highly productive, crop-based sedentary, rainfed (and also irrigated) agriculture. There is little or no vegetation growth until early summer due to droughts and drying winds, coinciding during springtime. The semiarid steppe has relatively high value for agriculture only on loess or loessial soils, especially for the production of annual crops such as grain. Thus, mainly the dark, deeper, loessial lowland soils have been converted to rainfed cropland, which was then subject to erosion by water received as runoff from overgrazed uplands. Deep erosion channels on rainfed croplands, however, reduce the area of cultivated land, contributing to overcultivation and further land reclamation from rangelands. An estimated 60% of the total area of the semi-arid to arid transition zone in northern China has been degraded according to this pattern, especially during the last 300 years (Sheehy, 1992; Ho, 2001).

According to ancient and historical cycles or alternations of grain farming and livestock husbandry as described above, desertification happens on an intermittent or periodic basis. The process continues up to present. For example, major emphasis of recent agricultural development in Inner Mongolia is the conversion of grazing land into sedentary agriculture (as well as increasing the number of domestic grazing animals). In Zhaowuda League, during most of the $2^{nd}$ half of the $20^{th}$ century, over 133,000 ha of grazing land was converted to cropland at a rate of over 4,000 ha annually. In the Kerquin area, sandy cover formed ca. 13% of all land in the early 1950s and grew to 78% of the entire area in 1989, with the Kerqin 'meadow' having turned into a desert within half a century due to land reclamation (Lin and Tang, 2002). In the Tibetan Autonomous Region, land reclamations for crops from rangelands were associated with severe land degradation especially in valley locations, such as the middle reaches of the Yarlung Zangbo, the Lhasa and the Nianchu Rivers (Liu and Zhao, 2001).

Semiarid steppe land has the capacity to recover ecological conditions between periods of disturbance, thus making possible large-scale alternations between grain farming and extensive grazing. In the arid parts, the amount of rainfall dominates the response in biological productivity, while in less arid parts, positive biomass trends could be due to desertification control measures such as planted protective farmland forests, shelterbelts, and aerially seeded rangelands, implemented on the Ordos Plateau, for example, since 40 years (Runnström, 2000). A measure to control the advancement of contemporary 'sandification' was the green wall project, the goal of which was to plant in the 1978-2000 period 10,000 hectares of trees per year over 7,000 kilometers, stretching from eastern Inner Mongolia to the Xinjiang Autonomous Region (Sheehy, 1992).

In contrast to steppe zones, land conversion in other steppe areas was reportedly associated with declining soil productivity, but less so with 'sandification'. For example, Ram, Tsunekawa, Sahad and Miyakazi (1999) as well as Khresat, Rawajfih and Mohammad (1998) mention crop production and livestock operations as well integrated land use systems which have sustained traditional farming practices and a predominantly settled population for millenia, in the Thar Desert and Eastern Mediterranean zone.

*Traditional Farming in Now Degraded River Ecosystems*

The second striking feature of Asian drylands is dry and hot lowland plains, depressions or basins which carry river, lake or delta ecosystems under (sub)desert conditions as well as long-settled land uses. This includes hyper-arid desert sites – such as in the Saudi, Thar and Taklimakhan deserts – with very hot, dry summers, but cool and sometimes wet winters, and average annual precipitation subject to considerable fluctuations. Basin climates are more pronounced than plain climates, because basin units span from dry and hot lowlying zones to much wetter and colder mountains, i.e., can be under the influence of nearby alpine glaciers and snowpack. Annual rainfall in basins is usually sparse and highly variable, ranging from as low as 30 mm (in the terminal area of basins) to as high as 800 mm per year (in upper reaches). Even the driest ecosystems still depend for their functioning upon ice and snowmelt waters from surrounding mountains, which is also the key factor for oasis agriculture there. Nonetheless, these climates have sustained (semi)nomadic grazing or settled farming, even, for centuries and millenia – such as nomadic pastoralism in the Saudi Desert, millet cropping in the Indian Thar Desert, and oasis agriculture in the Taklimakan Desert (Ram, Tsunekawa, Sahad and Miyazaki, 1999; Yang, 2001; Lin, Tang and Han, 2001; Feng, Endo and Cheng, 2001; Lin and Tang, 2002).

Vast regions were cited to be rich in groundwater resources, and water has been the key element for the formation of oases. Along ancient stream courses in the Taklimakan Desert, for example, constant river flow was reported from about 2000 AD (when more humid phenomena were also reported from northern Africa) well into the 16th and early 19th centuries (with the last two wet events happening simultaneously with the global ice expansion of the Little Ice Age). Reported evidence suggests that widespread water degradation such as salinization and chemical pollution became a contemporary phenomenon during the second half of the 20th century. For example, in the Tarim River Basin salinization, less (potable) water in rivers or boreholes, widespread drop in water tables and river discharge, pollution of surface and/or groundwater, and the shrinkage or shortening of rivers and lakes is widespread. Annual tree rings of *Populus*, indicating groundwater availability, were wider in 1700 to 1876 then for the years after 1877, and flood water had not reached the now dry delta of Keriya River since the end of the 1980s (Yang, 2001). Salt contents, mainly in groundwater, increased, and between 10 and 20% only of total surface runoff from the upper reaches of Tarim River was received by lower reaches in the 1990s as compared to the 1950s. Terminal lakes were reduced or disappeared (such as Lop Lake in 2000, the surface of which had been 3000 km$^2$ in 1950), and salinization in terminal areas proceeds at a rate of 100 ha of farmland groundwater per year. The water table of Tarim River fell from 2-3 to 4-10 meters between 1960-80, and continues to decline annually by 20 cm from 1980 to the present (Feng, Endo and Cheng, 2001). In the Kashi Plain, available water supply was lessened, high fluorine concentration in surface and ground water were observed, flood irrigation has salinized more than half of all arable land, and about one fifth of the salinized lands have turned into a salt desert, most widespread in the lower reaches of Kashi and Yerqiang Rivers (Lin, Tang and Han, 2001).

Similarly, river discharge in the Hei River Basin decreased and more than 30 tributaries dried up in the last 50 years, with salinized areas in the 1990s now covering 13% of the total cultivated land (Genxu and Guodong, 1999). Salinization is widespread in the Caspian Plain and adjacent Yergenin Uplands, with 100% of irrigated land now saline, and with almost 2,000 ha of the Tsaryn system canals in northern Kalmykia turning into salinized land in only 8 to 10 years' time (ca. 1980-89). In the Turkmenia Plain, saturation occurred along the Karakum Canal for 100 km (50 km on either side), totalling 10,700,000 hectares, and the annual average input of salts amounts to 9 to 10 million tons (Rozanov, 1991). In the Aral Sea Region, the sea area decreased from 66,900 to 32,000 km$^2$ in the 1960-95 period, the sea volume from 1064 to 310 km$^3$, and the sea level dropped from 53 to 37 m. The length of the shoreline shrank by 480 km, and sea salinity increased from 11-14 to 34 g/liter, while the salinity of groundwater (in irrigated lands) rose from 1-3 to 10-12 g/l. The inflow of fertile silts into rivers decreased, while the inflow of dissolved salts increased, and chemical pollution of groundwater rose from 20 to 80% of all irrigated land (Saiko and Zonn, 2000).

As with steppe ecosystems, the climate had changed several times during the Quaternary, and the current dry interval led to the development of mainly sandy soil properties. Different from other dryland areas in the world, sands, but also gravels, date back to Quaternary and Tertiary sediments, so that parent material or lithology remains virtually unchanged until present, with soil development being seriously impeded by arid conditions (Khresat, Rawajfih and Mohammad, 1998; Brown and Schoknecht, 2001). 'Sandification', or contemporary increase in sandy cover was reportedly widespread. Along the Tarim and Hei Rivers, it concentrated in reclaimed riverbeds that gradually dried up. Most of the (semi)fixed dunes had been activated and moved towards, or encroached even oases especially in the lower reaches (Genxu and Guodong, 1999; Lin, Tang and Han, 2001; Feng, Endo and Cheng, 2001). In the lower reaches of Hei River, 30,700 ha of cultivated land (in 1949) had been reduced to 3,000 ha (as by 1997), while the rest turned into desert. Desertified land expanded at a rate of ca. 11,000 to 13,000 ha per year in the 1960-90 period (Genxu and Guodong, 1999). In the Aral Sea Basin, a desert had formed on previous sea basin ground in the 1960-95 period, and fixed dunes started to move at rates of 900 to 1200 m annually in 1983-95, altering fundamentally the natural soil drainage conditions (Saiko and Zonn, 2000). The trend was less pronounced in the Caspian Sean Region, where 13% of the total area under study had converted from semi-desert and dry steppe soils to 'true' sandy desert as by 1989 (Rozanov, 1991).

What makes river and lake ecosystems different from steppe regions is that vegetation degradation is most widespread. For example, timber stock volume and shrub forests have badly been degraded in the Tarim River basin, namely, *Populus* forest and Tamarix (Yang, 2001). Hundreds of kilometers of poplar belts, especially along the lower reaches of rivers, have died due to frequent drying up of the rivers (Lin, Tang and Han, 2001). Annual loss rates of vegetation cover in the $2^{nd}$ half of the $20^{th}$ century range from 6.25 km$^2$ for shrub and meadow areas to 119 km$^2$ in the case of *P. euphratica* and *E. angustifolia*, with the latter being the main tree species in northwestern China (Feng, Endo and Cheng, 2001). In the Hei River Basin, dry river bed vegetation had been reduced, hydrophytes and swamp vegetation declined or dried out, natural grasslands of reduced productivity had been extending, thick stands of *Diversifolia schrenk* and Ulster wood as well as meadow vegetation and bush areas with high canopy coverage were reduced since about 1950, and the forest line in surrounding mountains had receded by several kilometers in the 1960-90 period (Genxu and Guodong, 1999). In the Aral Sea Basin, deltaic shrub, woodland and swamp vegetation had been reduced, and halophyte vegetation expanded, with grasses such as tamarik and saltwort replacing natural reed in the 1960-95 period (Saiko and Zonn, 2000). Decreases in herbaceous, grass, and scrub cover as well as increases of sparse, degraded and (for livestock) less palatable vegetation were mentioned for comparable sites in the Saudi and Thar Deserts as well as from the East Mediterranean Basin (Khresat, Rawajfih and Mohammad, 1998; Brown and Schoknecht, 2001; Weiss, Marsh and Pfirman, 2001).

*Expanding Irrigation Schemes, Booming Non-Agrarian Uses*

Oases or traditional irrigation agriculture were originally based on the use or extraction of surface and/or groundwater resources along ancient river courses or in delta or lake basin areas. They were cited to date back to historical and mostly ancient farming cultures or practices. Examples stem from the Yellow River, intersected into the Ordos Plateau, from the Tarim and Hei River Basins, and the Aral Sea Region (Sheehy, 1992; Genxu and Guodong, 1999; Runnström, 2000;

Saiko and Zonn, 2000; Yang, 2001; Ho, 2001; Feng, Endo and Cheng, 2001). Until the late 19$^{th}$ century, irrigation agriculture had a reportedly low impact on natural environments. It appears that the most fundamental pressures on dryland ecosystems through irrigation farming had been exerted during the second half of the 20$^{th}$ century. As a general trend across many areas of the Central Asian desert and steppe region, small-scale traditional oasis agriculture had been transformed into large-scale irrigation agriculture through advances in water technology and for economic and demographic reasons. Additional demands on surface water resources were created by the policy-driven influx of booming industries (such as the oil/gas industry), mining activities, and expanding settlements, in addition to a shift towards water-demanding cash crops. Cotton monocultures, in particular, became the key crops in drylands, but also vegetables, fruits, and grapevines, in addition to irrigated food crops such as grain and rice. The transition was described as gradual, having a low impact only on ecosystems until the 20$^{th}$ century.

Large-scale irrigation development in the 2$^{nd}$ half of the 20$^{th}$ century was made possible by the construction of hydrotechnical installations, which included dams (especially in the upper reaches of rivers), reservoirs (especially the build-up of plain reservoirs), irrigation canals, collectors and artificial drainage networks. These developments induced some fundamental and mostly irreversible changes to especially the natural hydrographic network, altering the hydrological cycle in most of the cases. While irrigated agriculture in arid lands was originally based on the use of surface and spring waters, with economic and technological advances in the last 30 to 40 years, canal systems and hydrotechnical installations have greatly increased water usage. They also triggered the expansion of farming land on to marginal, not easily irrigated sites (Rozanov, 1991; Genxu and Guodong, 1999; Saiko and Zonn, 2000; Feng, Endo and Cheng, 2001; Lin, Tang and Han, 2001). Two examples are given in the following, one from the Tarim River and one from the Hei River Basin.

In the Keriya River area of the Tarim River Basin, agricultural irrigation in artificial oases (around Yutian) and the natural oases in the dry delta areas (around Daliyaboyi) started in the early 10$^{th}$ century in small areas, and – due to immigration and increasing population numbers – expanded to a limited number of oases in the 16$^{th}$ century. Some more artificial oases formed around settlements ca. 250 years ago. Since 1949, irrigated land cultivation has accelerated and a huge oasis had been created (two thirds of the land which was cultivated since 1958 has become desertified). Expansion of irrigated land was accompanied by the shift to more water-demanding cash crops, and the major crop in many communes is now cotton. In Yutian, the production of cotton has increased continuously from 764 t (in 1986) to 9,323 t (in 1998). With the construction of several reservoirs in the middle reaches, the majority of water was channelled to agricultural fields outside of the traditional irrigation areas (Yang, 2001). In the total basin, the total length of canals increased by 2,000 km during the 1950-95 period, and the construction of reservoirs along the rivers has led to much of the existing water being cut off and diverted to remote farmlands, or wasted by evaporation. Wastage had also been due to poor water management legislation, uncoordinated distribution, and uncontrolled development (Feng, Endo and Cheng, 2001; Lin, Tang and Han, 2001).

In the Hei River Basin, which serves as one of the most important grain production bases of China, large-scale water development, including the construction of reservoirs, began as early as in 140 AD, but land use expanded and intensified only later in the 20th century. The gradual development of oasis crop production since the 2nd century AD left the hydrological status and river system largely unchanged until 1900 – despite of the scale of water development and population pressures which had continuously increased. Prior to 1900, the rivers and lakes were connected, there were abundant water resources as well as large areas of forest, grassland and farmland. In 1944, the first large reservoir was built, and from that time on, especially after 1949, the scale of water development increased significantly. Between 1949-78, 93 reservoirs have been built (excluding small water storage dams), the main and branch irrigation channels increased from 2,818 to 3,968 km in the 1974-94 period, and wells for groundwater pumping increased from 5,735 before 1985 to 9,650 by 1994. More and more oases were set up around reservoirs, and irrigated farmland area expanded drastically (Genxu and Guodong, 1999).

In addition to agricultural uses, several non-agrarian uses were reported to be located in irrigation farming areas. Oil-gas field complexes, such as the Astrakhan gas field in the Caspian Sea region or oil exploration in the Taklimakhan Desert, increased the water demand especially since the 1980s. The industries usually triggered immigration of workers and employees, mainly Han Chinese in the case of northwestern China). In the center of the desert, groundwater has been extracted for irrigating the many private, artificial gardens in the new settlements areas (Rozanov, 1991; Yang, 2001). Additional water demands arose from increased urbanization and industrialization what could be measured in the increase of the number of power dams, coal-burning power plants and chemical factories (Feng, Endo and Cheng, 2001). The continuous growth of population has led to the expansion of urban and rural settlements at the cost of cultivable land (Khresat, Rawajfih and Mohammad, 1998), thus increasing pressure on remaining crop (and grazing) land.

As per today, original oasis sites are often reported as the primary sites of contemporary desertification in river and delta ecosystems. Formerly productive land, mainly used for food production, got desertified due to destruction of irrigation systems, due to soil salinization and the advancement of surrounding desert sands. Different from the semi-arid steppe grazing lands, there seems little or no opportunity to restore land to a higher level due to the extreme aridity, as can be taken from the case of the Aral Sea Region.

The Aral Sea Basin is a region of ancient irrigation farming culture, the peak of which dates back to the 4th century BC to the 2nd century AD. The total irrigated land area of those times, located in the lower Amudarya and Syrdarya, exceeded the contemporary irrigated area by factor three. These lands were abandoned due to river and stream migration, war, and the undue practice of shifting cultivation. The latter created land which was regularly abandoned and left saline after one or two irrigation seasons, when farmers moved to another area. During the late Tsarist,

Soviet and post-Soviet periods, agricultural production of irrigated products in the basin, cotton in particular, was expanded for economic and demographic reasons, i.e., to assure national self-sufficiency. Irrigated expansion began as early as in the 1890s, but only from the 1960/70s onwards the area of irrigated land for the cotton industry developed rapidly. Volume and yield per hectare decreased after 1980 in all of the Central Asian republics involved (Kazakstan, Uzbekistan, Turkmenistan). Crop production had mainly been achieved by extension of land rather than by intensification. Nonetheless, the expansion of cotton monoculture was supported by the use of excessive amounts of fertilizers and pesticides which led to widespread chemical pollution of soils and surface as well as groundwater. Water-efficient methods were hardly developed and most commonly used practices of watering crops such as furrow irrigation continued to be used all over the periods of farmland expansion. As a consequence, considerable and increasing water losses occurred through evaporation and infiltration, triggering either water and soil mineralization, or leading to waterlogging and a decline in natural drainage conditions (Saiko and Zonn, 2000).

## African Drylands

Rather than generating one coherent land use and environmental history for all of the African drylands, again several major narratives have been identified from the African case studies of desertification:

- oscillations of climate, vegetation and land use along a distinct north-south gradient in West Africa, including the Republic of Sudan;
- topographies, soil differences and livestock herbivory overriding the impact of natural climate oscillations, especially in savannah ecosystems of East Africa;
- increased risk of overgrazing due to reduced flexibility (or sedentarization) of nomadic herders, intensification of livestock raising, and rising conflicts about land between farmers and herders;
- millenia-old rotational, mixed and livestock-integrated hand cultivation of traditional crops in West Africa, with multiple constraints on land productivity and low impact on degradation;
- vulnerable monocultures of mechanized maize and wheat farming in East and southern Africa, with low degrees of livestock integration and high impact on soil erosion; and
- millenia-old, widespread harvesting of (dry) natural ecosystem products and most recent, poorly performing rice irrigation schemes for remote, but growing urban demands.

*Natural Oscillations of Climate, Vegetation and Land Use in West Africa*

In West Africa, including the Republic of Sudan, differences in rainfall along a north-south gradient set the stage for a (sub)desert zone (<300 mm annual rainfall), a semi-desert zone (300-400 mm), and a semi-arid zone (>400 mm), fundamentally structuring in terms of cultivation potential the sequence of Sahelian pastures in the north to the Sudanian agricultural zones further south. The sub and semi-desert zones are able to support relatively large herds of camels and goats, and further south, relatively intensive stock grazing is possible, including more village-based goats and sheep, along with cattle. The northern limit of rainfed cultivation roughly runs along the northern fringe of the semi-arid zone (traditional grain-legume mixtures). It had been fluctuating according to most of the cases, but no major shifts occurred in the northern cultivation limit (same with the southern fringe of the Sahara). The Sudanian agricultural zones in the south were also important for seasonal movements (or transhumance) of livestock between these major zones. These oscillations make it especially difficult, if not impossible at all, to separate natural productivity changes from human-driven causation impact.

During pre-historic times, the climate in the West African Sudan-Sahel zone had changed from unusually wet conditions (9500 to 4500 BP) to more arid conditions (4500 to 3000 BP). Mean annual rainfall has been on a marked decrease since about the mid-$20^{th}$ century, accompanied by a southward movement of isohyetes in the 1951-89 period such as in Senegal. Rainfall appears to be consolidated since the 1990s. Fluctuations of precipitation usually have spatial and temporal variations with a considerable stochastic component (Venema, Schiller, Adamowski and Thizy, 1997; Gonzalez 2001; Rasmussen, Madsen and Fog, 2001).

Vegetation formation is governed by the rainfall gradient, i.e., increasing humidity and decreasing evapotranspiration from north to south. This leads to transitions from desert, semi- or subdesert vegetation formations to dry grasslands and (dry) steppe or savannah, to shrub or woody savannah with more or less thick stands of trees at same places, and further to open or closed woodlands or tree-shrub mosaics, with some more or less forest stands occasionally (Helldén, 1991; Ringrose and Matheson, 1992; Turner, 1999a). Within each broader vegetation unit, land use activities, including bush fires, led to characteristic bush and tree vegetation associations. Anthrophytes had been common such as trees and shrubs which owe their present distribution to the occurrence of past human activities. For example, the cultivation of fossil dunes, in particular, had been practiced since about ca. 3,000 years, with large populations of *Faidherbia albida* and *Adansonia digitata* (baobab) indicating former agricultural activities. As a consequence, mosaics were contemporarily dominant to such an extent that common zonal classifications of natural vegetation provide a poor guide only to what might be observed on the ground (Mortimore, Harris and Turner, 1999; Rasmussen, Madsen and Fog, 2001).

For the latter half of the $20^{th}$ century, considerable changes of natural vegetation had been reported in terms of plant cover decreases, reductions in tree density and biomass production, bush or scrub encroachments, and related alterations in floristic composition, with evidence indicating that the natural dynamics of vegetation change had been considerable. For example, dry periods of varying duration were reported to lead to some degree to 'deforestation' as part of the Sahelian ecosystem, since rainfall has been fluctuating (Benjaminsen, 1993; Gonzalez, 2001), and phases of herbaceous plant cover, i.e., duration of growth, density, and floristic composition, were reported to reflect local variations in geomorphic setting, precipitation and land use, thus varying from year to year (Helldén, 1991; Rasmussen, Madsen and Fog, 2001). Land management practices were also reported to contribute directly to high vegetation dynamics. For example, cropped fields were often abandoned and allowed to regenerate with secondary shrub species such as in central parts of the semi-desert and semi-arid zones of West Africa (Ringrose and Matheson 1992; Rasmussen, Madsen and Fog, 2001), and natural shrub stands were sometimes cleared and replanted along field boundaries (Gonzalez, 2001). Also, restoration of the vegetative cover was reportedly a common practice of farmers and herders. For example, and apart from the massive plantation of exotic species such as *Eucalyptus camaldulensins*, natural regeneration of shrubs and trees had traditionally been promoted, either by setting aside parcels of land for recovery, or by selecting small trees in the fields, protecting them, pruning them to promote raid growth, and raising them to maturity (Gonzalez, 2001). In the densely populated zone of northern Nigeria, for example, many trees were reportedly scattered in farms and on field boundaries which had been preserved, or planted, because they provide valuable fruit, medicines, building materials, browse and shade (Mortimore, Harris and Turner, 1999).

*Less Natural Oscillations in Savanna Ecosystems of East and Southern Africa*

Mean annual rainfall had reportedly been on a marked decrease since about the mid-20$^{th}$ century in other African dryland areas, too (e.g., 20% during the 1983-96 period in Botswana), and droughts or drought periods had been prolonged and recurrent, especially since 1970 in the Kalahari Sandveld (Keya, 1998; Dube and Pickup, 2001). Normal variations of rainfall on a year-to-year basis simply include drought periods, even in wetter dryland parts such as the northern zone of the Kalahari Sandveld. In southern Africa, the effects of both wet periods and prolonged droughts in the 20$^{th}$ century fell together with a phase of major land use change, so that in the end it is difficult to separate the two impacts when extent and rate of land degradation are to be assessed.

While gently undulating floodplains or flat plateaus, both of low relief energy, are common topogaphical features of most African drylands, the East African grassland zone is distinct. It has various, vast landscape units which, together with soil differences, govern vegetation formation to a considerably higher extent than in other parts of Africa. The Masailand (or Masai Steppe) in northwestern Tanzania, for example, comprises flatlands and undulating plains, rolling to moderately dissected (flat)lands, and hilly or mountainous units, each showing a distinct land cover dynamics (Mwalyosi, 1992).

In southern Africa, the order of broad transitions between (sub or semi) desert vegetation and more humid savannahs is still governed by the rainfall gradient to a certain degree, but the combined effect of orographic (topographical) and soil differences dominate in East Africa. For both regions, characteristic associations or mixtures of vegetation units were more typical than zonal classification of natural vegetation, as in West Africa. Examples are the dry savannah/grassland–cultivation mosaics such as in the Masailand (Mwalyosi, 1992), or a mixture of *Acacia* woodlands, (dwarf) shrublands and annual grasslands such as in the Turkana Steppe (Keya, 1998). Prolonged agricultural expansion during the 20$^{th}$ century had transformed large areas of natural savanna woodland into a mosaic of permanent or temporary fields and naturally regenerating fallows aside with undisturbed zones. Similar to anthrophytes in West Africa, herbivores in the savannas of East Africa are known to have altered ecosystems substantially, with domestic livestock contemporarily overriding the effects of indigenous species (Keya, 1998). The savannah landscapes are thought to be highly resilient due to their dynamic response to substantial short- and medium-term rainfall variability (Dube and Pickup, 2001), being the baseline of some contemporary vegetation changes. For example, in the Masailand of Tanzania, woody vegetation decreased from 630 to 144 km$^2$ in 1987 as relative to 1957, i.e., by 77% in total, or 2.1% yearly, with annual losses thus ranking far above the national average which had been 0.9%. On reverse, and to some degree due to deforestation, the area of grassland was found to have increased by 16% in the thirty years from 1957-87 (Mwalyosi 1992). In the Turkana Steppe, biomass production declined for grass, forb and shrub formations, and virtually no standing biomass existed on the grazed parts in the dry season. The productivity of palatable plant biomass for livestock had been on a decline, while invading species, undesirable to livestock, had increased such as *Heliotropium steudneri*, *Amaranthus sp.*, *Tribulus terrestris*, *Solanum dubium* and *Indigofera sp*. In the arid lands of northern Kenya, livestock herbivory had influenced biomass dynamics of the herbaceous layer claimedly to a considerably greater extent than abiotic factors such as rainfall did (Keya 1998). In the central Kalahari, ephemeral grasses such as *Urochloa trichopus*, which do not protect the soil in periods of droughts, dominated in the 1980-90 period. An estimated one third of natural savannah area had been invaded and replaced by woody weed species such as *Dichrostachys cinerea* and *Maytenus tenuispina*. The low-density cover of these species does not develop back into denser growth even during good rainfall. Also, in the southern Kalahari, grasses and palatable dwarf shrubs were replaced by woody shrubs and unpalatable dward shrubs (Ringrose, Vanderpost and Matheson, 1996; Palmer and Rooyen, 1998; Dube and Pickup, 2001).

Sandy soils dominate the savannahs of East and southern Africa, as they do in West Africa. The Kalahari sandveld is a paramount example of this, consisting of deep aeolian sand deposits, longitudinal dunes, pans and fossil valleys (Dube and Pickup, 2001). As in West Africa, despite of their highly weathered status and low inherent fertility, the high infiltration rates on especially fossil dune soils attracted considerable cultivation activities (Palmer and Rooyen, 1998). In the transitional zone from plains to pediplanes and mountains, mixtures of sandy plain and other mountain soils prevail, such as loamy or clay soils (Keya, 1998). Physical soil changes had been observed in present times. In parts of the Masailand, bare ground cover increased from 81 to 107 km$^2$ in 1987 as relative to 1957 (i.e., by 33%), and denudation rate in the Ardai Plains of the Masai Steppe rose from 5.3 to 8.0 t per ha and year when the two consecutive 10-year periods 1959-69 and 1969-71 were compared (Mwalyosi, 1992). In the central Kalahari, 5% of natural savannah had been converted into bare ground cover during the first half of the 1990s, which did not develop back to vegetational cover even during good rainfall (Ringrose, Vanderpost and Matheson, 1996).

*Sedentarization, Intensification and Rising Conflicts about Land*

Only few natural rangelands were reported to be solely used by nomadic households, such as in the (hyper)arid parts of the Republic of Sudan (Kordofan, Darfur), in the interior of the Gourma Region of the Niger River Delta, and in the Turkana Steppe of northern Kenya (Keya, 1998; Ayoub, 1998; Benjaminsen, 1993). Traditional nomadic grazing is century-old, but does hardly persist in undisturbed form until present. It was characterized by well-adapted species, such as camels and goats, and seasonal movements to transitional grazing land pastures. Although large numbers of livestock died, the system was adapted to erratic environmental conditions and low human population densities. Thus, the impact of pre-sedentarization land use on drylands had been rather low.

However, in most of the cases, traditional pastoralism was reported to have undergone changes in traditions and rights which already reduced the possibility of controlling natural rangelands. A characteristic sequence of land tenure transformation was that flexible modes of nomadic pastoralism were reduced or eliminated, and increasingly replaced by transhumant modes of livestock keeping. The latter could easily turn into extensive, village-based and open communal grazing systems, which occasionally shifted to fence-controlled ranching systems under individual management and ownership, mainly in East and southern Africa. Typically, the areas on which previous livestock uses got undermined and new, more intensified land uses were practised became subject to particularly high grazing pressure.

The settlement of former nomads was motivated out of a variety of reasons among which political motives ranked high. They were related to the needs of post-colonial nation building (e.g., control over territory, taxation, etc.). In addition, there are recent development pressures, and the constant food demands of growing human populations in the second half of the $20^{th}$ century. Sedentarization meant that the herds of nomads have been concentrated on grazing land around their new homes. This often implied that cattle industries, such as those of the Dinka and Nuer in Sudan, got located in areas where drinking water was available, or where it had been made available through ground-water extraction activities. Most commonly, increases in herd size and high grazing densities follow both the introduction of borehole technology, which started in the 1950s (or later), and improvements in veterinary services. The rise in livestock numbers went mostly hand in hand with changes in livestock composition. New, but also less well-adapted species such as cattle, donkeys, horses, and sheep were introduced. In particular in East African grasslands, they were added to indigenous herbivore species, overriding the impact of wild animals' grazing there. Examples are plentiful according to which quick overstocking occurred especially on permanent use rangeland while seasonal use rangeland got less affected. The intensive use of vegetation through permanent grazing, often in combination with fuelwood collection and digging for herbs, radiates from the village or water points on permanent rangelands. These were the points where people kept their livestock or small ruminants at night, and from where individuals dispersed daily to gather fuelwood and other products, or where large number of especially cattle gather daily and trampling effects were considerable. Thus, threats to dryland ecosystems were mainly reported from permanent rangelands, from around boreholes or watering points, or densely populated human settlement areas (Ringrose and Matheson, 1992; Ringrose, Vanderpost and Matheson, 1996; Keya, 1998; Turner, 1999a,b; Dube and Pickup, 2001).

Today, much more widespread than traditional nomadic pastoralism are the various modes of extensive grazing, including semi-nomadic pastoralism. For example, a rather homogenous mode of semi-nomadic pastoralism exists in the St. Louis and Louga area of Senegal. It is characterised by traditional live fencing and social fencing of village agreements, and allows natural regeneration of shrubs and trees, while uncontrolled browsing by livestock threatens trees and shrubs (Gonzalez, 2001). Similarly, in the Senegal River Basin, overgrazing by livestock was reported to occur only if deforestation reduced the remaining vegetated and highly resilient pasture (Venema, Schiller, Adamowski and Thizy, 1997). In the Ouadalan of northern Burkina Faso, livestock belongs to local farmers and to nomadic or transhumant pastoralists (or is herded by the latter group), and is concentrated around lakes if they hold water in the dry season (Rasmussen, Madsen and Fog, 2001).

Extensive grazing dominates African drylands, and cases in which sole activities of intensive livestock keeping were reported are few. Two rather divergent examples stem from the southern Okavango Delta of the Kalahari Sandveld in Botswana and the Kano Close-Settled Zone of the Sahel-Sudan zone in northeastern Nigeria. For example, beef production in Botswana was reported to be the second most important national economic activity (after mining). Cattle numbers had increased from ca. 670,000 in 1939 to more than 3.5 million in 1978, although numbers fluctuated with drought. The increases came after the introduction of borehole technology, improvements in veterinary services, and the development of export markets (servicing the European Union, in particular), accompanied by changes in land tenure which supported semi-commercial grazing operations under tribal law under the Tribal Grazing Land Policy (1975). In order to relieve grazing pressure from communal land, grazing land under village agreements was zoned into relatively small leasehold farms of 6,400 ha, centred on boreholes, i.e., ranches were leased to pastoralists for private use. There could be free movement of livestock between the ranches. Ranches were planned to carry a limited number of animals, but many exceeded this limit and were poorly managed (Dube and Pickup, 2001). In northern Nigeria, extremely high livestock densities were made possible through crop-livestock integration which is based on crop residues, under conditions even of high human population densities. Where grassland had been still available, it was common access rangeland, but the major trend was towards zoning of grazing areas, closing of the land frontier, and increased land use intensity achieved by mixed animal-agriculture-tree planting activities (Mortimore, Harris and Turner, 1999).

The sedentarization of nomads has been associated with rising competition and conflicts about land between farmers and herders. The reasons were that croplands expanded on transhumance corridors, herding flexibility as well as herding labour and skills were reduced, and livestock ownership increasingly shifted to settled farmers, merchants, and their urban lobby groups. Arenas of cooperation and mutual benefits between farmers and herders turned into arenas of conflict and tension (Mortimore, Harris and Turner, 1999; Turner, 1999b; Rasmussen, Madsen and Fog, 2001). In the rather short period from 1986-90, approximately 3,700 km$^2$ of natural rangeland were lost to millet clearing in the semi-desert zone (13,600 km$^2$ were lost to erosion and encroachment by sand dunes), and beginning millet encroachment on residual rangeland was also observed in the sub-desert zone. Similar changes occurred in the semi-arid zone further south, but on a more extensive scale. Increases in cultivated areas for both millet and sorghum at the expense of natural vegetation cover were at 16,600 km$^2$ (as for comparison: 116,000 km$^2$ were lost to erosion and sand encroachment), and the spatial pattern of land use in the late 1990s were large cropped fields with relatively small areas of extremely intensive stock grazing in between (Ringrose and Matheson, 1992).

*Traditional Hand Cultivation and Multiple Constraints in West Africa*

The cropping pattern of African drylands falls into two major categories, which both culturally (or institutionally) and agroecologically mean very different predispositions for vulnerability towards land degradation. It emerges that predisposition is lower in West than in East and southern Africa.

The major agricultural rainfed crops in the Sudan-Sahel of West Africa, and similarly in the Republic of Sudan, are pearl millet (*Pennisetum glaucum*), sorghum (*Sorghum bicolor*), cowpea (*Vigna unguiculata*), and groundnut (*Arachis hypogaea*). These are traditional crops which are mainly grown in grain-legume mixtures, and had thus been adapted not only to local conditions of soil and water, but also accepted in cultural terms. Opportunistic rain cropping of millet and sorghum had been practised, for example, in depressions and foothill zones of the more arid parts, which were also major cattle regions, and fossil dunes had been attractive for cultivation since centuries, or millennia even. The application of animal manures (though often in small amounts) was a major practice to maintain the fertility of cropped lands, while another strategy had been fallowing, or rotational farming at least. Some fields were temporarily abandoned due to low productivity or unavailability of household labour, and thus allowed to regenerate with secondary shrub species. Cash outlays to purchase inorganic fertilizers were reportedly not economical for most small farmers so that artificial inputs were low, or non-existent. The degree of mechanization was low, i.e., grain-legume mixtures were mainly grown on hand-cultivated fields (Ringrose and Matheson, 1992; Mortimore, Harris and Turner, 1999; Turner, 1999a,b; Rasmussen, Madsen and Fog, 2001).

Transhumance had historically played an integral role in sustaining rural livelihoods in the agricultural zone of West Africa. Cropping used to benefit from transhumance, not solely by animals moving out of agricultural areas during the growing season, but by well-fed animals supplied to the agricultural zone after crop harvest. This had been important not only because these animals are the wealth stores of agriculturalists and pastoralists alike, but also because animals in good physiological condition more efficiently provide manure to agricultural fields. Also, land productivity was multiply constrained (not only for ecological reasons of rainfall variability). Cultural constraints had mainly been due to nested systems of land rights, i.e., seasonally differentiated use rights to a piece of land by groups such as households, lineages, villages, groups of villages and pastoral clans. Therefore, no or hardly any syndromes of land enclosures were observed across all over the Sudan-Sahel, with just some increases in land sales in some heavily populated areas such as the Kano Close-Settled Zone. Due to the nested land use rights, investments of labour, manure, fallowing time, and fertilizers were intrinsically short term with limited productive benefit beyond five years.

Two examples of integrated crop and livestock management in the Sudan-Sahel are as follows. First, in the densely populated zones of northern Nigeria, some rural household members participate fully in cropping, others moved south with the cattle in the dry season. Farmers kept cattle for fattening, animal traction and transportation, with cattle either kept on mutually agreed and reserved common access grazing land, or tethered throughout the year consuming a large amount of stored hay, with straw produced from farmland. The principal livestock were small ruminants (goats and sheep), whose manure was systematically composted and distributed to the fields. Sorghum and millet stover was privately reserved on the farm, and scarce hay and straw kept to feed stock in the first part of the rainy season, before growth of farm weeds. Land scarcity had forced the pace of integration between crop and livestock husbandry, and current smallholders' farmland produces more biomass than, for example, reserved forests (Mortimore, Harris and Turner, 1999). Second, cultivation of paddy rice on flooded plots occurred on a rather smaller scale, and had mostly been combined with other activities such as fishing aside with pastoralism, for example, in the alluvial plain of the Niger river. There had either been mutual benefits for transhumant or nomadic pastoral groups (such as the Tuaregs) and sedentary millet cultivators (such as the Songhays), or groups had been involved in both land uses as semi-nomadic pastoralists and cultivators alike (such as the Fulani or Fulbe). At so-called 'fixation points', nomads had been granted rights to land where they could stay part of the year to cultivate paddy rice and the bourgou plant used for animal fodder. Although most livestock-rearing Fulbe also cultivated rice, the major means by which they subsisted was by bartering or selling milk to obtain grain (millet or rice) (Benjaminsen, 1993; Turner, 1999a).

The West African Sudan-Sahel had seen prolonged agricultural expansion, especially during the $20^{th}$ century when large areas of natural savanna woodland were transformed into a mosaic of permanent or temporary fields and naturally regenerating fallows. It clearly emerges though, that land degradation can hardly be associated with land use change. Rasmussen, Madsen and Fog (2001), for example, found that continuous cultivation of fossil dune sites of the Oudalan did not necessarily end up with signs of land degradation. Mortimore, Harris and Turner (1999) note that many trees were found to be scattered in farms and on field boundaries which had been preserved, or planted, because they provide valuable fruit, medicines, building materials, browse and shade, thus enabling higher biomass production than measured on non-cultivated sites. Also, crop failures were mainly attributed to droughts rather than to inappropriate land use in central Mali (Benjaminsen, 1993). As for central Sudan, Helldén (1991) reports a severe impact on crop yield during the 1965-74 drought, which was followed by a significant recovery as soon as the rains returned. Most of the annual deviations from the mean annual production of major rainfed agricultural production systems (as well as natural ecosystems) could be explained by climatic variation. Similarly, Olsson (1993) found that during the 1984-85 drought in the same area, crop yields of millet and sorghum dropped to about 20% of normal, and that the variations were explained by climatic factors to a fairly high degree.

*Vulnerability Towards Desertification in East and Southern Africa*

In contrast to the West African Sudan-Sahel, predisposition for land degradation is much higher in the savannah ecosystems of East and southern Africa. The key crops reported from cases located in these drylands were maize (*Zea mays*), wheat (*Triticum*) and seed beans (*Phaseolus vulgaris*), all of them carrying more options for mechanized farming than the traditional West African crops. Integrated livestock and cropping activities were found to be next to absent in the savannahs of East and southern Africa. It appears that subsistence shifting agriculture could hardly compete with livestock raising which expanded at the expense of subsistence croplands, due to the increasing demand for pastoral land triggered by land redistribution policies. No nested systems of land rights and no multiple constraints existed on land. Rather, the value of land got often linked to directional programmes such as the Kalahari West/East Rural Water Supply Scheme in South Africa, which was started in 1985 to develop surface water resources by piping water from the Orange River to the southern Kalahari (Palmer and Rooyen, 1998). While no cases of shifting cultivation were reported from West Africa, it appears as if dryland shifting cultivation in East and southern Africa had rapidly turned into sedentary, mechanized farming or intensified, specialised livestock raising, with both activities showing much higher degrees of vulnerability towards desertification than rotational and livestock-integrated farming systems in West Africa.

For example, in the Okavango Delta of Botswana, small-scale subsistence crop production had still been practiced aside with (more important) subsistence cattle grazing during the 1983-96 period, in two farming modes: floodplain farming, which depended on Delta floods, and dryland shifting cultivation in the more sandy areas. For a certain Kalahari Sandveld study area (Ngamiland District), a main permanent village center was typical with several semi-permanent settlements grouped around it. Only during recent times have political changes and development pressure, together with population increase, led to a more sedentary population, and also to land redistribution, increased land use intensity and the introduction of (semi)commercial grazing. Contemporary land use in combination with rising population densities, permanent settlements, and reduced flexibility of herds, limited the time for ecosystem recovery between periods of localised, more intensive uses. In the period under study, a series of droughts and the reduction of Okavango Delta floods had further constrained farming activities in the area (Dube and Pickup, 2001).

For example, in the Masailand of Tanzania, as in many parts of Tanzania prior to the national government's Villagization Programme of 1974-76, a 16-years' rotational fallow had been practiced under traditional smallholder farming. It was a long enough period to allow for the natural recovery of soil fertility, i.e., four years of continuous cropping followed by ten or fewer years of fallow. Villagization meant that former pastures had been reallocated to become intensively cultivated plots, with new villages established and some old ones relocated. Current national development priorities continue to favour cropping, promote the increase of cropped land, and transform pastoral into agro-pastoral or farming societies. Today's village agriculture is based on the continuous cropping of mainly maize, but using inputs such as manure or artificial fertilizer to restore soil fertility is not widespread. Hardly any soil erosion control measures were done, apart from some old worn-out contour ridges and intercropping maize with pigeon peas. To most smallholder farmers, artificial ferilitizer was too expensive and some did not even have livestock from which to get manure. High immigration of agriculturalists, with small average land holdings of about 2 ha, was characteristic, and rules of village agriculture applied to immigrants, too: all designated cropland was cultivated each year, and settlement had to occur in delimited villages. Increased smallholder production had mainly been achieved by putting more land under cultivation. Ox-drawn ploughs and, to a far lesser degree, tractors were the most commonly used farm implements, mainly for the primary treatment of soils. The number of ploughs increased from 195 in 1983 to 1,530 in 1989, while tractors increased from 89 to 196 during the same period. However, further crop management was not mechanized, and the introduction of ploughs and tractors did not result in increased yields, but in the extension of farmland only. The expansion of cultivation had reduced grazing land not only for livestock, but also for game animals, thus intensifying overgrazing. Small peasant agriculture had reportedly been practiced aside with large agricultural farms cultivating wheat on the most fertile areas of the region. This land got alienated for European settlers during the 1940/50s in a policy to eradicate tsetse, and reportedly had remained under leasehold to date. Since 1971, more large-scale farms, or estates, had been established, mainly for the production of seed beans. In particular, maize contributed directly to soil degradation through wide spacing of rows exposing a large surface area to raindrop impact and rillwash throughout the growing season. Also, mechanized production of wheat, although grown on a smaller scale, involved ploughing and harrowing twice, while the soil remained exposed to structural deterioration and wind-erosion. In wheat farming, the rate of mechanization had been very fast, especially due to disk ploughs used. Low rainfall usually limited crop productivity, as did the predominantly poor soils (except on European farms). Consequently, maize and seed bean yields had been on a decline in the time period studied (1957-87), and soil erosion was widespread (Mwalyosi, 1992).

*Harvesting of Ecosystem Products, Irrigation Schemes for Remote Demands*

In the following, two examples are given of harvesting dryland ecosystem products, and one example of irrigation measures in river and basin sites.

The collection of plant and/or animal products from African drylands has a long land use history, and it appears as if only long-lasting impacts of it constitute proximate causes of desertification. The open woodlands, for example, or semi-arid natural mountain stands of *Juniperus oxycedrus, J. phoenica* and *J. thurifera*, dating back to the Tertiary and occurring in isolated parts of the western Mediterranean basin had been used for several millennia. In the Middle and High Atlas mountains of Morocco, they have a marked open structure which classifies them as juniper savannas or a low-density open woodland steppe, and they are mainly found at the tree line between elevations of 2000 and 2500 meters. The species consist of fairly old trees, probably more than several hundred years old, which are often scarred, with a steppic shrub strata, and with evergreen oaks and cedar as two concomitant arboreal species. In Algeria, thuriferous juniper, mixed with cedar, is strictly limited to the Aurès Mountains with a number of scattered and often very large trees, probably the remains of more extensive Juniper stands. Leaves are collected to feed livestock (sheep and goats) and to produce a type of tar, along with oil distilled from *J. oxycedrus* in veterinary medicine. The contemporary stands of thuriferous juniper are heavily degraded in the Moroccan Atlas (ca. 30,000 ha), and only small stands exist in Algeria. Increasing human activities go in synergy with limited natural regeneration, leading to the die back of *J. thurifera*. However, arthropods attacking the cones and seeds as well as the possible influence of a change in climate cannot be ruled out (Gauquelin, Bertaudière, Montes, Badri and Asmode, 1999).

Another example is the gathering of dry natural vegetation, in particular fuelwood for household consumption. Where wood is easily accessible and means of transport are available, fuelwood tends to be collected more frequently. In the Gourma of central Mali, wood collection had been combined with the harvest of fonio (*Panicum laetum*) which is a wild grain almost exclusively harvested by wood and charcoal traders as a low status trade and seasonal coping strategy for women mainly. People were not cutting trees, but picking dry wood up from the ground. Since rainfall is fluctuating, dry periods of varying duration leading to some degree to deforestation are part of the Sahelian ecosystem. Therefore, the availability of large supplies of dry wood in the Gourma is a logical consequence of the ecological conditions. In particular, drought years in the 1970s and 1980s ravaged wide areas of woodland, so that 'cemeteries' of dead trees could be located where just a few years previously there had been forests. This was the case in places far from fixed habitations, which were very little exploited by people and livestock, and where causal factors other than drought had to be excluded (Benjaminsen, 1993). However, other examples tell different stories of a largely damaging impact of fuelwood gathering upon dryland ecosystems (Ringrose and Matheson, 1992; Mwalyosi, 1992; Ringrose, Vanderpost and Matheson, 1996; Venema, Schiller, Adamowski and Thizy, 1997; Gonzalez, 2001).

Irrigation agriculture in African drylands has reportedly been practised along rivers, and rice was the newly introduced, water-demanding key crop to be cultivated in large-scale as well as village-scale rice farms, for example, in the Senegal River Basin. Rice is commonly the primary food staple of urban consumers in West Africa, but countries such as Mauritania and Senegal suffer from deficits in the production of staple cereals crops, and have only limited zones where rice could naturally be grown. Thus, they are used to rely upon imports (of wheat and rice) which contributed greatly to foreign debt. In particular, the huge foreign exchange burden of rice importation dictated an agricultural development policy which stressed the establishment of large, state-managed rice plantations in order to maximise rice self-sufficiency. The flows of Senegal River were regularized and dams constructed, thereby permitting annual double cropping, with the expectation that productivity would rise. However, the large-scale (>1000 ha), irrigated and heavily subsidized rice farms in the Senegal River Basin had shown poor performance only. Thousands of hectares of former rice production areas had been abandoned because of salinization, due to the difficult and poor drainage maintenance of large volumes of water to be managed, due to high evaporation rates, and due to poor natural drainage conditions of especially the lower flood plain and delta. There is now a shift to private investments in still large irrigation projects, but also to smaller, village-scale irrigation using simple motor pumps instead of hydrotechnical installations, the outcomes of which have still to be assessed (Venema, Schiller, Adamowski and Thizy, 1997).

**European Drylands**

The European drylands are characterized by three major variants of one dominant narrative which captures the coupled land use and environmental history of the whole region:

- a millennia-old tradition of agro-pastoral land use (extensive grazing, cropping), with only contemporary mechanization and abandonment of farming land;
- land use overriding the impact of climate, and topography in combination with various lithologies exerting a strong influence on soil development; and
- fire in combination with land use practices shaping a highly resilient vegetation which in turn reflects various states of soil degradation.

*Millenia-Old Agropastoralism, Current Mechanization and Land Abandonments*

Throughout Mediterranean Europe, land had been subject to intense agro-pastoral land uses since about 7,000 years. Contemporarily, land abandonment as well as intensified farming is a striking feature, while the harvesting of products from either natural or managed dryland ecosystems (mainly forestry and fuelwood collection) has reportedly no or low importance. Cropping seems to slightly outweigh livestock activities, and infrastructural uses rank low if compared to other dryland regions.

Extensive grazing by goats and sheep is most typical for the Mediterranean drylands of southern Europe. It is done in a rather excessive manner, and with seemingly ever increasing numbers of livestock, especially in remote hilly or mountain environments. Taking the case of Greece, Hill, Hostert, Tsiourlis, Kasapidis, Udelhoven and Diemer (1998) report that since the country joined the European Community in 1981, grazing in mountainous regions has greatly increased due to subsidies that became available through the Common Agricultural Policy, and due to new access roads, supported by the community's regional development and cohesion funds. This brought more animals, food supplies and humans up to hitherto remote mountain zones. In most of the other cases, sheep pasture was either part of a rotational system – such as wheat production, fallow, and cork extraction in the Alentejo Region of southern Portugal (Seixas, 2000) – or followed the abandonment of previously cultivated fields. The latter situation had been the most widespread one. Examples are the hilly areas of the Greek island of Lesvos, where moderate grazing constituted the main post-abandonment land use, with growing shrubs occasionally cleared by using fire (Kosmas, Gerontidis and Marathianou, 2000), the Sierra de Gata in Almería Province of southeastern Spain where today's rangelands on hillslopes for sheep and goat had reportedly been agricultural land before (Vanderkerckhove, Poesen, Wijdenes, Nachtergaele, Kosmas, Roxo and Figueiredo, 2000), and the Guadalentin catchment in southeastern Spain where grazing by sheep and goats occurred mainly in the matorral as well as on abandoned agricultural fields (Wijdenes, Poesen, Vandekerckhove and Ghesquiere, 2000).

Rainfed production of annual crops and perennials is most typical for the region (while no cases of irrigation agriculture had been reported in the cases). The most widespread annuals had been cereals, mainly wheat, and, to a lesser degree, rye. The most common perennials had been vines, olives and almonds. Permanent cultivation was common, with some rotational farming mentioned for the Alentejo of southern Portugal (Seixas, 2000). Farming appears to be limited to the pediment and lower slope zones of landscapes, while mountain areas were mainly used by sheep and goats. The cases provide evidence that during the second half of the 20$^{th}$ century abandonment of cultivated fields, mostly on a permanent, but also on a temporary basis, was a striking feature. Post-abandonment uses commonly implied grazing and, occasionally, intensified cultivation on the few remaining, most fertile plots. Only Bragança of northern Portugal had been cited as an area where continuous cereal cultivation had occurred from the past until present (Vanderkerckhove, Poesen, Wijdenes, Nachtergaele, Kosmas, Roxo and de Figueiredo, 2000). On the hilly island of Lesvos, rainfed cropping of cereals, vines, and olives had been the major land use some 40 to 45 years ago, but the majority of cultivated sites had been abandoned due to low productivity. After abandonment, the area was moderately grazed, with fires occasionally used for clearing growing shrubs. On the few remaining cultivated plots, present cultivation practices had not significantly changed, if compared to the pre-abandonment period. There were some exceptions, though. Crop husbandry usually included animal labour, and additional use is now made of small, two-wheel tractors. Animal manure had been traditionally applied as the only fertiliser, and inorganic fertiliser, in combination with manure, is widespread now (the use of pesticides had been greatly restricted to protect grazing animals from poisoning). Full mechanization was not easy and did not happen on a large scale, because of steep slopes and narrow terraces restricting the use of heavy machinery. However, bulldozers had occasionally been used to remove rock fragments and boulders to support mechanical cultivation, such as on volcanic lavas. The consequence was reported as a mixing of the A-horizon with sub-surface soils, triggering a drop of organic matter content (Kosmas, Gerontidis and Marathianou, 2000). In the Sierra de Gata of Almería Province, today's hilly rangelands, especially the lower slopes, had been used to be cultivated for wheat production, with most of abandonment dating back at least 40 years. Some slopes were still cultivated until 10 to 20 years ago, with a thin plant cover having colonized all lands since abandonment (Vanderkerckhove, Poesen, Wijdenes, Nachtergaele, Kosmas, Roxo and Figueiredo, 2000). And, the pediment zone of the Guadalentin catchment area was mostly cultivated with wheat, almonds, and some vines, while some of the fields were (temporarily) abandoned. Almond tree plantations, locating most of active gully development, had been on an increase during last decades, mainly at the expense of matorral, with grazing done in the matorral and on abandoned fields (Wijdenes, Poesen, Vandekerckhove and Ghesquiere, 2000).

*No Climate-Driven Soil Erosion*

The Mediterranean climate covers semi-arid to (sub)humid conditions, and has typically a very hot and dry summer and a temperate winter. Characteristic for the more semiarid parts are low precipitation, but intense rainfall events which are mainly concentrated in spring and autumn. Spatial and temporal variability of precipitation is high. The subhumid parts are characterized by higher annual precipitation and a more even rainfall distribution. Throughout Mediterranean Europe, high and very likely increased interannual rainfall variability or oscillations have been observed (Wijdenes, Poesen, Vandekerckhove and Ghesquiere, 2000; Vanderkerckhove, Poesen, Wijdenes, Nachtergaele, Kosmas, Roxo and de Figueiredo, 2000; Basso, Bove, Dumontet, Ferrara, Pisante, Quaranta and Taberner, 2000; Seixas, 2000). The seasonal extremes were traced back to quaternary but also contemporary climate change (Basso, Bove, Dumontet, Ferrara, Pisante, Quaranta and Taberner, 2000; Gauquelin, Bertaudière, Montes, Badri and Asmode, 1999).

The major features of topography, as reported in the cases, included small-scale lowlands, rather narrow valley floor plains, and small-sized inter-mountain (sedimentary) basins, all of them with transitions into the lower and middle slope zones, but also rolling hills and mountains. Slopes had been a widespread feature of the typical Mediterranean relief. None of the landscape units as given above occurred on a large scale. Rather, most landscape units were relatively small in extent and linked to each other in nested geomorphic systems.

Throughout Mediterranean Europe, slopes had been subject to intense agro-pastoral land uses for millennia, which led to the formation of shallow, skeletal soils, and consequently soil erosion. The most widespread soils had been predominantly loamy, sandy, and gravelly. Loamy soils were reportedly well drained, but limited in depth. Sandy and gravelly soils were the second most represented soil types, but associations of various soil types had been common, too. Sandy or silt loams often showed a high rock fragment content or cover, and were formed on a variety of parent materials. Rather than soils *per se*, the soil parent material – i.e., lithology, geological setting, geomorphology – was cited to define plant growth and be critical for ecosystem resilience (Vanderkerckhove, Poesen, Wijdenes, Nachtergaele, Kosmas, Roxo and Figueiredo, 2000; Seixas, 2000) For example, Wijdenes, Poesen, Vandekerckhove and Ghesquiere (2000) mention highly erodible sedimentary deposits which include Tertiary marls and conglomerates as well as Quarternary fills. Locally, these deposits had been more sandy or gravelly, and land degradation such as gully erosion had a closer relation to lithology and land use than to precipitation. Similarly, Kosmas, Gerontidis and Marathianou (2000) found on hilly, cultivated sites that critical soil depths were defined by various lithological formations – rather than rainfall conditions – and constituted determinants of whether land change occurred towards irreversible degradation or potential restoration.

Soil erosion in many ways limited plant growth, and gully – rather than sheet – erosion had reportedly been the most widespread form of soil degradation. 'Badlands' with low, no or ancient human impact on their creation have to be separated from areas which degrade contemporarily. For example, Basso, Bove, Dumontet, Ferrara, Pisante, Quaranta and Taberner (2000) characterize the eroded middle reaches of the Agri valley in S-Italy as 'badlands' due to the combined effects of extreme seasonal rainfall events and geological setting. Differently, Hill, Hostert, Tsiourlis, Kasapidis, Udelhoven and Diemer (1998) observed that soil erosion on grazing land is most widespread, and that many areas existed in mountain zones of Crete which appeared irreversibly degraded or desertified due to heavy grazing pressure. Very active gully development had been observed in both the Bragança and Almería Provinces of northern Portugal and southeast Spain, respectively (Vanderkerckhove, Poesen, Wijdenes, Nachtergaele, Kosmas, Roxo and Figueiredo, 2000). It had reportedly been driven by the coupled effects of rainfall distribution (rather than amount), differences in soil structure and moisture conditions related to various lithologies, and land use change. Similarly, it was found that many banks of incised ephemeral streams in southeast Spain show widespread gullying due to the combined effects of land use, lithology and topography (Wijdenes, Poesen, Vandekerckhove and Ghesquiere, 2000). The enlargement of (wheat, almond) fields and introduction of different crop cover and vegetation types had reportedly changed the runoff potential dramatically over the last decades. The most active gully formation was observed in the cultivated fields, especially in almond groves, followed by those in wheat fields. Marls and almond groves located relatively the most active gully heads, while conglomerates and abandoned land had less. It was concluded therefore that almond tree plantations had been very important runoff generating areas, (re)activating bank gully heads, and that these gully heads were potentially important sediment sources in cultivated fields.

*Resilient Vegetation Reflecting Various States of Soil Degradation*

Microtopography, exposure, lithology and substrate were reportedly critical in creating typical, land use- and fire-driven vegetation mosaics. *Phrygana* had been the most commonly cited contemporary vegetation formation, i.e., a low shrub vegetation such as 'matorral' or 'garrigue' (shrubland, shrubby thornscrub), referring to extensive covers of flammable, aromatic and largely unpalatable plants. In open Mediterranean landscapes, *phrygana* could either exist aside with open woodlands (or forest relics), tree-shrub mosaics, or 'maquis' which are trees reduced to shrubby form, or be intersected with steppe-like grassland patches.

The millenia-old history of agropastoral land use and fire had created these vegetation mosaics, with extensive grazing by goats and sheep being of overriding importance. Not grazing, but intense fires were reportedly able to result in the complete removal of vegetation, while - except for tree stands – the adaption of plant species was reportedly high (Seixas, 2000). Frequent and often uncontrolled fires, either through lightning or set by humans – e.g., shepherds clearing growing shrubs on abandoned land for grazing – mean a considerable threat especially to the remaining woody cover (Hill, Hostert, Tsiourlis, Kasapidis, Udelhoven and Diemer, 1998; Gauquelin, Bertaudière, Montes, Badri and Asmode, 1999; Kosmas, Gerontidis and Marathianou, 2000). The impact of fire was reportedly greatest in those areas with the lowest fire frequencies, because an increase in fire frequency would lead to fewer plant species, caused by the loss of those which could not persist when fire became too frequent. The mosaics could thus be interpreted as different stages of a degraded, but highly dynamic and resilient vegetation. On fallow land, vegetation usually recovered quickly in the form of grasses, interspersed by colonizing Cistus plants, for example, which recovered also on abandoned fields in the form of semi-natural or secondary vegetation (Vanderkerckhove, Poesen, Wijdenes, Nachtergaele, Kosmas, Roxo and Figueiredo, 2000). Next to all of the 'natural' vegetation in the Mediterranean Basin, apart from a few forest stands, had been grazed to some extent, and episodic energy bursts of vegetation, linked to variable rainfall, were mentioned as important ecosystem features.

Natural sclerophyllous evergreen forests (such as pines and oaks) were found to be limited to some relict patches or strongly reduced areas of the indigenous cover, mainly on upper slopes, rolling hills, mountain ridges, and scattered between fields. Semiarid natural forest ecosystems – such as the mountain stands of thuriferous juniper in isolated parts of the western Mediterranean basin – could date back to the Tertiary, but had been altered since millenia by cultivation, grazing, uncontrolled fires, and product harvesting. In the Ebro basin, for example, juniper stands were abundant until the $18^{th}$ century, but today's trees remain only between fields. The natural tree stands had been extensively cleared for arable agriculture, construction (wooden beams, floors, cabinet making), and for fuelwood provision, in particular during the Spanish civil war, and were also used for producing liquor, as livestock fodder, or as ornamental trees (Gauquelin, Bertaudière, Montes, Badri and Asmode, 1999).

**Australian Drylands**

The Australian drylands are characterized by two major narratives, both dealing with features of coupled land use and environmental histories:

- European colonization overriding Aboriginal land use, and commercial pastoralism turning out to be increasingly less capable of coping with changing environmental and economic parameters; and
- episodic, decade-long shifts in rainfall triggering on-going vegetation and land use changes, land degradation and property restructuring due to economic misery.

*European Colonization, Commercial Pastoralism*

Prior to the arrival of Europeans, Australian drylands carried *spinifex* grasslands as the most extensive rangeland vegetation type, covering about one fifth of the Australian territory, which was used by hunter-gatherers and grazed by native herbivores. Aboriginal people had manipulated vegetation patterns and animal populations by burning, which created a mosaic of vegetation in various stages of recovery since burning. The mosaic was fine grained close to waters and other inhabited areas, and coarser further out, eventually merging with the very coarse wildfire-generated pattern in waterless, infertile areas. From about 1850 onwards, European pastoralists displaced the native people and introduced sheep and cattle. The fine-grained pattern disappeared as Aboriginal land use declined and most *spinifex* areas were then swept by large wildfires, with the loss of habitat triggering large-scale native animal extinctions. Be it the introduced exotic animal species or feral (wild) animals – e.g., horses and donkeys in northern Australia and, especially, the rabbit in the southern part of the continent – there is little contention that all the newly introduced European land uses were direct causes of dryland degradation, especially when poorly managed and in combination with climatic variability (Pickup, 1998).

From their introduction into the Australian dryland ecosystems, sheep and cattle were usually grazed around a limited number of natural waters, often in very large number by current standards. By the early 20$^{th}$ century, severe land degradation around many of the natural waters were common, and massive stock losses occurred during droughts. For example, in New South Wales sheep numbers reached 19 million in the 1890s, fell to 3.5 million in the drought of 1901-02, and never recovered subsequently. Pastoral development came later in Western Australia where the number of sheep increased from 1.6 to 4.8 million between 1900-30, and fell to 3 million in 1960, largely as a result of droughts and land degradation. Development reached central Australia later, and cattle numbers grew to a peak of 360,000 between 1900-58, crashing to 120,000 in the 1960s drought, and then rising to another peak during the record wet years of the 1970s, after which there was a decline. The increase in animal numbers was accompanied by an expansion of the grazed area due to the provision of artificial waters, and the area might have escaped much of the severe localised early degradation experienced elsewhere. Most pastoral development, following the initial degradation events, involved extension of the grazed area using artificial waters such as dams and boreholes, and the fencing of land into large paddocks. For example, extensive grazing for beef production in the Northern Territory is reportedly centred on artificial watering points, which restricts the distance over which animals range and which produces greater animal impact per unit area closer to water. Driven, among others, by the increased availability of earthmoving equipment for dam construction, the opening up or more land for grazing has continued until present. Another important development since World War II has been the expansion of the road network and improvements in road transport which allow for rapid removal of stock from drought-affected to other areas. However, despite continuing improvements in farm productivity, many of the 6,000 rangeland grazing enterprises in the arid and semi-arid parts of New South Wales, Queensland, South and Western Australia experience increasing economic pressure due to declining terms of trade. Many ranches are too small or carry too large a debt to be economically viable, especially since recent wool prices have collapsed. These enterprises have increasingly less capacity to cope with drought and to carry out grazing in a sustainable manner (Bastin, Pickup and Pearce, 1995; Pickup, 1998).

As per today, commercial pastoralism, mainly for export markets, is the principal land use on 60% of the Australian rangelands which occupy 70% of the continent, holding a vast arid and semi-arid zone. Pastoralism is carried out on an extensive basis with individual paddocks varying in size from about 1,500 ha such as in the southern sheep production areas, to 500 km$^2$ (or more) such as in the cattle grazing areas of arid central Australia. Individual properties, mainly run as restricted-purpose leaseholds, typically vary from 35,000 ha to 12,500 km$^2$ with 1 to 40 paddocks. Management intensities vary with relatively intensive sheep production and cases in remote areas where semi-feral cattle are tracked down and shipped out for slaughter once a year, or less (Pickup, 1998).

The impact of European colonization in turning *spinifex* grassland ecosystems into rangelands speeded up the episodic disappearance of vegetation due to natural oscillations, especially around artificial watering points, and/or once changes in fire regimes occurred. Fire frequency greatly increased after wet periods, but at the same time, fire frequency had been reduced over time because traditional burning practices by Aboriginal people had largely ceased in most areas. This had been due to the fact that Aboriginals were dispossessed and moved from their traditional lands, because herbage fuel had been reduced by grazing, and because pastoralists had excluded fire as a land management practice (Bastin, Pickup and Pearce, 1995; Pickup, 1998).

## Decade-Long Rainfall Shifts Triggering Land Change and Degradation

Australian drylands are mainly composed of alluvial and sandy plains with sand dunes. There had been major variations in climate throughout the Quaternary, and contemporary climatic conditions cover arid to semi-arid conditions. In the more arid areas of the continent, carrying shrubs and dry grasslands such as the hummock grasslands (*Triodia*), short term rainfall is subject to substantial variability. Differently, rainfall tends to be seasonal in the less arid parts, with southern Australia, for example, dominated by winter rainfalls, and with the semiarid fringe carrying associations of grasslands, tree-shrub mosaics and *Acacia* woodlands. There was no reported rise in rainfall variability, but some evidence was cited that extreme weather events had increased (Bastin, Pickup and Pearce, 1995; Pickup, 1998).

Most importantly, and different from other dryland areas experiencing variable rainfall as well, Australian drylands are characterized by shifts in rainfall which commonly extend over periods of several decades (rather than single or few years only). Droughts, for example, were experienced in large parts of the continent in the 1890s, 1920s and 1960s, with unusually wet, decade long periods in between, or after. Simple linkages of this pattern to global climate have not emerged. Rather, rainfall variability in the continent's drylands had anyhow been greater than that of comparable climates elsewhere in the world.

The shifts in rainfall over periods of several decades include not only droughts but also the creation of unusually long wet periods. Although Australian drylands had been virtually void of cropping – except on an intermittent basis in favoured locations on the semi-arid fringe – the occurrence of unusually long wet periods repeatedly leads to the expansion of cropping land into previously drier areas, with the result that land degradation and economic misery is triggered there once drought periods occur, and that, consequently, on-going changes in land use and property restructuring occur. This pattern is also true for livestock development, which typically has boom and bust cycles linked to the pattern of decadal rainfall variations. The expanding, contracting, expanding, etc. land uses, in combination with climatic variability as triggering mechanism which affect immediate changes in vegetation cover and land productivity, are the direct causes of dryland degradation, especially if they were poorly managed. Land degradation reportedly increased environmental patterning, i.e., the landscape was increasingly partitioned into areas of bare and eroding source zones and heavily vegetated sink zones (Bastin, Pickup and Pearce, 1995; Pickup, 1998).

The vegetation typically displays a highly dynamic response to rainfall and landscape features, with vegetation cover rapidly increasing after rainfall and then slowly decaying. Different landscape types show varying responses, though, such as fertile alluvial soils supporting higher total cover than impoverished stone-mantled soils. However, the disappearance of vegetation had been speeded up by land use impact such as grazing and/or changes in fire regimes. In the more arid areas of the continent, huge variations occurred. Over the last 120 years, Pickup (1998) identified three long periods of below-average growth of herbage cover in central Australia – resulting from dry periods starting in the late 1890s, the late

1920s, and the early 1960s – and two exceptional growth pulses – one during 1920-21 and the other during 1973-75 – with most vegetation cover change occurring with greater frequency and in response to smaller rainfall events. During the growth pulses, major recruitment of tree and woody shrub species occurred, which resulted in an increase of woody weeds that affected large areas, thus reducing pastoral productivity. The growth pulses also established herbage on areas where it had previously been lost due to grazing or drought. The response of degraded and undegraded cover to rainfall variability turned out to be similar, but that of degraded vegetation cover was much more dampened and herbage cover had been substantially reduced. It is important to make the point that the episodic, climate-driven response in vegetation and related changes in land productivity occur at a national rather than at a local scale (Bastin, Pickup and Pearce, 1995; Pickup, 1998).

## North American Drylands

North American Drylands, all of them situated in the Southwest of the United States, are characterized by a dominant narrative dealing with historical land use and environmental history, and one dealing with contemporary events:

- biophysical predisposition for desert formation, including severe droughts, and European colonization triggering vegetation change at historical frontiers leading to overgrazing, with only few farming land available; and
- irrigation farming and urban developments triggering water degradation and contemporary desertification, especially in ancient (pleistocene) inland basins.

### *Biophysical Predisposition and European Colonization*

Since at least 200 years, drylands of the US Southwest have been reportedly dominated by extensive and open grasslands (grassland-savanna, steppe), and, to a lesser degree, by shrublands turning into shrubby thornscrubs at some sites. The climate is arid to semi-arid, with annual rainfall ranging from 100 to 255 mm. Northerly winds in winter usually generate frequent dust storms, leading to the deposition of aeolian sediments (Brown, Valone and Curtin, 1997), while the semiarid parts commonly bear relatively wet cool-seasons and moist warm-seasons (Ludwig, Muldavin and Blanche, 2000). Arid to semiarid lands had frequent desertification events in the prehistoric past, and at some sites there is still long evidence of wind erosion as a geological process (Okin, Murray and Schlesinger, 2001). Landscape units bear mainly sandy deposits and clay soils, both fine-grained and shallow. Soil erositivity is mostly high, and there was indication of an increase in loosened soil and the destruction of soil crusts, as well as an increase in desert-like land cover such as sands and coppice dunes. The formation of (shrub, mesquite) coppice dunes in particular, could be seen as the first topographic and soil manifestation of increasing aridity and much more sand concentration to come

(Rango, Chopping, Ritchie, Havstad, Kustas and Schmugge, 2000; Okin, Murray and Schlesinger, 2001). Another predisposition for degradation stems from drought periods that have intensified over the past. For example, there were severe winters such as in 1885/86 as well as sequences of drought years which put severe stress on vegetation, such as 1891-94, 1901-04, and 1951-56. The latter period had been the most severe recorded during a 350-year period (Fredrickson, Havstad, Estell and Hyder, 1998). In most of the cases, aridity has reportedly increased, either in terms of on-going drought conditions – such as on the Colorado Plateau and in the Chihuahuan Desert (Rango, Chopping, Ritchie, Havstad, Kustas and Schmugge, 2000; Mouat, Lancaster, Wade, Wickham, Fox, Kepner and Ball, 1997) – or in terms of more sustained winds triggering dust and sand storms or emissions – such as in the Chihuahuan and Mojave Deserts (Brown, Valone and Curtin, 1997; Okin, Murray and Schlesinger, 2001).

What had been large areas of desert grassland became replaced by desert scrub mesquite to a large extent. Especially following droughts, grass cover could be severely reduced, while scrubs such as mesquite readily established after the event. Vegetation changes in grasslands have continued until present, resulting in an overall reduction of densities and productivity of grasses and herbaceous plants, and concomitant shrub invasion. For example, in the San Simon Valley of the Chihuahuan Desert of southern Arizona, total cover of woody shrubs had increased threefold as by the 1980s, and there was also greater shrub numbers and woody weed encroachment (Brown, Valone and Curtin, 1997). On the Colorado Plateau of southern Utah, Buffalo and blue gramme grass had been removed under heavy grazing in the late 1980s, even in years of above average rainfall (Mouat, Lancaster, Wade, Wickham, Fox, Kepner and Ball, 1997). In the Californian Mojave Desert, *A. polycarpa*, which is a perennial shrub, and annual exotic grasses such as *Schismus* dominate now areas of greatest human disturbance (Okin, Murray and Schlesinger, 2001). Shrubs such as the creosote bush (*Larrea spp.*) and mesquite (*Prosopis spp.*) displaced all valuable native grasses in the Jornada del Muerto Basin area of southern New Mexico, with native grass areas shrinking from 388 km$^2$ in 1858 (or 58% of the total area) to 25% in 1915, 23% in 1928, and 0% in 1963 (Rango, Chopping, Ritchie, Havstad, Kustas and Schmugge, 2000).

Such vegetation changes are generally considered to be undesirable for grazing livestock, because they reduced carrying capacities for livestock, contributed to soil erosion and reductions in stream flows, altered wildlife habitat, and threatened pastoralists or ecosystem sustainability. However, the directionality was not uniform. For example, wind – rather than water – erosion was cited as the major type of soil change with degrees ranging from light to strong. Soil surfaces could reportedly erode slightly at 0.4 mm/year on average in the 1983/84 to 1995 period (Ludwig, Muldavin and Blanche, 2000), and strong impact could lead to soil losses of 3 to 5 cm annually in mesquite dunes and grasses. However, there were also net soil gains of ca. 2 cm, though solely in mesquite dunes (Rango, Chopping, Ritchie, Havstad, Kustas and Schmugge, 2000).

Newly introduced animals during Spanish conquest were held at least somewhat responsible for vegetation change, but there seems to be no unidirectional cause stemming from land use alone (see also the following section on vegetation change in the adjacent drylands of northern Mexico). Ludwig, Muldavin and Blanche (2000), for example, document how desert grassland vegetation on the Otero Mesa of the New Mexican Chihuahuan Desert is very responsive to precipation change over relatively short time periods, showing increases of mesic grass cover and decline of desert shrubs and xeric grass cover in the 1982/83 to 1995 period, with an area grazed by cattle, goats, mules and horses from 1885-1954. Also, Brown, Valone and Curtin (1997) found that episodes of shrub increases in the Chihuahuan Deert region earlier in the 20$^{th}$ century occurred during periods of unusually high winter precipitation which favoured the establishment of cool-season active woody shrubs at the expense of warm-season active grasses. Thus, not the combination of drought and grazing, but increased winter precipitation seems to be the main explanatory variable.

Nonetheless, European immigration into the region, beginning in the mid 1500s, greatly affected the Southwest. Spaniards introduced cattle and horses along the Jornada del Muerto in New Mexico at the turn of the 16$^{th}$ and 17$^{th}$ century. However, indigenous nomads became quickly skillful horsemen, and, thus, prevented the large-scale colonization of the Southwest for some 250 years following. Especially the Navajo adopted many husbandry techniques of the Spanish extensive grazing system. All in all, and mainly because of the nomadic attacks, settlements were small and localized, as was the impact of new European land use on the land. However, what had been large areas of desert grassland became replaced by the desert scrub mesquite, and the newly introduced animals were at least somewhat responsible for it: fed to horses and cattle, it resulted in the spread of mesquite beans around settlements and campsites (Fredrickson, Havstad, Estell and Hyder, 1998).

Grazing animals were reportedly cattle as the principal animal, but also goats, sheep, mules, and horses. Most strikingly, present grazing reportedly exerts only light grazing pressure, while cases of overstocking due to drastic livestock increases were mainly due to past grazing, with a period of drastic overstocking reported from the 1880s until the mid 20$^{th}$ century (Brown, Valone and Curtin, 1997; Ludwig, Muldavin and Blanche, 2000). For example, Rango, Chopping, Ritchie, Havstad, Kustas and Schmugge (2000) claim that overstocking with cattle and sheep occurred in 1880 in New Mexico due to increases from approximately 0.7 million head of livestock in 1870 to about 5.8 million head by 1885, mainly sheep.

By mid-19$^{th}$ century, the United States had extracted Arizona, New Mexico and California from Mexico, but only after the Civil War of the 1860s, colonization of the Southwest became intensive. Soldiers were assigned to subdue raiding nomadic peoples, and large numbers of unemployed Americans sought their fortune in the West, where free grazing was offered to whoever reached it first. Since little land was suitable for farming, most newcomers raised livestock. The newly constructed railroads provided an effective means to transport cattle to the lucrative markets in the East. Cattle and sheep numbers escalated in the 2$^{nd}$ half of the 19$^{th}$ century. In those times, nomadic shepherds and better organized cattlemen competed for the same public forage, and violent land conflicts arose. Excessive grazing was perceived as a problem as early as the 1880s. This was partly due to limited land only (65 ha), given to settlers through the Homestead Act of 1862. The act adopted eastern United States standards, but was inadequate for pastoralism in the arid lands of the Southwest. The allotments were not increased until 1909 when the Enlarged Homestead Act (130 ha) was passed. Large livestock (and mining) corporations began to buy up land from settlers and obtain public land as well. One method included 'grazing out' so-called squatters by overgrazing land claimed by newcomers. The 1880s and 1890s were largely the period of struggle between large owners and the newcomers (Fredrickson, Havstad, Estell and Hyder, 1998).

A severe winter in 1885/86 and a sequence of drought years during 1886, 1891-94, and 1901-04 ended the cattle boom. Many cattle died of starvation, while rangelands were left severely degraded. Legal contracts did not exist, and little incentive was offered to improve the range. In the early 1900s, considerable efforts, including rangeland research, were started to conserve, for example, range and forest resources in the region, including grazing control and fire suppression. However, large tracts of chiefly less valuable arid lands were not protected, and overgrazing on the public domain continued until a more effective legal framework was created in the 1930s, for example, by creating grazing districts. Before the 1940s, management of public rangelands was largely custodial, and programs were started to help livestock producers maximise forage use. For another 30-year interval following this, improvements were made to increase forage production and stabilize the soil in seriously degraded areas. Drought was the pre-eminent event for much of the Southwest during the 1950s. In the Chihuahuan desert, grass cover was severely reduced, while scrubs such as mequite readily established after the drought. The resulting transition from grass to desert scrub now occurred on land grazed and not grazed by livestock (Fredrickson, Havstad, Estell and Hyder, 1998).

Reflecting the historic rural period, previous events unleashed a cascade of events in the Southwest that ultimately affect the current conditions. The reintroduction of the horse by the Spaniards, for example, greatly affected nomadic people, the lives of sedentary agriculturalists, the rate of the Spanish conquest, and finally the occupation and fundamental transformation of the area by the United States. Currently, total livestock numbers expressed as animal units are well the below the historic average. The efficiency of livestock production has improved to many more management options available. However, due to competition on global markets, prices offered to beef producers have fallen below a point that makes implementing new technologies economically feasible. Grazing issues continue to remain a public concern, less so a problem of desertification, and, in the $2^{nd}$ half of the $20^{th}$ century, rapid population growth and urbanization became the major problems (Fredrickson, Havstad, Estell and Hyder, 1998).

*Irrigation, Urban Growth, and Water Degradation*

From the 1950s onwards, human demography in the arid and semi-arid lands of the US Southwest changed from a predominantly rural society to urban populations becoming a dominant force. Land use shifted towards increased settlement activities, military uses, recreational activities, and other non-consumptive uses, a situation which was reflected in the Multiple Use Act of 1962. The following land laws aimed at improving human interactions with the land (Fredrickson, Havstad, Estell and Hyder, 1998).

Conflicts between rural agriculturalists and urban environmentalists were common, and citizen involvement evoked changes in land use which proceeded at a phenomenal rate. Driven by the rapid growth of predominantly urban populations, irrigated agriculture, recreational activities (vehicles), strip malls and housing developments, military manoeuvres, mining, pipeline, road and powerline construction bear the potential of damaging large areas of land, by destroying vegetation cover and exposing the soil to wind erosion (Okin, Murray and Schlesinger, 2001; Mouat, Lancaster, Wade, Wickham, Fox, Kepner and Ball, 1997). The main topographical features of the US Southwest – i.e., flat plains and flat or gently sloping upland plateaus (mesas) with some, mostly ancient inland basins, deltas, (dried up) river courses, and rolling dunelands – meant a predisposition for processes of desertification, too. It appears as if the most vulnerable or extremely fragile lands were topographically situated at the intersection of ancient (pleistocene) inland sites with human uses such as irrigation farming or mechanized agriculture. In the Manix Basin, for example, a palaeolake basin in the Californian Mojave Desert, ground water recharges were used for several phases of agriculture, with agricultural land use declining and desertification emerging in the most recent past. In the 1800s, the basin was used for dryland farming, and limited irrigation farming started in 1902. The acreage of irrigated land increased sharply after World War II. After the mid-1970s, central pivot agriculture became the dominant form of land use in the area, and *alfalfa* hay is now the major agricultural product. However, many fields have since been abandoned throughout the northern part of the basin due to increasing costs of ground-water pumping. While past agriculture demonstrated unregulated and unmanaged land use for short-term gain, as farming became less profitable, land was simply abandoned to natural degradation processes (i.e., sand blasting and burial of vegetation and equipment, dust emissions) without implementing long-term remediation strategies. The 1980s and 1990s were the period in which large areas of the Manix Basin were abandoned from agriculture and in which the greatest land degradation has been observed (Okin, Murray and Schlesinger, 2001).

Increased water requirements for growing cities and irrigated fields have significantly lowered water tables in the region, thus lowering the agricultural productivity of the land, but also threatening riparian habitat and species. Both upland and riparian species habitat underwent fundamental changes in the second half of the $20^{th}$ century. From the 1940s onwards, agricultural intensification programs lessened the competion for grass by other herbivores than domesticated livestock animals, and thus reduced the number of feral or wild grazing animals. The same program implied the elimination of livestock predators so that, as a consequence, free-ranging grizzly bears and gray wolves were eliminated, and prairie dogs reduced to a few very small populations. With an increasingly urban population from the 1950s onwards, the lowering of water tables and river discharges due to growing cities and agricultural irrigation began to threaten those riparian habitat and species. Urban developers started to buy up agricultural water

Initial Conditions 87

rights, in order to fuel urban growth, and left plowed fields to be eroded by wind and water. Land use laws addressing the shift from rural to urban uses turned out to be seemingly ineffective or insufficient to contend with the new critical drivers of land change (Fredrickson, Havstad, Estell and Hyder, 1998).

## Latin American Drylands

The three narratives, capturing coupled land use and environmental histories and stemming from various dryland areas of Latin America, are all rooted in historic to ancient processes with implications for contemporary desertification:

- contemporarily advancing desertification in northern Mexico through clearance of woody individuals, fire, chemicals and mechanical suppression for large-scale grazing land;
- Still high biotic diversity aside with impoverished rural livelihoods due to woodfuel harvesting and ancient to contemporary soil erosion, induced by farming practices in south-central Mexico; and
- intense and continuous grazing in the dry-cold Patagonian rangelands which turns *tussock* grass steppes into shrub-grass formations (with features of environmental patterning), occasionally even further into irreversibly damaged 'badlands' or deserts.

*Large-Scale Grazing and Advancing Desertification in Northern Mexico*

The northern territories of Mexico, spanning a range from semi-arid to extreme dry and hot climatic conditions, are typical rangelands. All of the land under available forage is used for livestock grazing, which is an estimated 70 million hectares of the Mexican drylands. Vegetation is as diverse as the rainfall and topographical conditions, including mainly dry grasslands (such as in the Chihuahuan and Sonoran Deserts) and shrublands (such as the Thamaulipan 'matorral'). Grazing exerts an important selection pressure in these ecosystems, and the frequency, intensity, extent and magnitude of it had been substantially changed with the introduction of livestock by Spaniards early in the $16^{th}$ century. In the northern part, abundant forage was also used by wild herbivores until mining activities were developed there, and subsequently numerous herds of cattle were moved from southern and central areas to the north. Since then, northern Mexican lands had been used essentially for livestock production. Mainly due to cultural and ecosystem properties, livestock populations show distribution patterns by animal type. Cattle is the most preferred domestic animal in the northern and northwestern states of Chihuahua, Durango, and Sonora, where extensive areas of natural, induced and cultivated grasslands are used to freely graze numerous herds, and bush encroachment appears to be the most important directional change in vegetation. In contrast, small ruminants such as goats are more abundant in the northeastern States of Coahuila, Nuevo León and Tamaulipas. Here, but also on both sides of the Rio Bravo/Rio Grande (i.e., in the coastal plain of northeastern

Mexico and in southern Texas), woody species of the Tamaulipan 'matorral' have been the dominant life form for centuries. Only since the early 1900s, large parts of the plant community had been cleared to seed artificial pastures, but also for cropland and expanding urban uses. Adaptation of cattle to the brush habitat very likely occurred as early as when Spaniards brought grazing herds to the region. Cattle used the shrubland for warmth and protection during winters and for calving in spring (Manzano, Návar, Pando-Moreno and Martinez, 2000).

Raising cattle for meat production is mostly done under an extensive free grazing management system where rangeland is usually managed collectively under the 'ejido' system, although some areas are used individually for intensive ranching. During colonial times, pastoralism shifted from nomadic pastoral systems to a system between transhumance and the intensive, heavily mechanized grazing domestic system. Livestock activities were carried out by private landowners who have developed range management techniques (originally adapted from North American ranchers) which aimed at controlling shrub invasion in rangelands. Clearly, this private grazing system had contributed to desertification in that large grazing areas in most of the northern region had been cleared of any woody individual through the use of fire, chemicals, and mechanical suppression, either to induce the growth of native grasses or to raise exotic grass species. Although the range management technique underwent some current changes, many private farm rangelands have continually been overused and shrub-cleared in the last few decades of the 20$^{th}$ century. This was especially true during and after the 'Green Revolution' in the 1960s, when sown pastures (and agriculture) started to expand dramatically. For example, agricultural areas in Tamaulipas expanded from 242,800 ha in 1953-54 to 1,310,000 ha in 1980-81, mainly at the expense of thornscrub. A total of 150,000 hectares of thornscrub were totally cleared in Nuevo León during a five-year period in the early 1980s (Manzano, Návar, Pando-Moreno and Martinez, 2000).

Since the Mexican Revolution, displaced private farms have been replaced by the communal or 'ejido' system, which is blamed for most of the severe overgrazing and consequent land degradation, too. While crop production is done on individual plots, rangelands as well as forests are common property lands which were collectively, freely and heavily grazed without charges or regulations for the use of community resources. There is some indication that, when cooperation among landholders in the management of communal ranges fails, more land gets allocated to crops and less to range, but, on the other hand, stocking rates increased drastically. Most recent government initiatives dealing with land issues led to the semi-privatization of communal land, and landholders learnt to make use of increasingly hardy and more agile types of domestic livestock that feed on a wider range of plant species. Also, the combined production of livestock and wild animals on diversified private ranches increased considerably in the last two decades, with more than 20 million hectares utilized for sport hunting activities in northern Mexico alone (Manzano, Návar, Pando-Moreno and Martinez, 2000).

Due to rainfall above longterm average in the 1965-80 period, and in addition to economic and political incentives for livestock expansion, the grazing stock in all of the country increased from 22 million grazing animals in the early 1950s (cattle, goats, sheep and hogs) to ca. 50 million in the mid-1980s. The sharp increases in livestock densities were then continuously reduced to 43 million grazing animals in the mid-1990s, leading to smaller stocks as by now. The drought spell, which had been hitting northern Mexico since the 1980s, might be partially responsible for the reduction in grazing animals, i.e., due to reduction of ecosystem productivity, namely soil erosion, degradation of plant communities, and henceforth reduction of rangeland carrying capacity. Mainly due to overgrazing, desertification is an advancing process. Heavy grazing pressure and mostly exceeded livestock carrying capacities emerged as problems of sustained livestock keeping, especially in the Sonora and Tamaulipas states. Widespread trampling, for example, causes soil compaction or loosening, leading to erosion in these zones (Manzano, Návar, Pando-Moreno and Martinez, 2000).

Vegetation changes in grasslands, as a consequence of livestock overgrazing, had resulted in a reduction of densities and productivity of grasses and herbaceous plants, and concomitant shrub invasion in most of the overgrazed areas of northern Mexico. These changes were considered to be undesirable because they had reduced carrying capacities for livestock, contributed to soil erosion and reductions in stream flows, altered wildlife habitat, and threatened pastoralists or ecosystem sustainability. At least 200 years ago, these areas had been dominated by grassland-savanna. However, directional change in vegetation through encroachment holds true only for arid grass ecosystems. The woody species of the shrubby thornscrub ecosystem, called Taumalipan 'matorral', shows a different degradation path under intense, frequent and heavy grazing which points just in the opposite direction of those of grasslands: from woody species to mainly annual grasses and herbs, leading to a patched plant community, simpler, and more prone to disturbances (Manzano, Návar, Pando-Moreno and Martinez, 2000).

*Impoverished Rural Livelihoods, Woodfuel Harvesting and Soil Erosion*

Different from the (semi)arid grazing lands of northern Mexico are the semiarid uplands with intersected drainage basins at 1,600 to 2,000 m, such as that of Rio Zapotitlán, in south-central Mexico. Here, many areas support a semi-natural, perennial vegetation, either in the form of broken canopies of small trees (e.g., *Lysiloma divaricata*) and arborescent cacti (e.g., *Stenocereus stellatus*), or in the form of a more extensive cover provided by a variety of shrubs and smaller cacti. Some of these areas had been pre-Columbian agricultural fields which were not cultivated any more since Spanish conquest, but left an impoverished 'natural' vegetation due to large-scale soil erosion in a zone which receives seasonal rainfall principally during the summer season from May to September. On remaining areas, extreme soil erosion on hillslopes, leaving only very thin soils on top of weathered rocky parent material, as well as the exposure of bedrock date back to agricultural practices which persist from ancient cultural practices until present. In some instances, pronounced hillside erosion, associated with deposition on adjacent valley floors, could also date back to the Holocene to Pleistocene climatic transition. Although contemporary erosional processes and soil losses are about to reduce species richness in the area, the woodland, tree-shrub mosaic, and shrublands in south-central Mexico still have a high biotic diversity. It corresponds to the nearby Tehuacán Valley which reportedly had an extremely rich flora that is cited to be the 'one of the most biologically rich semi-arid regions of the western hemisphere' (McAuliffe, Sundt, Valiente-Banuet, Casas and Viveros, 2001).

Subsistence agriculture played the most striking part in land degradation. Pre-Columbian agricultural societies modified valley and adjacent basin landscapes in many ways to provide the agricultural production that had both been required by and fostered a steadily increasing population. At the time of the Spanish conquest in 1521 AD, the Tehuacán Valley supported a population estimated at 100,000, which was then severely decimated, principally by the ravages of diseases to which indigenous people were immunologically defenseless. During the $20^{th}$ century, it had again experienced a growing population, which was estimated at about 882,000 in the 1990s. Maize was apparently domesticated in south-central Mexico, and the valley had yielded the oldest yet-discovered remains of maize dated at 4,700 before present. Evidence for increased land use on hillslopes stems from as early as 1300 to 1700 years BP. Maize and bean cultivation are reportedly the major land uses of today's population. The valley as well as the adjacent drainage basins show a mosaic of intensively cultivated and irrigated valley floors and seasonally farmed non-irrigated lands, aside with areas occupied by 'natural' vegetation. The almost ubiquitous presence of remnants of ancient canals and terraces indicate that many areas supporting natural vegetation had in fact been cultivated in pre-Conquest times. Extreme losses of soil, dating back even to the climatic transition period, caused the virtual collapse of significant contemporary hillslope agriculture, despite of continued tilling and planting (McAuliffe, Sundt, Valiente-Banuet, Casas and Viveros, 2001).

Hillslope farming was, and still is, completely dependent on incident summer rainfall, and the effects of drought were compounded by lack of sufficient soil volume for storage of plant-available moisture. Even in years of favourable rainfall, insufficient supply of moisture in extremely thin soils contributes to very poor yields, occasionally with no production at all during 'bad years'. Although maize is frequently planted in fields where soil thickness is 10 cm, or less, maize plants in shallow soils attained heights of only 50 cm, and failed to flower (reflecting the insufficient supply of soil water to enable growth). Some of the land had already been abandoned 45 years ago, and the recurrent failure of maize for the last several years caused hunger in parts of the local population. The currently cultivated lands are estimated to supply only about one fifth and one third of the community's subsistence needs for maize and beans, respectively. The shortage of arable land is acute, and as human population continues to grow, poverty is extreme and increasing. Sustaining rural livelihoods are further endangered by a ban which was recently imposed on cutting live trees within the communal land, so that from 1991 onwards most of the wood has to be obtained from travelling vendors. Before, harvesting of fuelwood for both domestic and commercial purposes (e.g., firing of ceramics) was a common practice. Dense woods of leguminous trees such as *Acacia cochleacantha* were preferred for domestic use in food preparation, and fuel that burns more slowly and at lower temperatures were used in firing dynamics, such as the small tree *Ipomea arborescens*, dried stems of various cactus species, trunks of Yucca, and stalks of Agave (McAuliffe, Sundt, Valiente-Banuet, Casas and Viveros, 2001).

*Patagonian Rangelands Turning into 'Badlands' or Deserts*

As in northern Mexico, the drylands of Patagonia in southern Argentina were dominated by extensive and open grasslands at least 200 years ago, but were altered fundamentally through heavy grazing. They had been (semi)arid grass ecosystems which were shrub-invaded mainly as a result of livestock overgrazing. Different from the northern grasslands, the climate in Patagonia is that of a dry, cold, and windy (semi)desert. It lacks a well-defined rainy season, has cool winters as opposed to mild summers, and is subject to a strong west to east precipitation gradient, i.e., with precipitation in the humid Andean region to the west (2,000 to 3,000 mm) decreasing to less than 150 mm in the extra-Andean region to the east, with hills and plateaus stretching between these two zones. Significant 'barrancas' (canyons, or incised stream channels) cut through alluvial deposits in the floors of basins, plains, and plateaus (Aagesen, 2000).

Along the strong gradient of decreasing precipitation, *tussock* grass steppes turned into shrub-grass steppes, and those to deserts, so that as per today hardly any relic areas of native plant communities dating back to the pre-European times can be found. The most important changes towards gradual and seemingly irreversible vegetation degradation – i.e., reduction of total cover and number of plants, disappearance of valuable fodder species and invasion of undesirable species – started in the Santa Cruz and Neuquén provinces between about 1885 and 1914, when sheep had been introduced on a large scale and livestock numbers reached peak values in those times. Vegetation degradation and sheep farming expanded later on to other Patagonian provinces (Aagesen, 2000).

Although it was first explored in 1520, Patagonia remained a source of basic provisions and shelter during inclement weather for Europeans, when they pursued trade and colonization elsewhere. Not much is known about indigenous settlements and native species, until a group of Welsh settlers started colonization on the coast and in the lower Chubut valley in 1865. As this group established itself successfully, Argentine authorities became interested in solidifying national sovereignty in the region, including territorial control over the natural resources. In 1879 to 1885, during a military campaign ('conquista del desierto') native hunters and gatherers were decimated. The territorial organization that followed favoured ranchers, settlers, traders, the military, and national as well as religious institutions. Indigenous land was reduced to small patches and the remaining native population was accommodated on reservations, which were located in isolated places and contained usually the least fertile land. Much of the conquered territory was divided amongst those who financed or led the military campaign in the 'desert conquest' (i.e., 35 million ha divided amongst 24 individuals). A population boom occurred once the indigenous population had been all but eliminated, and most of this growth was oriented towards the sheep industry, which spread rapidly along the coast, across the central plateau, and into the more accessible Andean valleys. The initial colonization between about 1880-1920 was impelled for extensive subsistence sheep farming, with settlers pushing into frontier lands with enough provisions from ecosystems to establish ranches ('estancias'). Those lands with the richest vegetation cover and adequate water supplies, which had not already been distributed after the 'desert conquest', were usually the first to be occupied. During those first decades, sheep ranching thrived, and wool began to make its way to the Atlantic ports, and eventually to Buenos Aires and Europe. National authorities provided substantial assistance such as tax releases to immigrants, with many products subsequently imported from England as ranchers invested in and improved their homes, canals, tools, and fences (Aagesen, 2000).

Further on, market growth for livestock products went hand in hand with technological improvements in the transport and processesing sector, and even when economic conditions worsened overstocking was rather the rule during the first half of the $20^{th}$ century. In 1894, the introduction of refrigerated ships for the transport of meat gave further impulse to an ever-growing sheep industry. Livestock numbers rose from 20,000 head of sheep in 1896 to about 5 million sheep by 1911. The increase in the quantity of sheep in Patagonia was sustained from the late 1880s until 1952, when their number peaked at about 22 million. In the initial colonization period, British entrepreneurs were among the first to invest into the booming livestock sector, by acquiring large parcels of best land with abundant forage resources in widely separated parts of the region, and establishing large-scale, London-based sheep farms there. Intra-regional specialization of livestock rearing occurred, i.e., Merino sheep, raised primarily for wool, became dominant in the arid north, Corriedale sheep (introduced from New Zealand in 1905) were raised for both wool and meat in the south, and half-breed black-faced sheep were introduced to coastal valleys to produce high quality lambs. During World War I, due to tripling wool prices and a heavy demand for meat, refrigerated meat plants were established in 1916, triggering a cultivation boom so that even the most marginal lands were used to raise livestock. With the end of World War I, wool prices returned to their former levels, and dwindling profits were felt especially by ranchers who had opened land on poorer sites (while ranchers with high quality rangelands survived). In addition, the economic situation worsened since import taxes were reinstalled, and legislation made it more expensive to lease public lands. 'In an effort to make ends meet, most ranchers responded by increasing the stocking rate on their rangelands' (Aagesen, 2000).

Since about the mid-1950s, the livestock industry in the region has been on a steady decline, with livestock numbers dropping to about 13 million in 1988. The decline could be attributed to both the appeal of synthetic materials (rather than wool) and a number of economic factors. Also, rangeland degradation was also cited as a causative factor, i.e., the substantial reduction in the quantity and quality of forage. In the meantime, extensive grazing had turned into more intensified uses, exerting very high grazing pressure all over the region. Valle, Elissalde, Gagliardini and Milovich (1998) estimate grazing pressure at 20 to 60 sheep per $km^2$, be these ranches situated in the sub- and extra-Andean region, Patagonian Monte, or the Magellanic Steppe. Intense and continuous grazing was, and reportedly still is the most widespread management strategy in Patagonia, which means that year-round grazing is practised in which animals roam the range unherded (Aagesen, 2000).

Only at few sites, an extensive (subsistence) mode had survived. An example is the Percey River Watershed in the Chubut Province of southern Argentina where subsistence sheep ranching was still practiced. As with intensive sheep ranching, livestock numbers decreased in the second half of the $20^{th}$ century. It appears that small ranches were grazed more heavily than large ones, and that despite of government incentives for diversification, all landholders still derived most income from livestock, with hardly any alternatives to livestock emerging. The highly skewed land-distribution, which operates as mediating factor of land change and was inherited from the post-'desert conquest' period – contemporarily, 10% of the ranches occupy 63% of the rangeland in all of Patagonia – blocks the majority of small farmers to have seizable and economically viable holdings, as Aagesen (2000, p. 212) notes.

> Those who have less land, or the vast majority of Patagonia's 13 000 sheep ranchers, often try to make ends meet by overstocking their holdings. Particularly hard hit are small-scale subsistence ranchers, who commonly respond to declines in productivity by stocking more heavily.

The process of desertification – namely, vegetation degradation, erosion, salinization, alkalinization, soil crusting and compaction – is seen to be irreversible in those parts of Patagonia where, long before sheep colonization, land had been clearly delimited 'badland' because of plateau building and erosional processes during geological times. An example is the Chubut Formation dating back to the Cretaceaous age. It makes land degradation in the Chubut Province more dynamic than elsewhere, i.e., land was more desertified, and 'natural' desert-type complexes were forming. In the whole of Patagonia, a fragmentation or patterning of what had previously been continuous vegetation habitats had been observed. This means that intensively grazed patches alternate with slightly grazed ones, within a matrix where bare soil is generally a constant, and land use impact causes instability in the system, triggering overgrazed patches of vegetation to increase in size under adverse natural conditions, such as drought periods (Valle, Elissalde, Gagliardini and Milovich, 1998; Aagesen, 2000).

In recent times, land degradation became also related to activities such as oil exploration, mineral extraction, and construction operations (road building, reclamation construction). Irrigated lands in the provinces of Chubut and Rio Negro were reportedly at various risks of desertification, and salinization as well as alkalinization were associated with both irrigated and non-irrigated land (Valle, Elissalde, Gagliardini and Milovich, 1998). Nonetheless, there is agreement that the most significant impact on the region's native or Pre-European vegetation – which had disappeared as by now – occurred in the very early years of colonization, when millions of sheep were so swiftly introduced to Patagonian ecosystems that did not know this land use before (Aagesen, 2000).

Chapter 4

# Causes and System Properties

**Introduction**

Results are summarized on the frequency of occurrence of proximate and underlying causes of desertification (broad clusters, detailed activities). The mode of causation and the type of interaction between causative variables is explored, addressing hierarchical scales and mediating factors. Finally, system properties of desertification are presented (feedbacks, thresholds, control points).

**Proximate Causes**

*Broad Clusters and their Regional Variation*

At the proximate level, desertification is best explained by the combination of broad categories of factors such as agricultural activities (i.e., extension of livestock and crop production), increased aridity, extension of infrastructure (e.g., irrigation techniques, settlement), and wood extraction, or related extractional activities (e.g., herbs) – see Table 4.1. These combinations explain 90% of the cases, while single explanatory variables were found in 10% of the cases only. Both extractional activities and infrastructure expansion do not operate as individual causes leading directly to desertification, but are always associated with agricultural activities. Increased aridity appears in more than four fifths of the cases as a proximate causation. Dominating the proximate factor combinations is the term 'agricultural activities – increased aridity' (26% as a sole combination), which is inherent to all-important combinations explaining four fifths of the cases.

Some regional variations of these broad causal patterns are obvious. Desertification in the arid and semi-arid zones of Australia, for example, is solely caused by agricultural activities in combination with infrastructure extension under the condition of increased aridity. This combination is important, too, in the Asian and African cases, but notably less so in cases from Europe and Latin America. As another example, the combination of agricultural activities, increased aridity and extractional activities (such as wood harvesting) do best characterize the cases from the West African Sudan-Sahel zone, East African grasslands and the Kalahari Sandveld of Southern Africa. The combination is considerably less important, or non-existent, in other parts of the world's drylands. The full and complex interplay of all broad clusters is particularly prevalent in cases from Asia, and does not matter at all in cases reported from Europe, Australia, and Latin America.

**Table 4.1  Frequency of Broad Clusters of Proximate Causes of Desertification***

| | All Cases (N=132) | | | Asia (n=51) | | Africa (n=42) | | Europe (n=13) | | Australia (n=6) | | N-America (n=6) | | L-America (n=14) | |
|---|---|---|---|---|---|---|---|---|---|---|---|---|---|---|---|
| | abs | rel | cum | abs | rel | abs | rel | abs | rel | abs | rel | abs | rel | abs | rel |
| *Single factor causation* | | | | | | | | | | | | | | | |
| Agro (agricultural activities) | 7 | 5 | 5 | 6 | 12 | 1 | 2 | 0 | | 0 | | 0 | | 0 | |
| Arid (increased aridity) | 7 | 5 | 10 | 0 | 0 | 6 | 14 | 1 | 8 | 0 | | 0 | | 0 | |
| Infra (infrastructure extension) | 0 | | | 0 | 0 | 0 | 0 | 0 | | 0 | | 0 | | 0 | |
| Wood (wood extraction) | 0 | | | 0 | 0 | 0 | 0 | 0 | | 0 | | 0 | | 0 | |
| *Two-factor causation* | | | | | | | | | | | | | | | |
| Agro-arid | 34 | 26 | 36 | 6 | 12 | 10 | 24 | 8 | 62 | 0 | | 3 | 50 | 7 | 50 |
| Agro-infra | 4 | 3 | 39 | 2 | 4 | 2 | 5 | 0 | | 0 | | 0 | | 0 | |
| *Three-factor causation* | | | | | | | | | | | | | | | |
| Agro-infra-arid | 27 | 21 | 60 | 9 | 18 | 9 | 21 | 1 | 8 | 6 | 100 | 1 | 17 | 1 | 7 |
| Agro-wood-arid | 15 | 11 | 71 | 3 | 6 | 10 | 24 | 1 | 8 | 0 | | 0 | | 1 | 7 |
| Agro-wood-infra | 7 | 5 | 76 | 1 | 2 | 0 | 0 | 1 | 8 | 0 | | 0 | | 5 | 36 |
| *Four-factor causation* | | | | | | | | | | | | | | | |
| All | 31 | 24 | 100 | 24 | 47 | 4 | 10 | 1 | 8 | 0 | | 2 | 33 | 0 | |
| Total | 132 | 100 | | 51 | 100 | 42 | 100 | 13 | 100 | 6 | 100 | 6 | 100 | 14 | 100 |

* Abs = absolute number; rel = relative percentages; cum = cumulative percentages. Relative percentages may not total 100 because of rounding.

The following tables provide a breakdown of causative variables, both proximate and underlying, by broad geographical regions. They show the absolute number as well as the relative percentages of the frequency of causative variables reported in the case studies. Tables 4.1 and 4.6 give only the broad clusters of proximate causes (and underlying driving forces), while tables 4.2 to 4.5 (on proximate causes) and 4.7 to 4.11 (on underlying causes) provide a detailed breakdown of the broad clusters by specific factors. Only the frequency of the modes of causation (single or multifactorial) by broad clusters show relative percentages of cases which add up to 100% (Tables 4.1 and 4.2), while the relative percentages of the frequency of occurrence of specific factors do not add up to 100%, as multiple counts exist because of causal factor synergies (discussed later).

*Specific Causes*

The specific proximate causes are given here in the order of decreasing importance as found in the cases, i.e., agricultural activities, increased aridity, extension of infrastructure, and extension of wood (or related) extractional activities.

*Agricultural activities* Agricultural activities, or productional activities related to managed ecosystems such as livestock and crop production, are by far the leading land uses associated with nearly all desertification cases. They are inherent to nearly all reported cases of desertification. (n=125, or 95%) – see Tale 4.2. Livestock production slightly outweighs crop production, but in most of the cases both activities are intricately interlinked with immediate causative consequences for degradation. This means that cropland expansion on areas previously used for pastoral activities leads to overstocking on remaining, reduced rangeland, triggering soil mining at sites not suitable in particular for permanent agriculture. It appears that this causative interlinkage is not valid for cases from North and Latin America, i.e., rangelands of the US Southwest and Patagonia. Nomadic grazing is reported to be far less important in causing desertification than extensive grazing, the various forms of which include sedentary as well as transhumance activities. Overgrazing by nomads is reported mainly from Asian cases. It appears that only in the Asian cases, nomadic grazing has higher impact than extensive grazing. Wild animals' (feral) overgrazing has never been reported as a cause of desertification for itself, but becomes important only when linked to the extension of extensive grazing (e.g., on wilderness areas, leading to overstocking of wild animals on remaining untouched land). In all combinations, extensive grazing is associated with three fifths of all desertification cases. To a lesser degree than livestock activities, crop production has been associated with more than half of all desertification cases. Crop production mainly implies rainfed-production of annual crops, but is not limited to it. Important exceptions are 'wetland' cases of irrigation and oasis agriculture (14%) leading to dryland degradation, especially in the Central Asian desert and steppe region.

**Table 4.2** Frequency of Specific Agricultural Activities Causing Desertification[a]

| | All Cases (N=132) | | Asia (n=51) | | Africa (n=42) | | Europe (n=13) | | Australia (n=6) | | N-America (n=6) | | L-America (n=14) | |
|---|---|---|---|---|---|---|---|---|---|---|---|---|---|---|
| Livestock production | 98 | 74% | 36 | 71% | 32 | 76% | 7 | 54% | 6 | 100% | 5 | 83% | 12 | 86% |
| ... nomadic grazing | 16 | 12% | 13 | 26% | 3 | 7% | 0 | | 0 | | 0 | | 0 | |
| ... extensive grazing[b] | 45 | 34% | 8 | 16% | 19 | 45% | 7 | 54% | 1 | 17% | 3 | 50% | 7 | 50% |
| ... nomadic & extensive | 22 | 17% | 15 | 29% | 7 | 17% | 0 | | 0 | | 0 | | 0 | |
| ... intensive production | 8 | 6% | 0 | | 3 | 7% | 0 | | 0 | | 0 | | 5 | 36% |
| ... intensive & extensive[b] | 1 | 1% | 0 | | 0 | | 0 | | 0 | | 1 | 17% | 0 | |
| ... wild animals' grazing | 0 | | 0 | | 0 | | 0 | | 0 | | 0 | | 0 | |
| ... wild & extensive[b] | 6 | 5% | 0 | | 0 | | 0 | | 5 | 83% | 1 | 17% | 0 | |
| Crop production | 78 | 59% | 40 | 78% | 20 | 48% | 10 | 77% | 5[c] | 83% | 1 | 17% | 2 | 14% |
| ... annual crops | 46 | 35% | 23 | 45% | 14 | 33% | 2 | 15% | 5[c] | 83% | 1 | 17% | 1 | 7% |
| ... perennial crops | 2 | 2% | 1 | 2% | 0 | | 1 | 8% | 0 | | 0 | | 0 | |
| ... annual & perennial | 11 | 8% | 0 | | 3 | 7% | 7 | 54% | 0 | | 0 | | 1 | 7% |
| ... wetland cropping[c] | 18 | 14% | 15 | 29% | 3 | 7% | 0 | | 0 | | 0 | | 0 | |
| ... annual, perennial, wet[c] | 1 | 1% | 1 | 2% | 0 | | 0 | | 0 | | 0 | | 0 | |
| Total | 125 | 95% | 51 | 100% | 36 | 86% | 13 | 100% | 6 | 100% | 5 | 83% | 14 | 100% |

[a] Multiple counts possible; percentages relate to the total of all cases for each category; relative percentages may not total 100 because of rounding.
[b] Sedentary as well as transhumance.
[c] Including irrigation and oasis agriculture.
[d] On an intermittent basis at the semi-arid fringe only.

*Increased aridity* Increased aridity turns out to be a robust factor of proximate causation of desertification in more than four fifths of the cases (n=114, or 86%) – see Table 4.3. A rough breakdown into two groups shows that factors related to the indirect impact of climate variability and related aspects of the climate which directly influence surface vegetation tend to be equally important. There are regional distinctions, though. Climatic variability associated indirectly with increased aridity seems less prominent in the Australian and Latin American cases, while climatic aspects of aridity with direct influence on land cover are found to be most pronounced in the Australian and North American cases. A further breakdown of these two, still broad categories reveals a striking difference between the Asian and African cases. Decrease in annual precipitation due to increased rainfall variability features more cases in Africa (and Europe), but strikingly less so in Asia. In contrast, increased aridity associated with warmer and drier climate conditions (including more droughts and sand storms, for example), and very likely due to climate change, appears to be more important in cases from Asia than in any other cases. Aridity increases associated with changes in fire regimes – such as increased fire frequency, increased fuel load, and more uncontrolled fires (lightning and set by humans) – have a relatively low frequency of occurrence, and do not matter at all in the Asian and Latin American cases. Similarly, cases which show a rise in aridity due to pronounced oscillations of warmer and drier climatic conditions (including droughts and sand storms, for example) with wetter and more humid conditions (with the consequence of increased erosion) have a relatively low frequency of occurrence too. They seem to matter in Australia and Latin America only. An increase in aridity related to the occurrence of mainly prolonged droughts (which cannot be attributed to either climate change or variability increases) are frequent in one third of the cases.

*Extension of infrastructure* The extension of infrastructure is associated with more than half of all cases (n=73, or 55%), mainly from Australia and Asia – see Table 4.4. There are no dominant individual causes, but cause combinations of various infrastructure elements which occur in four fifths of the cases. Most of these combinations are reportedly grouped around or linked to the development of water-related infrastructure, mainly for cropland irrigation and pasture development (such as reservoirs, dams, canals, boreholes, pump stations, etc.). The establishment of irrigation infrastructure is reported to occur in combination with settlement expansion, road network extension, and commercial as well as industrial developments such as the build-up of oil/gas industry facilities, mining, quarrying, and touristic facilities. Most of the combinations reveal no distinct pattern. Only the Australian cases could be characterized as driven predominantly by an industry-water-road complex.

Table 4.3  Frequency of Increased Aridity Causing Desertification[a]

| | All Cases (N=132) | | Asia (n=51) | | Africa (n=42) | | Europe (n=13) | | Australia (n=6) | | N-America (n=6) | | L-America (n=14) | |
|---|---|---|---|---|---|---|---|---|---|---|---|---|---|---|
| Climate variability[b] | 63 | 48% | 25 | 49% | 27 | 64% | 7 | 54% | 1 | 17% | 2 | 33% | 1 | 7% |
| ... higher rainfall deficit | 37 | 28% | 3 | 6% | 24 | 57% | 7 | 54% | 1 | 17% | 1 | 17% | 1 | 7% |
| ... warmer, drier (due to climate change)[c] | 34 | 26% | 23 | 45% | 5 | 12% | 4 | 31% | 0 | | 2 | 34% | 0 | |
| Aspects of climate with direct influence on surface vegetation | 62 | 47% | 17 | 33% | 21 | 50% | 6 | 46% | 5 | 83% | 5 | 83% | 8 | 57% |
| ... changes in fire regime[d] | 16 | 12% | 0 | | 3 | 7% | 5 | 39% | 5 | 83% | 3 | 51% | 0 | |
| ... more oscillations of drier/wetter conditions | 12 | 9% | 0 | | 0 | | 1 | 8% | 5 | 83% | 0 | | 6 | 43% |
| ... simple occurrence of (prolonged) droughts[e] | 42 | 32% | 17 | 33% | 20 | 47% | 0 | | 0 | | 3 | 50% | 2 | 14% |
| Total | 114 | 86% | 42 | 82% | 39 | 93% | 12 | 93% | 6 | 100% | 6 | 100% | 9 | 64% |

[a] Multiple counts possible; percentages relate to the total of all cases for each category; relative percentages may not total 100 because of rounding.
[b] Indirect impact, including accompanying feedbacks to the atmosphere.
[c] Including shift of winter rainfall season (in low-latitude drylands) and less snow cover, withdrawing glaciers and alteration of freeze-thaw soil processes (in high-latitude drylands).
[d] Increased fire frequency, increased fuel load, more uncontrolled fires (lightning and set by humans).
[e] No mention whether long-term climate change or short-term oscillations were responsible.

**Table 4.4  Frequency of Specific Infrastructure Activities Causing Desertification[a]**

| | All Cases (N=132) | | Asia (n=51) | | Africa (n=42) | | Europe (n=13) | | Australia (n=6) | | N-America (n=6) | | L-America (n=14) | |
|---|---|---|---|---|---|---|---|---|---|---|---|---|---|---|
| Water/irrigation | 14 | 11% | 7 | 14% | 6 | 14% | 0 | | 1 | 17% | 0 | | 0 | |
| Residential/settlement | 13 | 10% | 9 | 18% | 3 | 7% | 1 | 8% | 0 | | 0 | | 0 | |
| Water & residential | 13 | 10% | 2 | 4% | 11 | 26% | 0 | | 0 | | 0 | | 0 | |
| Water, residential, road, mineral[b] & industrial[c] | 8 | 6% | 8 | 16% | 0 | | 0 | | 0 | | 0 | | 0 | |
| Water, road & industrial[c] | 5 | 4% | 0 | | 0 | | 0 | | 5 | 83% | 0 | | 0 | |
| Water, road, mineral[b] & industrial[c] | 5 | 4% | 0 | | 0 | | 0 | | 0 | | 0 | | 5 | 36% |
| Water, res. & industrial[c] | 5 | 4% | 5 | 10% | 0 | | 0 | | 0 | | 0 | | 0 | |
| Roads | 2 | 2% | 0 | | 0 | | 2 | 15% | 0 | | 0 | | 0 | |
| Resid., road & mineral[b] | 2 | 2% | 2 | 4% | 0 | | 0 | | 0 | | 0 | | 0 | |
| Water & road | 1 | 1% | 1 | 2% | 0 | | 0 | | 0 | | 0 | | 0 | |
| Water, rd., ind.[c] & others | 1 | 1% | 1 | 2% | 0 | | 0 | | 0 | | 0 | | 0 | |
| Residential & touristic | 1 | 1% | 0 | | 0 | | 0 | | 0 | | 1 | 17% | 0 | |
| Mineral[b] & others | 1 | 1% | 1 | 2% | 0 | | 0 | | 0 | | 0 | | 0 | |
| Residential, water & road | 1 | 1% | 0 | | 0 | | 0 | | 0 | | 1 | 17% | 0 | |
| Industrial[c] & road | 1 | 1% | 0 | | 0 | | 0 | | 0 | | 0 | | 1 | 7% |
| Total | 73 | 55% | 36 | 71% | 20 | 48% | 3 | 23% | 6 | 100% | 2 | 33% | 6 | 43% |

[a] Multiple counts possible; percentages relate to the total of all cases for each category; relative percentages may not total 100 because of rounding.
[b] Mineral extraction like mining and quarrying.
[c] Industrial, including commercial infrastructure such as for oil/gas development.

*Wood extraction and related activities* The extraction of wood, mainly from natural forests or woodlands, and related extractional activities such as digging for medicinal herbs and the collection of plant and/or animal products in natural ecosystems are reported to be frequent in the drylands of Asia, Africa, Europe and Latin America in about half of the cases (n=49, or 45%) – see Table 4.5. (e.g., fuelwood, polewood, charcoal) These activities seem to be of lower importance in cases from North America, and have apparently no importance at all in cases from Australia.

**Table 4.5   Frequency of Specific Extractional Activities[a] Causing Desertification[b]**

|  | Wood extraction from natural/planted forests | | Digging for medicinal herbs | | Other collection of plant and/or animal products | | Total | |
|---|---|---|---|---|---|---|---|---|
| Asia | 29 | 57% | 10 | 20% | 5 | 10% | 31 | 61% |
| Africa | 17 | 41% | 1 | 2% | 2 | 5% | 17 | 41% |
| Europe | 3 | 23% | 0 |  | 1 | 8% | 4 | 31% |
| Australia | 0 |  | 0 |  | 0 |  | 0 |  |
| N-America | 1 | 17% | 0 |  | 0 |  | 1 | 17% |
| L-America | 6 | 43% | 0 |  | 0 |  | 6 | 43 |
| All cases | 56 | 42% | 11 | 8% | 8 | 6% | 59 | 45% |

[a] Wood extraction and related activities.
[b] Multiple counts possible; percentages relate to the total of all cases for each category; relative percentages may not total 100 because of rounding.

### Underlying Driving Forces

*Broad Clusters and their Regional Variation*

At the underlying level, there is no robust factor or factor combination underpinning the proximate causes to explain desertification. Desertification is best explained by regionally distinct combinations of multiple and coupled socio-economic and climatic underlying factors or driving forces. Factor combinations explain more than 90% of the cases, while single variables were found in 10% of the cases only, with two thirds of the cases being driven even by three up to six factors each – see Table 4.6.

*Individual or single key factors* Economic, technological, socio-cultural and demographic forces do not operate as individual or single key factors driving desertification, but are always associated with climatic and institutional or policy factors, and mostly with each other. Climatic forces, or meteorological factors, were found to be causative to increased aridity associated with desertification as a sole force in merely 5% of the cases. These cases, in which no adverse human or land use influences were reported, stem overwhelmingly from the Sudano-Sahelian zone in Africa. Another regional cluster of single factor causation (again low, at the 4% level) relates to institutional or policy factors underpinning livestock activities, wood extraction and especially market infrastructure extension associated with desertification. These cases stem exclusively from the Patagonian rangelands where the long-term effect of former European colonization policies – such as the introduction of sheep, a pronounced export production rather than subsistence orientation of the economy, ever increased taxation, etc. – contributed to continuous pressures upon land and water resources in a dryland ecosystem that had not experienced these land uses before.

*Factor combinations* Factor combinations dominate the causative pattern of underlying driving forces. It appears that demographic factors are increasingly associated with desertification once the complexity of factors involved is increasing, i.e., no single factor causation of human population dynamics, one dominant two-factor occurrence (in African cases only, together with climatic factors), involved in 3 out of 6 three-factor terms, in 5 out of 8 four-factor terms, and in 4 out of 5 six-factor terms. This pattern, i.e., generally low impact of demographic factors, but growing in importance in multi-factorial combinations, is not valid for other drivers such as climatic factors. The latter are inherent to factor combinations of more than four fifths of the cases regardless of the complexity of the multi-factorial term. It further appears that, in particular, cases from Asia are driven by the full interplay of all factors involved. Therefore, it could be stated that complex multi-factorial combinations of underlying forces, mainly coupled socio-economic and climatic factors to varying degrees underpin the proximate pattern of causation. From the results on specific causes (to be discussed in the following) it could be drawn that robust broad factor combinations – though differing widely in the range of specific factors involved – imply climatic factors leading to increased aridity, agricultural growth policies, newly introduced land use technologies, and malfunct traditional land laws or tenure arrangements not suited to dryland ecosystem management.

*Specific Causes*

As in the case of proximate causes, the specific underlying driving forces are given here in the order of decreasing importance as found in the cases, i.e., climatic factors, technological factors, institutional or policy factors, economic factors, demographic factors, and cultural or socio-political factors.

**Table 4.6** Frequency of Broad Clusters of Underlying Driving Forces of Desertification*

| | All Cases (N=132) | | | Asia (n=51) | | Africa (n=42) | | Europe (n=13) | | Australia (n=6) | | N-America (n=6) | | L-America (n=14) | |
|---|---|---|---|---|---|---|---|---|---|---|---|---|---|---|---|
| | abs | rel | cum | abs | rel | abs | rel | abs | rel | abs | rel | abs | rel | abs | rel |
| *Single factor causation* | | | | | | | | | | | | | | | |
| Climatic (clim) | 7 | 5 | 5 | 0 | 0 | 6 | 14 | 1 | 8 | 0 | 0 | 0 | 0 | 0 | 0 |
| Institutional (inst) | 5 | 4 | 9 | 0 | 0 | 0 | 0 | 0 | 0 | 0 | 0 | 0 | 0 | 5 | 36 |
| Economic (econ) | 0 | 0 | | 0 | 0 | 0 | 0 | 0 | 0 | 0 | 0 | 0 | 0 | 0 | 0 |
| Technological (tech) | 0 | 0 | | 0 | 0 | 0 | 0 | 0 | 0 | 0 | 0 | 0 | 0 | 0 | 0 |
| Cultural (cult) | 0 | 0 | | 0 | 0 | 0 | 0 | 0 | 0 | 0 | 0 | 0 | 0 | 0 | 0 |
| Demographic (pop) | 0 | 0 | | 0 | 0 | 0 | 0 | 0 | 0 | 0 | 0 | 0 | 0 | 0 | 0 |
| *Two-factor causation* | | | | | | | | | | | | | | | |
| Pop-clim | 10 | 8 | 17 | 0 | 0 | 10 | 24 | 0 | 0 | 0 | 0 | 0 | 0 | 0 | 0 |
| Tech-clim | 9 | 7 | 24 | 0 | 0 | 0 | 0 | 6 | 46 | 1 | 17 | 2 | 33 | 0 | 0 |
| Econ-tech | 2 | 1 | 25 | 2 | 4 | 0 | 0 | 0 | 0 | 0 | 0 | 0 | 0 | 0 | 0 |
| Econ-inst | 2 | 1 | 26 | 0 | 0 | 2 | 5 | 0 | 0 | 0 | 0 | 0 | 0 | 0 | 0 |
| Cult-clim | 1 | 1 | 27 | 1 | 2 | 0 | 0 | 0 | 0 | 0 | 0 | 0 | 0 | 0 | 0 |
| *Three-factor causation* | | | | | | | | | | | | | | | |
| Tech-inst-clim | 4 | 3 | 30 | 0 | 0 | 4 | 10 | 0 | 0 | 0 | 0 | 0 | 0 | 0 | 0 |
| Pop.tech-inst | 4 | 3 | 33 | 3 | 6 | 0 | 0 | 1 | 8 | 0 | 0 | 0 | 0 | 0 | 0 |
| Pop-econ-clim | 2 | 1 | 34 | 0 | 0 | 2 | 5 | 0 | 0 | 0 | 0 | 0 | 0 | 0 | 0 |
| Econ-tech-clim | 2 | 1 | 35 | 0 | 0 | 1 | 2 | 0 | 0 | 0 | 0 | 1 | 17 | 0 | 0 |
| Tech-cult-clim | 2 | 1 | 36 | 0 | 0 | 2 | 5 | 0 | 0 | 0 | 0 | 0 | 0 | 0 | 0 |
| Pop-cult-clim | 3 | 2 | 38 | 1 | 2 | 0 | 0 | 1 | 8 | 0 | 0 | 1 | 17 | 0 | 0 |

**Table 4.6  Continued**

| | Abs | Rel | Cum | | | | | | | | | | |
|---|---|---|---|---|---|---|---|---|---|---|---|---|---|
| **Four-factor causation** | | | | | | | | | | | | | |
| Econ-tech-inst-clim | 6 | 5 | 43 | 3 | 6 | 0 | 1 | 2 | 0 | 0 | 0 | 3 | 21 |
| Pop-econ-inst-clim | 4 | 3 | 46 | 3 | 6 | 1 | 2 | 5 | 0 | 0 | 0 | 0 | |
| Tech-inst-cult-clim | 3 | 2 | 48 | 0 | | 2 | 0 | | | 1 | 17 | 0 | |
| Pop-tech-inst-clim | 5 | 4 | 52 | 1 | 2 | 3 | 1 | 7 | 0 | 0 | | 2 | |
| Pop-econ-cult-clim | 2 | 1 | 53 | 0 | | 0 | | 8 | 0 | 0 | | 2 | 14 |
| Econ-tech-cult-clim | 1 | 1 | 54 | 1 | 2 | | 0 | | | 0 | | 0 | |
| Pop-econ-tech-clim | 1 | 1 | 55 | 1 | 2 | | 0 | | | 0 | | 0 | |
| Pop-econ-tech-inst | 1 | 1 | 56 | 1 | 2 | | 0 | | | 0 | | 0 | |
| **Five-factor causation** | | | | | | | | | | | | | |
| Econ-tech-inst-cult-clim | 15 | 12 | 68 | 1 | 2 | 5 | 10 | 0 | 5 | 0 | 0 | 4 | 29 |
| Pop-econ-tech-inst-clim | 12 | 9 | 77 | 11 | 22 | 1 | 2 | 0 | 0 | 0 | 0 | 0 | |
| Pop-econ-inst-cult-clim | 5 | 4 | 81 | 4 | 8 | 1 | 2 | 0 | 0 | 0 | 0 | 0 | |
| Pop-econ-tech-inst-cult | 4 | 3 | 84 | 3 | 6 | 1 | 2 | 0 | 0 | 0 | 0 | 0 | |
| Pop-econ-tech-cult-clim | 4 | 3 | 87 | 0 | | 1 | 2 | 3 | 23 | 0 | 0 | 0 | |
| **Six-factor causation** | | | | | | | | | | | | | |
| Pop-inst-econ-tech-cult-clim | 17 | 13 | 100 | 15 | 29 | 1 | 2 | 0 | | 0 | | 1 | 17 |
| Total | 132 | 100 | | 51 | 100 | 42 | 100 | 13 | 100 | 6 | 100 | 6 | 100 |
| | | | | | | | | | | | | 14 | 100 |

\* Abs = absolute number; rel = relative percentages; cum = cumulative percentages; relative percentages may not total 100 because of rounding.

*Climatic factors* Climatic factors leading to increased aridity, i.e., mainly rainfall decrease, are important and robust driving forces which have been reported in 114 (or 86% of the) cases – see Table 4.7. However, the specific impact of climate is reportedly unclear in about one third of the cases (n=46, or 35%). Cases in which climatic conditions operate in concomitant occurrence with other, socio-economic driving forces, and cases in which they operate as synergistic forces are more or less equally widespread, i.e., in about one fifth of all cases. In the Australian cases, mainly droughts were reported to occur aside with other socio-economic drivers (and not in synergy), while the reverse seems true at sites in southern Europe. Climate, or meteorological factors, were found to underlie increased aridity as a sole force in 5% of the cases (n=7), which mainly stem from the Sudano-Sahelian zone of Africa.

Table 4.7    Frequency and Mode of Climatic Factors Driving Desertification[a]

|  | [b] | | [c] | | [d] | | [e] | | Unclear impact | | Total | |
|---|---|---|---|---|---|---|---|---|---|---|---|---|
|  | abs | rel | abs | rel | abs | rel | abs | rel | abs | rel | abs | rel |
| Asia | 0 |  | 11 | 22 | 10 | 20 | 6 | 12 | 15 | 29 | 42 | 82 |
| Africa | 6 | 14 | 7 | 17 | 14 | 33 | 0 |  | 12 | 29 | 39 | 93 |
| Europe | 1 | 8 | 0 |  | 4 | 31 | 0 |  | 7 | 54 | 12 | 92 |
| Australia | 0 |  | 5 | 83 | 0 |  | 0 |  | 1 | 17 | 6 | 100 |
| N-America | 0 |  | 0 |  | 1 | 17 | 0 |  | 5 | 83 | 6 | 100 |
| L-America | 0 |  | 2 | 14 | 1 | 7 | 0 |  | 6 | 43 | 9 | 64 |
| All cases | 7 | 5 | 25 | 19 | 30 | 23 | 6 | 5 | 46 | 35 | 114 | 86 |

[a] Multiple counts possible; percentages relate to the total of all cases for each category; abs = absolute number; rel = relative perentages; relative percentages may not total 100 because of rounding.
[b] Main, single, key driver without human impact (natural hazard).
[c] Concomitant occurrence with other (socioeconomic) drivers.
[d] In synergy or causal chains with other (socioeconomic) drivers.
[e] Operating (more) as a threshold than as a driver.

*Technological factors* Technological factors are prominent underlying driving forces of desertification (69%) which show low geographical variation as a broad group – see Table 4.8. New technological innovations (56%) are reported to be associated with as many cases of desertifications as are deficiencies (52%).

Table 4.8  Frequency of Specific Technological Factors Driving Desertification[a]

| | All Cases (N=132) | | Asia (n=51) | | Africa (n=42) | | Europe (n=13) | | Australia (n=6) | | N-America (n=6) | | L-America (n=14) | |
|---|---|---|---|---|---|---|---|---|---|---|---|---|---|---|
| Introduction of | | | | | | | | | | | | | | |
| ... water management[b] | 17 | 13% | 13 | 26% | 3 | 7% | 0 | | 1 | 17% | 0 | | 0 | |
| ... transport & earth[c] | 16 | 12% | 7 | 14% | 2 | 5% | 0 | | 5 | 83% | 1 | 17% | 1 | 7% |
| ... agriculture[d] | 27 | 21% | 7 | 14% | 6 | 14% | 0 | | 5 | 83% | 3 | 50% | 6 | 43% |
| ... veterinary services[e] | 5 | 4% | 3 | 6% | 1 | 2% | 0 | | 0 | | 1 | 17% | 0 | |
| ... control measures[f] | 8 | 6% | 5 | 10% | 0 | | 3 | 23% | 0 | | 0 | | 0 | |
| Deficiencies of | | | | | | | | | | | | | | |
| ... resource management[g] | 39 | 30% | 16 | 31% | 12 | 29% | 2 | 15% | 5 | 83% | 1 | 17% | 3 | 21% |
| ... transport system | 2 | 2% | 0 | | 2 | 5% | 0 | | 0 | | 0 | | 0 | |
| ... watering technology[h] | 26 | 20% | 15 | 29% | 3 | 7% | 6 | 46% | 0 | | 2 | 33% | 0 | |
| Total | 91 | 69% | 42 | 82% | 20 | 48% | 11 | 85% | 6 | 100% | 5 | 83% | 7 | 50% |

a  Multiple counts possible; percentages relate to the total of all cases for each category; relative percentages may not total 100 because of rounding.
b  New water management technology such as motor pumps and boreholes.
c  Application of new transport and earth moving technologies such as trucks, tractors and carter pillars, including new transport-related processing and storage technology (e.g., refrigeration containers on ships and trucks for the transport of livestock products).
d  Agricultural innovations such as new crop varieties, fertilizer, and insecticides, and application of machinery.
e  Improvements in veterinary services, including research, leading to a reduction of tsetse, for example.
f  Desertification control measures such as shelter belts, afforestation, aerial seeds, and erosion barriers.
g  Unchanged management techniques despite of changing social and environmental conditions (e.g., fragmentation of land).
h  For example, water losses and poor drainage maintenance.

Deficiencies of technological applications often relate to the introduction of the very same new and innovative technology. In the Asian cases, land management technologies, e.g., for oasis and large-scale irrigation schemes, show reportedly low efficiency: they have reportedly high water losses and/or are characterized by poor drainage maintenance. Also, in the Australian and American cases desertification is mainly related to the introduction of new agricultural technologies (21% in total) which are not suited to, or not favourable for the sustained management of dryland sites. Examples are the rapid increase of irrigated garden products for distant markets due to road access, drastic increases in sheep number once transport technology gets improved (refrigeration chains, for example), and/or rapid increases in crop or livestock production due to market liberalization measures. Deficient or inappropriate land and water management techniques rank prominent as individual factors (30%), but show low geographical variations. This implies often standard applications of land use tools, which are recommended either by development authorities or practiced for traditional reasons by in-migrants, which, however, are not suited to ecological properties of dryland environments.

*Institutional and policy factors* Institutional and policy factors are prominent underlying driving forces of desertification (65%) – see Table 4.9. As with technological factors, modern as well as traditional policy or institutional aspects are equally involved in desertification cases. For example, the most important factor groups are both measures of growth-oriented agricultural policy (22%) - such as land (re)distribution, agrarian reforms, modern sector development projects and/or propagation of intensification – and institutional aspects of traditional law (20%) - such as the consequences of succession law upon land management, i.e., equal sharing of land and splintering of herds, reducing flexibility in management and increasing the pressure upon constant land unit. The introduction of new land tenure is another important factor group (19%). However, as a general rule no difference is apparent between private and state management impact. Policy and institutional factors show high geographical variation. They reportedly underpin more of the Asian, Latin American and Australian cases than other regional cases. In comparison with Latin America and Australia, mainly cases from the Central Asian steppe and desert region show a different and much more complex causative pattern. Especially economic policy measures, strongly directed towards modernization and growth of agriculture, in conjunction with some market liberalization policies, and land zoning measures (e.g., closing of frontiers, reservation of grazing areas, establishment of administrative boundaries) make them different from other cases. On the other hand, cases from the arid and semi-arid Australian and Patagonian rangelands are the only ones where the impact of former European colonization policies has been cited to fundamentally underlie desertification, i.e., the introduction of sheep and cattle with clear commercial rather than subsistence orientation, and modes of increased taxation contributing to continuous pressure upon land and water resources.

Table 4.9  Frequency of Specific Institutional and Policy Factors Driving Desertification[a]

| | All Cases (N=132) | | Asia (n=51) | | Africa (n=42) | | Europe (n=13) | | Australia (n=6) | | N-America (n=6) | | L-America (n=14) | |
|---|---|---|---|---|---|---|---|---|---|---|---|---|---|---|
| Agricultural development policies (reform, growth) | 29 | 22% | 21 | 41% | 7 | 17% | 2 | 15% | 0 | | 1 | 17% | 0 | |
| Market liberalization | 6 | 5% | 6 | 12% | 0 | | 0 | | 0 | | 0 | | 0 | |
| Sedentarization policy | 4 | 3% | 1 | 2% | 3 | 7% | 0 | | 0 | | 0 | | 0 | |
| Free grazing[b] | 4 | 3% | 0 | | 4 | 10% | 0 | | 0 | | 0 | | 0 | |
| Land zoning[c] | 12 | 9% | 8 | 16% | 4 | 10% | 0 | | 0 | | 0 | | 0 | |
| Lack of effective rule[d] | 16 | 12% | 8 | 16% | 1 | 2% | 0 | | 0 | | 1 | 17% | 6 | 43% |
| State management[e] | 9 | 7% | 5 | 10% | 2 | 5% | 0 | | 0 | | 1 | 17% | 1 | 7% |
| Private, commercial[f] | 16 | 12% | 6 | 12% | 5 | 12% | 0 | | 5 | 83% | 0 | | 0 | |
| Subsidies, incentives[g] | 17 | 13% | 6 | 12% | 3 | 7% | 2 | 15% | 5 | 83% | 0 | | 1 | 7% |
| Succession law[h] | 26 | 20% | 13 | 26% | 7 | 17% | 0 | | 0 | | 0 | | 6 | 43% |
| European colonization[i] | 11 | 8% | 0 | | 0 | | 0 | | 5 | 83% | 0 | | 6 | 43% |
| Total | 86 | 65% | 45 | 88% | 20 | 48% | 2 | 15% | 5 | 83% | 2 | 33% | 12 | 86% |

[a] Multiple counts possible; %s relate to the total of all cases for each category; relative %s may not total 100 because of rounding.
[b] 'Open access' situations and races for best grassland.
[c] Closing of frontiers, reservation of grazing areas, and establishment of political (administrative) boundaries.
[d] With particular reference to common property, including inequalities in access to land.
[e] Introduction of new land tenure, including a policy bias towards large landholdings.
[f] Introduction of new land tenure, including contracting-out of rangelands to households (e.g., private fencing).
[g] Direct or indirect farm or housing subsidies and 'grain for feed' or drought assistance programmes by the state.
[h] Equal sharing of land and splintering of herds.
[i] Introduction of colonial policy measures such as production orientation, taxation and application of new technologies.

The policy impact in African cases is distinct from other cases, in that sedentarization policies (directed towards nomads) and issues of free grazing (such as races for best grasslands or 'open access' situations) are prevalent. To a far lesser degree, European and North American cases are reported to be policy-driven. In both cases, socio-economic change – i.e., a shift from a predominantly rural towards an increasingly elderly, wealthier and urban population – seems to outweigh, or mitigate, the direct impact of institutional or policy factors leading to dryland degradation. Since about 1950, policies were implemented in the US Southwest for an increasingly urban population that had interest in land conservation and recreation, with frontier- and later market-driven ranching on a decline since long, and abandonment of especially marginal farmland due to low productivity and outmigration of working population from rural areas has widespread in southern Europe.

*Economic factors* Economic factors are prominent underlying driving forces of desertification (60%), with considerable regional variations – see Table 4.10. The leading factors are agricultural intensification, either land- or labour-scarcity driven intensification (21%), associated with desertification, and the increased demand for land to raise livestock or produce grain (20%). Concerning product price changes (10%), low prices leading to unsustained land management practices are reported to impact in more cases of desertification than do high prices. Economic factors underlie predominantly the Asian, Australian and Latin American cases. However, when it comes to specific economic factors the causative patterns vary per region. Asian cases appear to be mainly driven by remote influences such as industrialization, urbanization and commercialization (including export orientation and market competition), and by local factors such as agricultural intensification and land demand for crops and livestock which reportedly are responses of locals to distant market signals. Australian cases do more reflect local farmers' economic situation which relates to unfavourable prices in the export-oriented sheep sector leading to indebtedness, which altogether are seen to have induced the overuse of rare natural resources. Similarly, half of the Latin American cases of desertification are driven by increased land demand, however with considerably lower degrees of commercialization impact, product price or debt impact, but partly related to agricultural intensification (as in the case of Asia). Reduced economic vitality due to large-scale retirements in the agricultural sector is reportedly the single economic factor associated with desertification in the semi-arid European Mediterranean Basin, and limited to rugged hilly or mountain sites there.

*Demographic factors* Human population dynamics is an important underlying driving force of desertification (55%), with distinct regional variations – see Table 4.11. Asian and African cases of desertification are most commonly cited to relate to human population dynamics, while the impact of demographic factors seems of low importance in North and Latin America and is reportedly nil in the Australian rangelands.

Table 4.10  Frequency of Specific Economic Factors Driving Desertification[a]

| | All Cases (N=132) | | Asia (n=51) | | Africa (n=42) | | Europe (n=13) | | Australia (n=6) | | N-America (n=6) | | L-America (n=14) | |
|---|---|---|---|---|---|---|---|---|---|---|---|---|---|---|
| Land/labour-scarcity driven intensification | 27 | 21% | 21 | 41% | 3 | 7% | 0 | | 0 | | 0 | | 3 | 21% |
| Export orientation, commercialization | 18 | 14% | 9 | 18% | 2 | 5% | 0 | | 5 | 83% | 1 | 17% | 1 | 7% |
| Low prices[b] | 11 | 8% | 5 | 10% | 0 | | 0 | | 5 | 83% | 1 | 17% | 0 | |
| High prices | 3 | 2% | 0 | | 2 | 5% | 0 | | 0 | | 0 | | 1 | 7% |
| External demand[c] | 7 | 5% | 2 | 4% | 3 | 7% | 0 | | 0 | | 1 | 17% | 1 | 7% |
| Increased land demand[e] | 26 | 20% | 14 | 28% | 3 | 7% | 0 | | 0 | | 2 | 33% | 7 | 50% |
| Urbanization[f] | 5 | 4% | 4 | 8% | 0 | | 0 | | 0 | | 1 | 17% | 0 | |
| Industrialization | 8 | 6% | 8 | 16% | 0 | | 0 | | 0 | | 0 | | 0 | |
| Market failure[g] | 5 | 4% | 2 | 4% | 2 | 5% | 0 | | 0 | | 0 | | 1 | 7% |
| (Economic) poverty | 4 | 3% | 0 | | 2 | 5% | 0 | | 0 | | 0 | | 2 | 14% |
| Poverty-related features[h] | 7 | 5% | 1 | 2% | 1 | 2% | 3 | 23% | 0 | | 1 | 17% | 1 | 7% |
| Total | 79 | 60% | 45 | 88% | 15 | 36% | 3 | 23% | 5 | 83% | 2 | 33% | 9 | 64% |

[a] Multiple counts possible; percentages relate to the total of all cases for each category; relative percentages may not total 100 because of rounding.
[b] Including price decrease and collapse, for example, in the water, wool and beef sector.
[c] For example, for cotton, oil, beef, rice, peat, turf and bricks for house construction.
[d] With particular reference to common property, including inequalities in access to land.
[e] In particular, for raising livestock and producing grain.
[f] Including the loss of cultivated land for agricultural production, thus pushing farmers and herders onto marginal sites.
[g] For example, unjust credit system, poor distribution system, and excessive subsidation.
[h] Indebtedness, low or nil investments, low labour availability, and no longer viable farm sizes.

Table 4.11  Frequency of Specific Demographic Factors Driving Desertification[a]

| | All Cases (N=132) | | Asia (n=51) | | Africa (n=42) | | Europe (n=13) | | Australia (n=6) | | N-America (n=6) | | L-America (n=14) | |
|---|---|---|---|---|---|---|---|---|---|---|---|---|---|---|
| Increased population size | 28 | 21% | 21 | 41% | 3 | 7% | 2 | 15% | 0 | | 1 | 17% | 2 | 14% |
| ... unspecified ('more') | 26 | 20% | 20 | 39% | 3 | 7% | 2 | 15% | 0 | | 1 | 17% | 0 | |
| ... increase in last 50 yrs[b] | 3 | 2% | 1 | 2% | 0 | | 0 | | 0 | | 0 | | 2 | 14% |
| Population growth | 14 | 11% | 10 | 20% | 1 | 2% | 0 | | 0 | | 0 | | 0 | |
| ... unspecified | 1 | 1% | 1 | 2% | 0 | | 0 | | 0 | | 0 | | 0 | |
| ... fast, rapid | 4 | 3% | 3 | 6% | 1 | 2% | 0 | | 0 | | 0 | | 0 | |
| ... remote populations[c] | 6 | 5% | 6 | 12% | 0 | | 0 | | 0 | | 0 | | 0 | |
| ... 'overpopulation'[d] | 3 | 2% | 3 | 6% | 0 | | 0 | | 0 | | 0 | | 0 | |
| Migration | 15 | 11% | 12 | 24% | 2 | 5% | 0 | | 0 | | 1 | 17% | 0 | |
| ... unspecified | 3 | 2% | 1 | 2% | 1 | 2% | 0 | | 0 | | 1 | 17% | 0 | |
| ... forced migration[e] | 4 | 3% | 4 | 8% | 0 | | 0 | | 0 | | 0 | | 0 | |
| ... increased mobility[f] | 4 | 3% | 4 | 8% | 0 | | 0 | | 0 | | 0 | | 0 | |
| ... no emigration of males | 3 | 2% | 3 | 6% | 0 | | 0 | | 0 | | 0 | | 0 | |
| ... labour outmigration | 1 | 1% | 0 | | 1 | 2% | 0 | | 0 | | 0 | | 0 | |
| Natural increase | 4 | 3% | 4 | 8% | 0 | | 0 | | 0 | | 0 | | 0 | |
| ... unspecified | 1 | 1% | 1 | 2% | 0 | | 0 | | 0 | | 0 | | 0 | |
| ... high fertility increase | 3 | 2% | 3 | 6% | 0 | | 0 | | 0 | | 0 | | 0 | |

**Table 4.11    Continued**

| Population densities | | | | | | | | | | | | |
|---|---|---|---|---|---|---|---|---|---|---|---|---|
| | 18 | 14% | 3 | 6% | 12 | 29% | 3 | 23% | 0 | 0 | 0 | 0 |
| ... low[g] | 1 | 1% | 0 | 0 | 0 | 0 | 1 | 8% | 0 | 0 | 0 | 0 |
| ... high[h] | 6 | 5% | 0 | 0 | 6 | 14% | 0 | 0 | 0 | 0 | 0 | 0 |
| ... both low & high | 2 | 2% | 0 | 0 | 0 | 0 | 2 | 15% | 0 | 0 | 0 | 0 |
| ... increasing | 9 | 7% | 3 | 6% | 6 | 14% | 0 | 0 | 0 | 0 | 0 | 0 |
| Others | 5 | 4% | 0 | 0 | 1 | 2% | 4 | 31% | 0 | 0 | 0 | 0 |
| ... unequal age structure[g] | 3 | 2% | 0 | 0 | 0 | 0 | 3 | 23% | 0 | 0 | 0 | 0 |
| ... life cycle change[i] | 2 | 2% | 0 | 0 | 1 | 2% | 1 | 8% | 0 | 0 | 0 | 0 |
| Total | 73 | 55% | 43 | 84% | 21 | 50% | 6 | 46% | 1 | 17% | 2 | 14% |

[a] Multiple counts possible; percentages relate to the total of all cases for each category; relative percentages may not total 100 because of rounding.
[b] Mainly, a doubling of population numbers.
[c] Growth of remote rural as well as urban populations, due to natural growth and/or immigration; triggering the loss of cultivated land in (peri)urban zones and pushing cultivators and herders onto marginal sites.
[d] In remote, settled agricultural areas.
[e] Forced migration of herders onto marginal sites and/or of sedentary cultivators into drylands.
[f] For example, of nomadic households due to transport improvements.
[g] With the consequence that the stability of productive land is not ensured.
[h] With the consequence that the pressure on resources endangers the stability of land.
[i] Changes in the composition of households such as the shift from extended to nuclear families.

Increases in the size of population are the most important factor group associated with desertification (21%), but the specific components of it remain unclear (since they were hardly specified). However, there is indication that natural increases due to high fertility of local populations are of less importance (and limited to Asian cases), if compared to fast, rapid population growth as linked to migration and, thus, rising population densities (11% and 14%, respectively). Migrational aspects associated with desertification include the forced migration of herders to marginal sites or the shift of cultivators onto (marginal) drylands, as mainly reported from various sites of the Central Asian steppe and desert region. Aspects of migration, spatial as well as social, also include the increased mobility of nomadic people due to transport improvements, such as in the case of grazing in the Saudi Desert. Cases of in-migration seem to dominate the causative pattern, but it has also been reported, for example, that blocked outmigration of males and a consecutive high proportion of male population leads to overuse of Thar desert lands, and that labour outmigration, often in conjunction with a shift from extended to nuclear families, leads to lower livestock mobility and consecutive overstocking. Both low and high population densities are reported to be associated with desertification, with high and increasing densities being the most important factor. However, most of the cases in which population growth has been reported do not relate to fast, rapid growth of local populations. Rather, these are cases in which overpopulation or population pressure of distant urban populations or in remote settled agricultural areas impact upon local land use, in both situations leading to loss of cultivated land and thus triggering off migration of cultivators and/or herders onto marginal sites. Different from the Asian and African cases, demography-driven cases of desertification in the European Mediterranean Basin relate to imbalanced age structure (more elderly, less children) and a shift from high to low population densities at which the maintenance of landscapes is no longer ensured, land is abandoned, and the stability of agroecosystems gets easily jeopardized.

(Notes related to Table 4.12)

[a] Multiple counts possible; percentages relate to the total of all cases for each category; relative percentages may not total 100 because of rounding.
[b] Exceptional treatment of ethnic minorities; for example, they are allowed to have more children.
[c] Towards urban consumers and larger holdings.
[d] In terms of better living standards and self-sufficiency in food.
[e] High expectation in land settlement, consideration of water as 'free good' and of grazing as an 'inefficient land use practice'.
[f] Indifference, for example, towards efforts of remediation, restoration and conservation of agro-ecological environments, including low skills and energy applied to livestock husbandry. Unconcern and no or low interest in cooperation due to low remuneration.
[g] Land transfer and/or acquisition through inheritance rather than through markets, for example.

**Table 4.12  Frequency of Specific Cultural or Socio-Political Factors Driving Desertification[a]**

| | All Cases (N=132) | | Asia (n=51) | | Africa (n=42) | | Europe (n=13) | | Australia (n=6) | | N-America (n=6) | | L-America (n=14) | |
|---|---|---|---|---|---|---|---|---|---|---|---|---|---|---|
| Public attitudes/values | 52 | 39% | 30 | 59% | 10 | 24% | 1 | 8% | 5 | 83% | 1 | 17% | 5 | 36% |
| ...frontier mentality[b] | 8 | 6% | 7 | 14% | 0 | | 0 | | 0 | | 0 | | 1 | 7% |
| ...ethnic minorities[b] | 1 | 1% | 1 | 2% | 0 | | 0 | | 0 | | 0 | | 0 | |
| ...violent conflicts, war | 10 | 8% | 3 | 6% | 6 | 14% | 1 | 8% | 0 | | 0 | | 0 | |
| ...policy bias[c] | 8 | 6% | 0 | | 4 | 10% | 0 | | 0 | | 0 | | 4 | 10% |
| ...goal of 'catching up'[d] | 12 | 9% | 12 | 24% | 0 | | 0 | | 0 | | 0 | | 0 | |
| ...perceptional issues[e] | 12 | 9% | 7 | 14% | 0 | | 0 | | 5 | 83% | 0 | | 0 | |
| ...concern, response | 1 | 1% | 0 | | 0 | | 0 | | 0 | | 1 | 17% | 0 | |
| Individual & household behaviour | 53 | 40% | 18 | 35% | 10 | 24% | 3 | 23% | 5 | 83% | 3 | 50% | 9 | 64% |
| ...indifference[f] | 13 | 10% | 4 | 8% | 5 | 12% | 0 | | 0 | | 1 | 17% | 3 | 21% |
| ...life style change | 7 | 5% | 3 | 6% | 1 | 2% | 0 | | 0 | | 0 | | 3 | 21% |
| ...short-term gain | 4 | 3% | 0 | | 0 | | 0 | | 0 | | 1 | 17% | 3 | 21% |
| ...frontier mentality | 8 | 6% | 3 | 6% | 0 | | 0 | | 5 | 83% | 0 | | 0 | |
| ...traditional behaviour[g] | 3 | 2% | 3 | 6% | 0 | | 0 | | 0 | | 0 | | 0 | |
| ...breakdown of tradition | 2 | 2% | 0 | | 2 | 5% | 0 | | 0 | | 0 | | 0 | |
| ...poor, low education | 6 | 5% | 1 | 2% | 2 | 5% | 3 | 23% | 0 | | 0 | | 0 | |
| ...unawareness | 1 | 1% | 0 | | 0 | | 0 | | 0 | | 1 | 17% | 0 | |
| ...perceptional issues[e] | 9 | 7% | 4 | 8% | 0 | | 0 | | 5 | 83% | 0 | | 0 | |
| Total | 55 | 42% | 26 | 51% | 12 | 29% | 4 | 31% | 5 | 83% | 3 | 50% | 6 | 43% |

*Cultural or socio-political factors* Cultural or sociopolitical factors are important underlying driving forces which relate to less than half of the desertification cases (42%), and which show considerable regional variations – see Table 4.12. Both public attitudes, values and beliefs, and individual or household behaviour are associated with cases of dryland degradation to about equal parts, however with pronounced geographical distinctions. In the Asian cases, desertification has been cited to be more strongly driven by public than by private responses or motivations, which had mainly been outcomes of frontier mentality such as public desires for land consolidation and for establishing military strongholds in the northwester territories of China, for example. Frontier mentality reportedly operated in close conjunction with public desires to 'catch up' in terms of improving living standards or attaining self-sufficiency in food. It was accompanied by issues of perception such as that water is a 'free good' and that grazing – if compared to grain production – is 'inefficient'. In contrast, the Latin American cases are predominantly driven by individual responses or motivations, to a large part including indifference towards environmental degradation, low interest in cooperation, or aspirations towards independence (a 'better living'). The Australian cases are unique in that frontier mentality associated with desertification pervades the private livestock businesses of individuals (rather than being a state concept), and issues of environmental perception are reported to exist in both the public domain (i.e., high expectations in land settlement by individuals, regardless of environmental concerns) and on the side of individuals (i.e., water is considered as 'free good' and droughts are not perceived as risks). The impact of wars, civil wars, insurgency and violent conflicts about land upon the disruption of land management, leading to land degradation, is reported to be most pronounced in cases from West and East African rangelands. In addition, these are the sites where the collapse or breakdown of traditional systems (of pastoral rules, for example) has been cited to be directly associated with desertification. African as well as European cases appear to be less driven by cultural or socio-political factors than other regional cases, though. Poor or low levels of education (or agricultural training) among the rural farming population, for example, are the only factors noteworthy in European cases of desertification such as from the Basilicata region of southern Italy.

## Mode and Type of Variable Interactions

Not only are multiple causal factors at work, but their interactions also lead to desertification, which is why it is important to understand the system dynamics. The analysis here reveals that, in most cases, three to five underlying causes are driving two to three proximate causes, with agricultural activities and/or climatic factors inherent to all of the causes, not including feedbacks (which are considered later).

*Causes and System Properties* 117

*Interlinkages between Proximate Causes*

*Modes of interaction* The interlinkages between causative factors at the proximate level are made up of two predominant modes of interaction which feature nearly three fourths of the cases. These are factors that occur concomitantly (n=55, or 42% of the cases) – i.e., independent but synchronous operation of individual proximate factors leading to desertification – and factors that intervene in synergetic factor combinations (n=37, or 28%) – i.e., several mutually interacting proximate causes. The most important examples of the first mode are increased aridity and agricultural activities operating aside with each other (in 26 cases, or in combination with infrastructure extension in another 14 cases). Examples of synergistic operation commonly involve all proximate factors, i.e., agricultural activities, wood extraction, infrastructure extension, and increased aridity mutually interacting with each other. Cases of pure chain-logical connection are low in number (n=5, or 4%) – i.e., proximate factors connected as causal chain – with infrastructure extension triggering agricultural activities, and agricultural activities either triggering wood extraction or inducing, for example, a change in fire regimes associated with desertification (but no climatic factors unilaterally triggering change in another factor). However, there are more cases (n=21, or 16%) in which proximate causes are featured by both concomitant occurrence and chain-logical connection (with all chain-logically connected cases considered below in more detail). A pattern of single factor causation was found in 10% of the cases (n=14) – i.e., one proximate cause completely dominating other causes – with increased aridity and agricultural activities being reported as causative to desertification (to equal parts).

*Chain-logical connection* Interacting variables with chain-logical connections were involved in one third of all cases (n=41, or 31%), either as pure causes or as causative connections in concomitant occurrence with other proximate factors. If all connections in the cases are broken down by their most simple form which is a tandem (or two-factor chain), a total of 48 tandems emerges at the broad aggregate level – see Table 4.13. Clearly, the most frequent two-factor chain is infrastructure extension impacting upon agricultural activities in about one fifth of the cases (22%). Mainly the establishment of water-related infrastructure – in its various forms such as dams, reservoirs, canals, boreholes, and pump stations – is reported to lead to an expansion of irrigated cropland and pasture. As a characteristic of most of these cases, irrigation infrastructure extends in combination with settlement expansion, road network extension, and commercial as well as industrial developments. Thus, one might identify an 'industrial-water-road complex' driving agricultural expansion, what is especially true for the Australian cases. The role of increased aridity in causal chains is low, since climatic factors operate with land use activities mainly in a synergetic manner. As an addition to the infrastructure-agriculture tandem, in most of the infrastructure-driven cases where agricultural activities are associated with desertification, cropland extension and livestock development appear to be interlinked. In other words, it has been reported that both land uses are involved and mutually causing land degradation once infrastructure expansion happens. This means that cropland expansion on land previously used for pastoral activities reportedly leads to overstocking on the remaining, reduced rangeland, and accelerated land degradation happens at cropland sites now under rain-fed farming, which had previously been extensive grazing land under flexible modes of management.

Table 4.13  Chain-Logical Connection of Broad Proximate Causes (n=41)[a]

|  | Agricultural activities[b] | | Wood extraction[c] | | Infrastructure extension | Increased Aridity | | Total row |
|---|---|---|---|---|---|---|---|---|
| Agricultural activities[b] | NA | | 3 | 2% | 0 | 4 | 3% | 7 |
| Wood extraction[c] | 1 | 1% | NA | | 0 | 0 | | 1 |
| Infrastructure extension | 29 | 22% | 3 | 2% | NA | 0 | | 32 |
| Increased Aridity | 7 | 5% | 1 | 1% | 0 | NA | | 8 |
| Total column | 37 | | 7 | | 0 | 4 | | 48 |

[a] Row causes column; percentages relate to all cases of desertification (N=132).
[b] Livestock raising and crop production.
[c] Including related extractional activities (e.g., herbs).

*Interlinkages between Underlying Driving Forces*

*Modes of interaction* The interlinkages between causative factors at the underlying level are made up of three predominant modes of interaction, which feature 90% of the cases. In the order of decreasing importance, these are concomitant occurrence, chain-logical connection, and synergistic manner.

First, in nearly two thirds of the cases factors occur concomitantly, i.e., independent but synchronous operation of individual underlying forces. Drivers which operate in a fully synchronous manner are less frequent though (n=29, or 22% of the cases) than concomitant drivers which operate in cases that are also featured by chain-logical factor combinations (n=54, or 41% of the cases). Drivers which operate in a fully synchronous manner involve mainly two-factor combinations, i.e., climatic factors with human population dynamics, and climatic factors with technological factors. Drivers which operate concomitantly, but are part of a factor combination in which chain-logical connections are at work, too, involve mainly five-factor combinations of economic, technological, institutional and climatic factors in combination with either demographic or socio-cultural variables.

Second, underlying drivers connected as causal chain occur in three fifths of the cases. Cases of pure chain-logical connection are low in number, though (n=7, or 5%). They are chiefly made up of three-factor combinations including human population dynamics, technological changes, and institutional factors or policy impact (no climatic factors are involved in cases of pure causal chains). However, causal chains feature also cases in which concomitant drivers operate (n=54, or 41% of the cases) – mainly, five-factor combinations of economic, technological, institutional and climatic factors in combination with either population or socio-cultural variables – and in which several, but not all drivers interact in a synergistic manner (n=19, or 14%). As for the latter, the full interplay of all underlying driving forces is characteristic.

Third, factors intervene in synergetic factor combinations in about one third of the cases (n=37, or 28%). Cases of pure causal synergies are less frequent though (n=11, or 8%) than cases in which several mutually interacting drivers operate aside with factors that are also linked up in causal chains underpinning land use activities that lead to desertification (n=19, or 14%). In cases where pure causal synergies are prevalent, no characteristic factor combinations emerge. In cases where causal synergies and chain-logical connection occur, the full interplay of all underlying driving forces is most typical. A pattern of single factor causation was found in 9% of the cases (n=12) – i.e., one underlying driver completely dominating other causes – with climatic and institutional (or policy) factors involved to about equal parts.

*Chain-logical connection* Interacting variables with chain-logical connections were found to be involved in three fifths of all cases (n=75, or 57%), either as pure causal chains or, chiefly, together with drivers operating concomitantly or in a synergetic manner. If all connections in the cases are broken down by their most simple form which is a tandem (or two-factor chain), a total of 170 tandems emerges at the broad aggregate level – see Table 4.14.

The most frequent two-factor chains appear to be those in which policy or institutional factors are causative at first instance, and driving, for example, technological factors (in 22% of the cases) and economic factors (in 15% of the cases), underpinning proximate factors leading to desertification. Similarly, important are those two-factor chains in which economic factors are involved as driven by other underlying forces such as institutional or policy factors (15%), or, equally important, by human population dynamics (in 17% of the cases). The role of climatic factors as being causative to other drivers is nil. No climatic factors are impacted directly by other, socio-economic drivers, because human impact with climate happens in the form of synergy or concomitant occurrence at the proximate level, i.e., between (human) land use activities and increased aridity.

The most important chain-logical interaction is that of policy or institutional factors driving technological developments leading to desertification (22% of the cases), the policy/institution-technology tandem. The pattern, not including cases from Europe, Australia and North America, has recurrent and straightforward linkages. Agricultural, growth-oriented development policies – directed towards land (re)distribution, agrarian reforms, and modern development projects including intensification measures – are often coupled with the introduction of new, private land tenure. These measures are implemented against the background of lacking effective rules or regulation concerning common property resources, and against some inherited institutional barriers. In specific, traditional succession law, according to which land is equally shared among family members and herds are splintered, reduces flexibility in management and increases the pressure upon constant land units (even without new policy impacts). Once land zoning occurs (e.g., reservation of grazing areas, establishment of irrigation zones) in combination with the introduction of new land and water management techniques (canals, dams, reservoirs, boreholes, pump stations, or tractors, pesticides, insecticides, exotic crop varieties such as irrigated garden products), the reported outcomes of such tandems are always cited to be inappropriate land use with high losses especially in water-based cultivation, i.e., straightforward water losses or poor drainage maintenance leading to the spread of salinized land, for example. These, or similar linkages associated with desertification – i.e., salinization, reactivation of sand dunes, removal of vegetated cover and increases in bare, eroded ground – are reported, for example, from various sites in northern China, northwestern Mexico, the Sudan-Sahel, and East Africa. The sole exception were not private land tenure but state management and collective property was involved, were sites in southern Russia from which large-scale introduction of new irrigation technologies was reported.

Table 4.14  Chain-Logical Connection of Broad Underlying Causes (n=75)[a]

|   | A Demographic factors | | B Economic factors | | C Technological factors | | D Policy institutional factors | | E Cultural or sociopolitical factors | | F Climatic factors | | [c] |
|---|---|---|---|---|---|---|---|---|---|---|---|---|---|
| A | NA | | 22 | 17% | 0 | | 6 | 5% | 1 | 1% | 0 | | 29 |
| B | 2 | 2% | NA | | 3 | 2% | 6 | 5% | 8 | 6% | 0 | | 19 |
| C | 4 | 3% | 7 | 5% | NA | | 6 | 5% | 0 | | 0 | | 17 |
| D | 9 | 7% | 20 | 15% | 29 | 22% | NA | | 6 | 5% | 0 | | 64 |
| E | 3 | 2% | 10 | 8% | 13 | 10% | 15 | 11% | NA | | 0 | | 41 |
| F | 0 | | 0 | | 0 | | 0 | | 0 | | NA | | 0 |
| [b] | 18 | | 59 | | 45 | | 33 | | 15 | | 0 | | 170 |

[a] Row causes column; percentages relate to all cases of desertification (N=132).
[b] Total column.
[c] Total row.

Another important chain-logical interaction is human population dynamics driving economic factors leading to desertification (17% of the cases), the demography-economy tandem. The most common pattern is that increases in the size or density of local populations, mainly driven by in-migration, are at very first instance translated into growing demand for agricultural land to raise livestock or produce grain. Besides meeting the food (and shelter) needs of local populations, external economic demands are powerful signals to which growing, local populations respond, e.g., growing irrigated rice or cotton, raising cattle for beef, producing wool from sheep ranching, and digging peat and turf for bricks. These cases were not only limited to northern China, but were also typical for the US American Southwest, for example. Often, remote demographic impacts such as the growth of distant rural or urban populations, mostly in conjunction with the expansion of commercial activities, led to the loss of cultivated land at the previous homeland sites of farmers who moved on then (deliberately or forced) to marginal dryland sites, taking up commercial farm production. Such cases were reported from various East Mediterranean steppe sites, from the Sonora Desert, and from various sites of the Central Asian steppe region. Contrasting this prevailing pattern of population growth-driven land use are cases in which outmigration, often in conjunction with changes in household composition (i.e., the shift from extended to nuclear families) impacts upon the local economy, underlying land degradation. As a consequence, low investments in human capital (herding labour, for example) occurred, thus reducing livestock mobility and adaptive capacity of herding populations to oscillating rainfall, for example, in the West African Sudan-Sahel.

Another important chain-logical connection is that of policy or institutional factors driving economic factors leading to desertification (15% of the cases), the policy/institution-economy tandem. Two major patterns are evident from the cases. First, market liberalization and modern sector-oriented agricultural policies drive market competition, commercialization and export orientation of production leading to overuse of resources and land degradation. These, or similar market policy-driven examples are manifold, and characterize, for example, various sites in the Central Asian steppe and desert region. In some of these cases, government subsidies or incentives directed towards agricultural and infrastructure expansion are involved such as subsidies for house construction and 'grain for feed'. Another pattern, often overlaying the free or subsidized market policy-driven cases, is that both traditional elements of land use institutions (e.g., successional law practices) and modern measures of private property regimes (e.g., private fencing, contracting out rangelands to private households) lead to increased land demand, and henceforth to the expansion of cultivated land, often at the expenses of rangeland, thus increasing the pressure upon remaining grazing land. Most of these cases stem from the Great Plains of northern China, but are not limited to them.

*Interlinkages between Underlying and Proximate Causes*

Chain-logical connection pervades the causative pattern of all interlinkages between the underlying and proximate levels in all cases (n=132), i.e., one or more underlying factors driving one or more proximate causes leading to desertification (including here cases of single-factor causation). If all causative (chain-logical) connections are broken down by their most simple form which is a tandem (or two-factor chain), a total of 585 causative factor interlinkages emerges – see Table 4.15. The most frequent two-factor chains appear to be those in which demographic factors (129 tandems), economic and technological factors (107 tandems each), and policy or institutional factors (90 tandems) are causative at first instance, i.e., mainly driving agricultural activities (in 40 to 50% of the cases), infrastructure extension (20 to 30%), wood extraction (5 to 30%), and climatic factors (120 tandems) associated with increased aridity (in 86% of the cases), leading to desertification.

**Table 4.15    Chain-Logical Connection of Broad Underlying Forces Driving Broad Proximate Factors Leading to Desertification (N=132)[a]**

|  | Agricultural activities[b] | | Wood extraction[c] | | Infrastructure extension | | Increased Aridity | | Total row |
|---|---|---|---|---|---|---|---|---|---|
| Demographic factors | 55 | 42% | 39 | 30% | 35 | 27% | 0 | | 129 |
| Economic factors | 65 | 49% | 15 | 11% | 27 | 21% | 0 | | 107 |
| Technological factors | 69 | 52% | 7 | 5% | 31 | 24% | 0 | | 107 |
| Policy factors | 49 | 37% | 11 | 8% | 30 | 23% | 0 | | 90 |
| Cultural factors | 20 | 15% | 7 | 5% | 5 | 4% | 0 | | 32 |
| Climatic factors | 6 | 5% | 0 | | 0 | | 114 | 86% | 120 |
| Total column | 264 | | 79 | | 128 | | 114 | | 585 |

[a]   Row causes column; percentages relate to all cases of desertification (N=132).
[b]   Livestock raising and crop production.
[c]   Including related extractional activities (e.g., herbs).

**Interacting Hierarchical Scales**

Most of the underlying causes of desertification could be attributed to individual or multiple scales at which they operate, except for demographic and climatic factors – see Table 4.16. As for the latter, uncertainty was reported in up to one half of the cases where these causative factors actually originate. It was unclear, for example, what are the exact components which cause human populations to increase (e.g., fertility, reduced mortality, migration, or all of them?). Similarly, it remained unclear from the cases whether droughts are local phenomena or whether they are linked to wider changes in metrological conditions.

Table 4.16  Driving Forces of Desertification by Scale of Influence

| | Demographic factors[a] | | Economic factors | | Technological factors | | Policy factors | | Cultural factors | | Climatic factors | |
|---|---|---|---|---|---|---|---|---|---|---|---|---|
| | \multicolumn{12}{c}{Individual scales} |
| Local[b] | 17 | 23% | 14 | 18% | 26 | 29% | 10 | 12% | 1 | 2% | 0 | |
| Regional[c] | 0 | | 0 | | 0 | | 0 | | 8 | 14% | 0 | |
| National | 0 | | 10 | 13% | 0 | | 17 | 20% | 2 | 4% | 0 | |
| Global | 0 | | 3 | 4% | 0 | | 5 | 6% | 0 | | 14 | 12% |
| Total | 17 | 23% | 27 | 34% | 26 | 29% | 32 | 37% | 11 | 20% | 14 | 12% |
| | \multicolumn{12}{c}{Cross-scalar interactions ('interplays')} |
| Local-regional | 8 | 11% | 0 | | 10 | 11% | 21 | 24% | 16 | 29% | 14 | 35% |
| Local-national | 13 | 18% | 34 | 43% | 41 | 45% | 15 | 17% | 17 | 31% | 17 | 15% |
| Local-global | 0 | | 9 | 11% | 14 | 15% | 4 | 5% | 0 | | 11 | 10% |
| Regional-national | 0 | | 0 | | 0 | | 4 | 5% | 11 | 20% | 0 | |
| Regional-global | 0 | | 0 | | 0 | | 0 | | 0 | | 0 | |
| National-global | 0 | | 9 | 11% | 0 | | 10 | 12% | 0 | | 0 | |
| Total | 21 | 29% | 52 | 66% | 65 | 71% | 54 | 63% | 44 | 80% | 68 | 60% |
| | \multicolumn{12}{c}{No data, unknown (scale not specified)} |
| Total | 35 | 48% | 0 | | 0 | | 0 | | 0 | | 32 | 28% |

[a] 'Rising population density' was coded as local, but 'increase in population' was coded as unspecified.
[b] Farm, household, society, community, or small ecosystem.
[c] District level.

As a result, cross-scalar interactions of driving forces are much more important than causative factors operating at individual scales. Individual scales matter in between roughly 10 and 35% of the cases, while multiple scales matter in between 30 and 80% of the cases.

As for individual scales, the importance of either the local, regional, national or global scale varies by the type of underlying driving force involved. For example, causative factors which arise exclusively at the local scale are human population dynamics (natural population increases, mainly) and technological factors (inappropriate land and water management techniques, mainly). Economic and policy (or institutional) factors also originate from the local level – e.g., land or labour-driven agricultural intensification, rural poverty, lack of effective common property rules – but the national scale is important as well – i.e., external economic demands from remote urban populations elsewhere in the country, introduction of new land tenure, implementation of development projects, subsidies, (fiscal) incentives, etc. Differently, cultural or socio-political causative factors stem mainly from the regional (or district) level – i.e., traditional rules of land transfer, and regional conflicts about land – while climatic factors are reportedly important at the global scale of climate change.

As for multiple scales, or cross-scalar interactions, interplays of scales in which the local level is involved are much more common than any other type of interaction. For example, economic and technological factors interact mainly at the local to national level. Land demand for livestock and grain production, for instance, reportedly materializes at the local (and district) level, but intervenes also as a remote impact from remote urban centres; industrialization as a declared government policy programme triggers the establishment of oil-gas fields, mining enterprises, and processing factories at local sites in arid frontier regions of nation states; and the introduction at the local level of new land and water management technology such as hydrotechnical installations or improved seed varieties, fertilizer, etc. follows mainly development and extension programmes implemented by state rather than local authorities. In contrast, scalar interplays in which the global domain is involved, in particular local-global interplays, are much less common. This holds even true for climatic factors. This means that regional feedback loops, for example, between land use, albedo change, and worsening climatic conditions are reportedly more prevalent than cases in which global climate change was related to local rainfall deficits.

**Intermediate Factors**

Intermediate or mediating factors which shape, or intervene into the interplay of underlying driving forces and proximate causes, have been reported in about one fifth of the cases (26%) – see Table 4.17. It appears that they can take the form of single factors (in 16% of the cases), or be multi-factorial combinations (in 10% of the cases) to about equal parts. The only factor group worth to be considered due to its extent is that of wealth conditions versus poverty conditions in about one fifth of the cases (n=24, or 18%). The wealth/poverty divide appears to be a robust feature, i.e., geographical variations are low. It shows the following distinct patterns.

**Table 4.17    Mediating Factors in Desertification (n=34 cases)***

|  | abs | rel |
|---|---|---|
| Wealth/poverty (e.g., rich *versus* poor land users, smallholders *versus* large farmers) | 24 | 18% |
| Type of land management (soil, tree, water) (e.g., agroforestry *versus* monocultural, large-scale *versus* small-scale) | 9 | 7% |
| Access to resources (e.g., cropping land, pastures, cattle) | 8 | 6% |
| Ethnical affiliation | 7 | 5% |
| Other factors (e.g., quality of social relations, age structure, degree of mobility, property type) | 7 | 5% |
| Total | 55 |  |

\*    Multiple counts possible; percentages relate to the total of all 132 cases.

First, impoverished individuals, groups or larger parts of the population are most vulnerable to the outcomes of dryland degradation, especially in the course of drought-induced crop failures. Such cases were mainly reported from the West African Sudan-Sahel. Secondly, the actions of impoverished or deprived individuals, groups or larger parts of the population have been reported as causes of desertification. For example, at various Thar Desert sites decreases in soil fertility were reportedly due to continuous cultivation by impoverished farm holdings, and clashes between rich and poor farmers were cited as underpinning several cases of desertification in the US-American Southwest. Thirdly, impoverished or deprived individuals, groups or larger parts of the population have been reported to be both most vulnerable and directly causative to desertification. Such cases stem exclusively from rangelands in Australia and Patagonia. Many of the 6,000 rangeland grazing enterprises in the arid and semi-arid parts of Australia reportedly belong to this group, as well as the majority of small farmers in Patagonia.

## System Properties

*Amplifying and Mitigating Feedbacks*

In 40 or 30 of the cases, respectively, amplifying and mitigating feedbacks were specified, with both types overlapping occasionally. These cases include feedbacks – mainly positive or mutually self-enforcing – from proximate factors upon underlying driving forces (13%), feedbacks – mainly mitigating – from underlying demographic, economic, technological and institutional factors upon proximate causes (7%), feedbacks – mainly amplifying – from meteorological conditions upon proximate and underlying causes (9%), and other feedback loops in which long-term climate-driven change has been moulded into a process which operates more and more in causal synergy with local land use activities (5%).

*Amplifying feedbacks from proximate causes* In 12 cases (9%), proximate factors – mainly agricultural activities, to a lesser degree infrastructure extension, but not wood extraction or related activities – had a positive, amplifying feedback upon underlying driving forces. The most common pattern is a self-perpetuating process made up of expanding rangeland and cropland, followed by overstocking and soil mining, respectively, leading to increased demand for pasture or cropland, having as a consequence further expanding rangeland and cropland with continued overstocking and soil mining, etc. Related rangeland cases were reported, for example, from various steppe and desert sites in Northern China (Sheehy, 1992) as well as from East African rangelands such as the N-Tanzanian Masai steppe (Mwalyosi, 1992). Related cropland cases were reported, among others, from two sites in the Kano Close-Settled Zone of northern Nigeria, for example. As in many other cases, the expansion of cropland is negatively linked to grazing land development (i.e., reservation of grazing areas, closing of land frontier), leading to problems of overstocking and land degradation (Mortimore, Harris and Turner, 1999). Similarly, it was reported from the Maasina Region of central Mali that cropland expansion in the alluvial Niger River delta inland plain had a feedback upon the social organization of production in so far as less labour was available for, and consequently lower investments were made into, ensuring mobility and flexibility of livestock, thus leading to degradation (Turner, 1999a). Amplifying feedbacks of proximate factors upon human population dynamics have been described, for example, from remote ('matorral', 'garrigue', shrubland) mountain sites on Crete, where new, EC policy-driven access roads led to increasing human population numbers through better accessibility and secured food supply, thus raising the land use pressure upon fragile mountain ecosystems (Hill, Hostert, Tsiourlis, Kasapidis, Udelhoven and Diemer, 1998). Similar cases of a self-perpetuating relationship between local populations and land use systems which get adopted to demographic change stem, from the SW-Russian Caspian Plain and adjacent Yergenin Uplands (Rozanov, 1991) and the Tehuacán Valley in south-central Mexico (McAuliffe, Sundt, Valiente-Banuet, Casas and Viveros, 2001).

In five cases (4%), proximate factors or land use activities were clearly named to be part of a regional feedback loop made up between high and increasing rainfall variability and prolonged droughts (as driven by climate change) and agricultural as well as infrastructure extension leading to desertification and mutually impacting re-enforcing upon climate change. The Okavango Delta of the Kalahari Sandveld was cited an example (Dube and Pickup, 2001). Another example of natural phenomena operating in synergy with land use activities is a special variant of the desertification-land use-desertification-etc. spiral reported from the Qinghai-Xizang (Tibet) Plateau. There, wood extraction led to sand encroachment and consequent invasion of *Sophora moocroftiana*, a fuel plant which triggered continued wood extraction, and thus further land degradation (Liu and Zhao,

2001). Drought conditions and the establishment of infrastructure nuclei were described as a mutual relationship for the semi-arid parts of West and Central Sudan, i.e., enforcing the concentration of people and livestock around infrastructure nuclei which became initial conditions for desertification (Ayoub, 1998). Also, it was reported from the Mu Us Region (desert, 'meadow') of the Ordos Plateau in north-central China that humans through their land use activities recently enhanced the stabilization of previously oscillating desert margins, namely, dominant winter monsoons associated with dust storms and aeolian deposits. What had been long-term change not moulded by human impact operates now in synergy with land use activities, namely agricultural activities, wood extraction, and infrastructure development (Zhou, Dodson, Head, Li, Hou, Lu, Donahue and Jull, 2002).

*Mitigating feedbacks from underlying driving forces* There are few cases only (n=9, or 7%) in which social underlying forces – besides driving proximate factors – were reported to exert a dampening or mitigating impact upon proximate causes. As for the US-American Southwest, Fredrickson, Havstad, Estell and Hyder (1998) pointed out the 'importance of citizen involvement in evoking land use changes': while public interest was low and land degradation proceeding during the cattle boom of the 1880s, changes began once public interests other than ranching were aroused in the early 1900s (regulations of the Taylor Grazing Act addressed further erosion of the forage basis, and interests in recreation and land quality of growing urban populations in the 1960s and 1970s were responsible for Multiple Use Act and later laws meant to improve human interactions with the land). Wijdenes, Poesen, Vandekerckhove and Ghesquiere (2000) report from the Guadalentin catchment area in SE-Spain that minor technological changes such as small earth dams were able to dampen degradation (here: river bank gully erosion). Similarly, the introduction of desertification control measures (e.g., afforestation, green shelters, erosion control, aerial seed of rangelands) was reported to correlate with low degrees of desertification, though the control measures were not explicitly cited as mitigating feedbacks. Examples of the latter stem from two sites of the Agri Valley in the Basilicata Region of southern Italy (Basso, Bove, Dumontet, Ferrara, Pisante, Quaranta and Taberner, 2000), and five sites from the Ordos Plateau in the arid great plains of north-central China (Runnström, 2000).

As discussed earlier, there are cases in which regional amplifying feedbacks made up between high and increasing rainfall variability (prolonged droughts) and land use activities, mutually re-enforcing climate degradation and desertification (n=5, or 4%). However, in some cases (n=7, or 5%) droughts are reported to exert a distinct, mitigating impact, especially upon increasing livestock numbers. Examples stem from the arid US-American Southwest (Fredrickson, Havstad, Estell and Hyder, 1998), where (severe) droughts were cited to generally dampen livestock expansion. Similarly, in the rangelands of (semi)arid northern Mexico, the dampening impact of droughts since the late 1980s balanced sharp livestock increases which occurred in the 1965-80 period and which were favoured by good rainfall conditions (Manzano, Návar, Pando-Moreno and Martinez, 2000).

*Other feedback loops* Other feedbacks than reported above do not relate to specific proximate or underlying causes but are made up between the process of desertification itself and mainly meteorological components (n=7, or 5%). In the case of the Guadiana Valley in the Alentejo Region of southern Portugal, Seixas (2000) reports about positive feedback from the desertification process upon vegetation change as an increase of spatial heterogeneity and 'greenness' of the landscape, as a response to hydrological stress. Revealing the ecosystem's ability to adapt to adverse conditions by species replacement, thus, supporting the positive feedback cycle described by Schlesinger, Reynolds, Cunnigham, Huenneke, Jarrell, Virginia and Whitford (1990), changes in the floristic composition of scrubs reportedly led to increasing biomass, namely, rock-rose replacing sargasso. Hydrological stress had been linked to an increasingly drier climate due to global climate change, namely cyclical precipitation with a marked decline in the 1960-90 period. Thus, the 'greenness' level of the landscape increased despite of reduced precipitation, and although the total area of scrub and dense vegetation had been on a decline.

Similarly, i.e., not relating to specific land use activities or underlying forces, a feedback loop, linked to other feedbacks from remote areas, has been reported from northwestern Senegal: local land degradation, local albedo effects and precipitation effects in and outside the area, including negative feedbacks from distant rainforest zones (and deforestation there) are coupled in a self-perpetuating downward spiral, in which declining rainfall due to global climate effects (driven by changes in sea surface temperature) and the degradation-albedo-precipitation feedback loop overrides the impact from human activities (Gonzalez, 2001). Another case of climatically induced desiccation or aridization due to high rainfall variability, driving the whole of the process of desertification rather than impacting upon specific causative factors, has been reported from the Circum Aral Region, which renders desertification – regardless of various human activities involved - very dynamics (Saiko and Zonn, 2000). From two large sites in the west African Sahel, crossing the national boundaries of Mali, Niger and Burkina Faso, it has been reported that drought is responsible for increases in eroded ground and encroachment by wind-blown sand (dunes), without claimed human impact (Ringrose and Matheson, 1992). In the case of the Manix Basin, situated in the Mojave Desert of SE-California, the proliferation of exotic annual plants has been cited to increase fire frequency and fuel load, thus raising the vulnerability of burnt areas to wind erosion and amplifying desertification (Okin, Murray and Schlesinger, 2001). Finally, there is indication from the Qinghai-Xizang (Tibet) Plateau that human-induced land degradation – as evidenced through the removal of vegetation and soil cover – turns step-wise into irreversible desertification since it is aggravated by increased aridity driven by global climate change, not in causal synergy with local land use activities (Holzner and Kriechbaum, 2001).

Rainfall variability and amount of precipitation were also cited not to operate as drivers with possible feedbacks, but to be thresholds for (un)sustainable pathways of land change (discussed later).

## Thresholds

Thresholds and control points of desertification were specified in 47 cases (36%), with some overlaps between the two types. It appears that thresholds (hidden points or 'break points') were mainly inherent to Central Asian cases, while more cases of land use activities in African dryland ecosystems had reportedly been shifted beyond control points (or switch and choke points) than cases from other regions.

**Table 4.18    Thresholds of Desertification (n=47 cases)***

|  | abs | rel |
|---|---|---|
| Biophysical thresholds | 34 | 26% |
| ... dry-climate dependent rainfall limits water provision | 12 | 9% |
| ... Germination conditions driven by climatic variability | 9 | 7% |
| ... crucial minimum soil depth | 3 | 2% |
| ... low regenerative capability of vegetation | 3 | 2% |
| ... low temperature and short vegetation period | 2 | 2% |
| ... topographical threshold | 2 | 2% |
| ... crucial minimum vegetation cover | 2 | 2% |
| ... wind friction velocity (at 15-50 cm/sec) | 1 | 1% |
| Socio-economic thresholds | 18 | 14% |
| ... degree of flexibility for informal arrangements, etc. | 5 | 4% |
| ... critical wood collection distance | 5 | 4% |
| ... money spent on fuel wood | 5 | 4% |
| ... limited opportunities for hay provision | 2 | 2% |
| ... ratio of children (at <6) and elderly (>65) per 100 persons | 1 | 1% |

* Multiple counts possible; percentages relate to the total of all 132 cases.

Biophysical as well as socio-economic thresholds of desertification were specified in 34 cases (26%) – see Table 4.18. In most of these cases, there are single, mainly biophysical thresholds inherent to each case, but in eleven cases, several (mainly two) thresholds were specified. In cases where multiple thresholds were found, they incorporated both biophysical and socioeconomic aspects. Common examples of thresholds had been dry climate as limiting factor to water provision, germination conditions as driven by climatic variability and amount of precipitation, the degree of flexibility among rural societies for informal arrangements, etc., critical wood collection distance, the amount of money spent on wood, critical minimum soil depths, and the (low) regenerative capability of vegetation to develop back to dense growth.

*Climate dependent rainfall as limiting factor for water provision* The threshold mentioned most widely in the cases was climate dependent rainfall as a limiting factor for water provision (9%). Except for two sites from Mexico, all examples stem from the Central Asian steppe and desert region.

For the Circum Aral-Region of Kazakstan, Uzbekistan, and Turkmenistan, Saiko and Zonn (2000), for example, state a uniform, unidirectional, and irreversible process of desertification, because the expansion of the area under irrigation for cotton within the Aral Sea basin – since the 1930s, but speeding up from 1975 onwards – occurred despite the fact that the availability of water resources, under conditions of desiccation, was the main limiting factor on economic development. Small changes of underlying factors – such as the continued use of furrow irrigation which was not superseded by more water-efficient methods, even in post-Soviet times – pushed the system, and accordingly led to tremendous water losses in transport through infiltration and evaporation.

Similarly, Genxu and Guodong (1999) report from six sites situated at the lower to upper reaches of Hei River Basin in the Gansu Province of NW-China, with the middle reaches serving as one of the most important national grain production bases, that the river system changed markedly in the latter part of the $19^{th}$ century due to water resource development (and misuse) for irrigation. Small changes – such as water-saving irrigation techniques and better water distribution regulations – could have prevented soil salinization, formation and encroachment of sand dunes, even under an expanding economy, leading to inevitable land degradation with a step-wise trend to (potentially) irreversible desertification. Likewise are the case artificial oases around Yutian and natural oases along Keriya River (and in the now dried up delta area around Daliyaboyi) of the Tarim River Basin of the Taklimakan Desert in NW-China (Yang, 2001). In addition to the expansion of irrigated land, small changes pushed the system beyond the break point of limited water resources, namely the shift to economic crops during the period of market liberalization and the exploration of oil in the 1980s, not only increasing water demand, but also triggering immigration into desert regions.

Another water-related hidden point, inherited from pre-historic times, had been named for two semi-arid sites in the Tehuacan Valley of south-central Mexico. In a currently semi-arid and water-limited ecosystem, soil losses from pre-Columbian times had persisted to date, with the consequence that the recovery of adequate water-storage capacity of soils operated as the main limitation on primary production for local populations. As a consequence, there existed extreme and even growing poverty which was held responsible for driving continued short-term exploitation of the land (McAuliffe, Sundt, Valiente-Banuet, Casas and Viveros, 2001).

*Germination conditions* Germination conditions, controlled by climatic variability and the amount of precipitation, have been reported in 7% of the cases as thresholds of dryland degradation.

Taking the case of the Manix Basin in southeastern California, Okin, Murray and Schlesinger (2001) argue that in the arid shrublands of the Mojave Desert stable state plant communities were pushed beyond their threshold of resilience by land use or 'anthropogenic disturbance', and that it will take centuries for recovery, if this would be possible at all. The critical threshold was cited to be natural germination of native perennial vegetation which had been rare, and which was linked to interannual climate variability and long-term regional climatic conditions.

Taking evidence from eight sites at the Ordos Plateau of the N-Chinese steppe and desert region, Runnström (2000) argues that in the arid western part of the plateau the amount of rainfall dominated the response in biological productivity to desertification, while in the eastern part, which showed increases in biological productivity, positive biomass trends could be the result of desertification measures that had been applied over the last 40 years.

*Crucial minimum soil depth and vegetation cover* Related to the quality of germination conditions are thresholds which had been identified in terms of crucial minimum soil depth (2%) and critical vegetation cover (2%), necessary for the regeneration of (semi)natural vegetation.

Throughout cases from the Mediterranean Basin, parent material was characterized a soil-forming factor that not only affects soil properties, but also plant growth, soil erosion, and ecosystem resilience. The nature of the parent material becomes increasingly important in vegetation establishment and land protection once soil depth has been reduced due to erosion. As a quantitative measure, Kosmas, Gerontidis and Marathianou (2000) found that under soil and climatic conditions as given on the Greek island of Lesvos a cultivated hilly landscape should be abandoned before the soil has been depleted to a critical depth of 25 to 30 cm. The recovery of natural vegetation would be very low in soils of lesser depth, and erosional processes were found to be very active. If a soil had been eroded to a depth of about 10 cm or less, depending on the parent material, the perennial vegetation could not be supported and the remaining soil was rapidly removed by wind or water erosion, thus triggering an ultimately irreversible process of land degradation and desertification. In addition to the crucial soil depth of 10 cm (or less), a value of 40% vegetation was cited and considered critical, below which accelerated erosion dominated on sloping land. If the vegetation covered amounted to a greater area than 40%, then it would act as a factor of resilience or protection of the land (Kosmas, Gerontidis and Marathianou, 2000).

Such thresholds appear to depend on several factors, with the nature of the parent material, or various lithologies, ranking high. For example, soils formed on pyroclastics had a reportedly lower capacity to regenerate natural vegetation, thus leading to higher erosion, than soils formed on shale, ignimbrite, schist-marble or volcanic lava, which had a higher capacity for at least partial regeneration of natural vegetation (Kosmas, Gerontidis and Marathianou, 2000).

*Control Points*

In 11 cases (8%), control or switch and choke points were specified. For six sites on the fossil Bidi and Oursi dunes, part of the Ouadalan of northern Burkina Faso, Rasmussen, Fog and Madsen (2001) argue that droughts were not only trigger events leading to abrupt changes in the human-environment condition, but operated as control points for some parts of the area, at least. Cases of continuous cultivation on the dunes without significant land degradation were found as well as clear evidence of cultivation causing local degradation on the dunes. The picture of a uniform and unidirectional process of desertification emerges as a simplification of more complex realities, even in a rather small study region of a few hundreds of $km^2$. Taking the case of the Senegal River Basin, Venema, Schiller, Adamowski and Thizy (1997) specify water-stress, i.e., the mainly climate change-driven incapacity to let soils regenerate naturally, as control point which pushed the irrigation/river ecosystem into a state of criticality. Citing – as a biophysical threshold (not control point) – the low regenerative capability of dry savanna vegetation to develop back to dense growth at a Kalahari Sandveld site in south-central Botswana, Ringrose, Vanderpost and Matheson (1996) argue that the combined, sustained and effective human-induced pressure (i.e., fuelwood gathering, goat and cattle grazing, cropland and settlement expansion, and borehole installation, driven by demographic, economic and institutional factors) constitutes a control point which pushed the human-environment system towards irreversible desertification. Gauquelin, Bertaudière, Montes, Badri and Asmode (1999), taking the case of the Middle and High Atlas Mountain region of Morocco, argue that the low regenerative capability of thuriferous juniper stands to develop back to dense growth not only forms a threshold, but that the scarcity of regeneration and the decline of natural germination operated already as switch and choke point of a step-wise trend towards (potentially) irreversible desertification. From two arid to alpine grassland sites at the Qinghai-Xizang (Tibet) Plateau, Holzner and Kriechbaum (2001) report that changes in waterlogged soils and frozen underground, with limiting agroclimatic conditions for hay provision operating as threshold (i.e., short vegetation period and low temperature), pushed the highly fragile mountain ecosystem towards a step-wise trend to (potentially) irreversible desertification. Overgrazing, driven by cultural, social and economic variables, and increased cutting of turf and peat were reported to be combined with vegetation removal and drying out of the waterlogged surface, leading to complete decomposition of peat, soil losses due to erosion and desertification. An intermediate state was semi-desert vegetation, with desert, stone or gravel pavement being the final result.

# Chapter 5

# Syndromes and Process Rates

## Introduction

Syndromes are based upon the insight that, for any given human-environment system in drylands, a limited number of causes are essential to predict the general trend in land use. Different, but related to other approaches (explained in chapters 1 and 2), a syndrome of land change constitutes here the particular combination of specific causal conditions, involving both proximate and underlying factors, and rates of change, i.e., slow and fast causative variables. The latter need to be held apart from process rates of change. They indicate the speed at which desertification occurs. Again, a difference has been made between slow and fast rates of change in indicators of desertification.

## Syndromes

In nearly all of the desertification cases (n=126, or 96%), syndromes of change were held apart, as defined by specific causal combinations, involving both proximate causes, underlying driving forces and qualitatively derived (slow, fast) rates of change. Most of the cases are driven by a complex of both slow and fast variables (n=97, or 77%). Fast variables are causative to a mere proportion of the cases only (n=7, or 6%), and slow variables alone drive three times more of the cases (n=22, or 18%). Tables 5.1 and 5.2 provide examples of this pattern, taking selected cases from various landscapes of the Asian drylands. Table 5.1 shows that there is only one case in the northcentral section of the Chinese steppe and desert region which is characterized by slow variables alone, which is the Yellow River Valley crossing the Ordos Plateau. All the other cases, covering various upland steppe and lowland desert ecosystems in Asia, are driven by a complex of causative variables at different speed. Asian drylands are representative for a pattern which combines both fast and slow variables to varying degrees, and which has been found in many other dryland areas, too.

*Most Important Syndromes*

There are large differences, however, in the nature of syndromes per various dryland regions of the world. They relate to the frequency of occurrence of factors and processes as well as to their regional distribution – see Tables 5.3 to 5.7. In the order of diminishing frequency, most syndromes, inherent to three fifths up to three quarters of the cases, are those which are associated with resource scarcity causing a pressure of production on resources (n=98, or 74%), with slow factors or gradual processes being clearly more important than fast or rapid ones. Similarly important are syndromes such as changing opportunities created by markets (n=91, or 69%), be they rapidly or gradually changing. Syndromes associated with outside policy interventions are also important (n=79, or 60%), predominantly in the form of factors and processes occurring at slow speed.

**Table 5.1    Slow/Fast Rates of Desertification Causes in Central Asia[a]**

| Slow change | Fast change |
|---|---|
| *North West China: Taklimakan Desert, Qinghai-Xizang (Tibet) Plateau* | |
| 1a-1c-2c-3e-4c-5b-5e[b] | 1b-2a-2c-5a[b] |
| 1a-1c-2c-4c-5b-5e | 1b-2a-2c-4c-5a |
| 1a-1c-2c-4c | 1b-2a-2c-5a-5b-5e |
| 1c-1e-2c | 4b-4c |
| 1a-1c-2a[c] | 1b-2a-2c[c] |
| 2a-3c-4d-5b | 5a |
| 1e | 4c |
| *North central China: Loess/Ordos Plateau* | |
| 2a-3a | 2a-2c-3a-4c |
| 2a-3a[b] | 2a-2c-3a[b] |
| 1c-1e-2a-3a-3c-5a-5b-5d | 3a-3c-4c |
| 2a-3a-5b-5d | 1b-2a |
| 1a-1c-2c-4c | 1b-2a-2c-4c-5a-5b-5e |
| 2a-3a-5b-5d | 2a |
| 2a-3a-5b-5d[b] | 1b-2a[b] |
| 2a-3a-5b-5d[b] | 1b[b] |
| 2a-3a-5b-5d | |
| 1c-1e-2a-3a-3c-5b-5d-5e | 1a-2a-2c-5a |
| 3c | 4b |
| *North East China: Inner Mongolian Plateau* | |
| 1c-1e-2a-3a-5a | 3a-3c-4c |
| 1c-1e-2a-3a-3c-5a-5b-5d | 3a-3c-4c |
| 1c-1e-2a-3a-3c-5b-5d-5e[b] | 1a-2a-2c-5a[b] |

[a]    Codes of slow and fast causative variables are same as in Tables 5.3 to 5.7.
[b]    Two cases with same coding per area.

**Table 5.2**  **Slow/Fast Rates of Desertification Causes in Tibet, the former Soviet Union, India, Jordan and the Arabian Peninsula**[a]

| Slow change | Fast change |
|---|---|
| *Tibet Autonomous Republic* | |
| 1a-2a-3b-5b-5e | 2a-2c-4c |
| 1a-2a-3b-5b-5e | 2a-2c |
| 1c-1e-2a-2d-5e | 3a-4c |
| *Southern Russia, Kazakstan, Uzbekistan, Turkmenistan* | |
| 1c-1e-2a-3a-5a-5b-5d | 2a-2c-4b-4c |
| 2a-5a-5b | 1a-2a-2c-4a-5a |
| 1c-2a | 2a-2c |
| *East Mediterranean steppe zone (Jordan)* | |
| 1a-1d-2a-3e[b] | 1b-5a[b] |
| 1a-1d-2a | 1b-5a |
| 1a-1c-1d-1e-2a-3e[c] | 1b-5a[c] |
| *Saudi Desert (Kuwait, Saudi Arabia)* | |
| 2a | 2a-2c |
| 3b[d] | 1a-2a-2c[d] |
| *Thar Desert (northwestern India)* | |
| 1a-3a[e] | 2a-2c[e] |

[a] Codes of slow and fast causative variables are same as in Tables 5.3 to 5.7.
[b] Two cases with same coding per area.
[c] Six cases with same coding per area.
[d] Four cases with same coding per area.
[e] Three cases with same coding per area.

*Resource scarcity causing a gradual pressure of production on resources* Slow factors and gradual processes associated with resource scarcity causing a pressure of production on resources form the most important syndrome (n=86, or 65%). They are robust syndromes which have low geographical variation. This is especially true for the loss of land productivity on sensitive areas following excessive or inappropriate use, and the failure to restore or maintain protective work. Other factors or processes associated with resource scarcity and production pressure are less important and show distinct regional variations.

**Table 5.3  Frequency of Syndromes Associated with Slow Land Change – Resource Scarcity**[a]

| | All Cases (N=132) | | Asia (n=51) | | Africa (n=42) | | Europe (n=13) | | Australia (n=6) | | N-America (n=6) | | L-America (n=14) | |
|---|---|---|---|---|---|---|---|---|---|---|---|---|---|---|
| | abs | rel | abs | rel | abs | rel | abs | rel | abs | rel | abs | rel | abs | rel |
| 1. Resource scarcity causing a production pressure | 86 | 65 | 33 | 65 | 25 | 60 | 11 | 85 | 5 | 83 | 3 | 50 | 9 | 64 |
| a. Division of land parcels & shrinking land availability due to natural population growth | 35 | 27 | 22 | 43 | 7 | 17 | 0 | | 0 | | 0 | | 6 | 43 |
| b. Domestic life cycles leading to changes in labour availability | 4 | 3 | 0 | | 1 | 2 | 3 | 23 | 0 | | 0 | | 0 | |
| c. Loss of land productivity on sensitive areas following excessive of inappropriate use | 58 | 44 | 24 | 47 | 16 | 38 | 7 | 54 | 5 | 83 | 3 | 50 | 3 | 21 |
| d. Reduction of land productivity due to weathering or soil nutrient decline[b] | 14 | 11 | 6 | 12 | 6 | 14 | 0 | | 0 | | 0 | | 2 | 14 |
| e. Failure to restore or maintain protective works of resources | 44 | 33 | 13 | 25 | 12 | 29 | 8 | 62 | 5 | 83 | 3 | 50 | 3 | 21 |
| f. Heavy surplus extraction away from the land manager | 12 | 9 | 0 | | 4 | 10 | 0 | | 5 | 83 | 0 | | 3 | 21 |

[a] Multiple counts possible; abs = absolute number; rel = relative percentages; cum = cumulative percentages; percentages relate to the total of all cases for each category (column).
[b] In general terms, deterioration of the climate-soil-vegetation complex.

Table 5.4  Frequency of Syndromes Associated with Slow Land Change – Changing Markets and Policy Interventions[a]

| | All Cases (N=132) | | Asia (n=51) | | Africa (n=42) | | Europe (n=13) | | Australia (n=6) | | N-America (n=6) | | L-America (n=14) | |
|---|---|---|---|---|---|---|---|---|---|---|---|---|---|---|
| | abs | rel | abs | rel | abs | rel | abs | rel | abs | rel | abs | rel | abs | rel |
| 2. Changing market opportunities | 73 | 55 | 42 | 82 | 8 | 19 | 4 | 31 | 5 | 83 | 2 | 33 | 12 | 86 |
| a. Increase in commercialization[b] | 55 | 42 | 28 | 55 | 8 | 19 | 0 | | 5 | 83 | 2 | 33 | 12 | 86 |
| b. Improvement in accessibility through road construction | 24 | 18 | 11 | 22 | 0 | | 2 | 15 | 5 | 83 | 1 | 17 | 5 | 36 |
| c. Changes in market prices for inputs & outputs[c] | 10 | 8 | 4 | 8 | 0 | | 0 | | 5 | 83 | 1 | 17 | 0 | |
| d. Off-farm wages and employment opportunities | 7 | 5 | 1 | 2 | 1 | 2 | 3 | 23 | 0 | | 1 | 17 | 1 | 7 |
| 3. Outside policy interventions | 70 | 53 | 35 | 69 | 18 | 43 | 2 | 15 | 5 | 83 | 3 | 50 | 7 | 50 |
| a. Economic development programmes | 31 | 23 | 20 | 39 | 7 | 17 | 2 | 15 | 0 | | 1 | 17 | 1 | 7 |
| b. Perverse subsidies, policy-induced price distortions & fiscal incentives | 14 | 11 | 6 | 12 | 0 | | 2 | 15 | 5 | 83 | 0 | | 1 | 7 |
| c. Frontier development[d] | 19 | 14 | 8 | 16 | 5 | 12 | 0 | | 5 | 83 | 0 | | 1 | 7 |
| d. Poor governance & corruption | 9 | 7 | 0 | | 4 | 10 | 0 | | 0 | | 1 | 17 | 4 | 29 |
| e. Insecurity in land tenure | 20 | 15 | 8 | 16 | 4 | 10 | 0 | | 0 | | 2 | 33 | 6 | 43 |

[a] Multiple counts possible; abs = absolute number; rel = relative percentages; cum = cumulative percentages; percentages relate to the total of all cases for each category (column).
[b] Including an increase in agro-industrialization; for example, land devoted to cash crops.
[c] For example, erosion of primary production prices and unfavourable global or rural-urban terms of trade.
[d] For geopolitical reasons or to promote interest groups, for example.

A particular feature of the Asian, African and Latin American cases is both the division of land parcels and shrinking land availability due to (natural) population growth, and loss of land productivity due to the gradual deterioration of the climate-soil-vegetation complex, especially soil nutrient decline. However, only few cases provide sufficiently detailed data to come to broad conclusions about the speed of change. In cases from the Indian Thar desert, for example, it has been reported that a threefold increase of population density (i.e., from 24 inhabitants per $km^2$ in 1951 to 76 in 1991) happened together with subdivision and fragmentation of land holdings, under traditional land inheritance by division of parcels. This meant a decrease in the land available for an average rural household from 27 to 9 hectares, which caused a pressure of production on resources with partly drastic crop yield declines. Cluster beans had not been cultivated over the last ten years, and yields of pearl millet had been reduced from 190-220 to 60-90 kg/ha over the last 30 to 35 years, despite of the adoption of high yielding varieties and biotechnological manipulations (Ram, Tsunekawa, Sahad and Miyakazi, 1999). Similarly, a 5.3-fold increase of human population in the 1950-98 period was reported for the middle and lower reaches of Tarim River in the Taklimakan Desert. It caused a division of land parcels and triggered, in combination with the reservation of grazing areas, a decrease in land availability that impaired livestock carrying capacity (Feng, Endo and Cheng, 2001).

In contrast, fast variables associated with resource scarcity and production pressure are far less important (n=37, or 28%), and they all show distinct regional differences. For example, Asian cases are typical for their rapid in-migration, forced population displacement, and increased mobility, as well as a decrease in land availability due to encroaching other uses, creating a 'tragedy of enclosure'. These features make Asian cases almost completely different from other regions. It appears that cases of production pressure with fast demographic variables – such as immigration – and slow demographic change – through natural increment – do not coincide (with a sole exception).

*Changing opportunities created by markets* Syndromes related to changing opportunities created by markets are important, be they made up of slow (n=73, or 55%) or fast variables (n=65, or 45%). They differ from one broad geographical region to another, though. Examples of slowly developing market opportunities are the gradual increase in commercialization and agro-industrialization, which underlies more Latin American, Australian and Asian desertification cases than found in other regions (and which seems not important at all in Europe). Another example is improved market access as facilitated by road construction, which features more of the Latin American and Australian cases than found in other regions (and which seems to play no role at all in the African cases). Further examples of rapidly developing market opportunities are capital investments associated with desertification which appear to be more important in Asia and Australia than, in particular, in Europe and Africa. The same trend among fast variables seems to hold true for the quick spread of new technologies for intensification of resource use, i.e., a significant feature in Asia and Australia, while of lower importance in Europe and Africa.

Table 5.5  Frequency of Syndromes Associated with Slow Land Change – Adaptive Capacities and Social Organization[a]

| | All Cases (N=132) | | Asia (n=51) | | Africa (n=42) | | Europe (n=13) | | Australia (n=6) | | N-America (n=6) | | L-America (n=14) | |
|---|---|---|---|---|---|---|---|---|---|---|---|---|---|---|
| | abs | rel | abs | rel | abs | rel | abs | rel | abs | rel | abs | rel | abs | rel |
| 4. Loss of adaptive capacity, increased vulnerability | 31 | 23 | 6 | 12 | 13 | 31 | 0 | | 5 | 83 | 1 | 17 | 6 | 43 |
| a. Impoverishment[b] | 19 | 14 | 0 | | 7 | 17 | 0 | | 5 | 83 | 1 | 17 | 6 | 43 |
| b. Breakdown of informal social security networks | 7 | 5 | 0 | | 3 | 7 | 0 | | 0 | | 1 | 17 | 3 | 21 |
| c. Dependence on external resources or assistance | 10 | 8 | 5 | 10 | 0 | | 0 | | 5 | 83 | 0 | | 0 | |
| d. Social discrimination[c] | 13 | 10 | 1 | 2 | 7 | 17 | 0 | | 0 | | 1 | 17 | 4 | 43 |
| 5. Changes in social organization, resource access & attitudes | 51 | 39 | 21 | 41 | 10 | 24 | 4 | 31 | 5 | 83 | 2 | 33 | 9 | 64 |
| a. Changes in institutions governing access to resources[d] | 15 | 11 | 5 | 10 | 5 | 12 | 0 | | 5 | 83 | 0 | | 0 | |
| b. Growth of urban aspirations | 33 | 25 | 23 | 45 | 5 | 12 | 0 | | 0 | | 1 | 17 | 4 | 29 |
| c. Breakdown of extended families | 4 | 3 | 0 | | 1 | 2 | 3 | 23 | 0 | | 0 | | 0 | |
| d. Growth of individualism & materialism | 21 | 16 | 14 | 27 | 0 | | 0 | | 0 | | 1 | 17 | 6 | 43 |
| e. Lack of training, poor information flow | 27 | 20 | 11 | 22 | 2 | 5 | 3 | 25 | 5 | 83 | 0 | | 6 | 43 |
| Slow variables, total | 119 | 90 | 51 | 100 | 32 | 76 | 11 | 85 | 5 | 83 | 6 | 100 | 14 | 100 |

[a] Multiple counts possible; abs = absolute number; rel = relative percentages; cum = cumulative percentages; percentages relate to the total of all cases for each category (column).
[b] For example, creeping household debts, no access to credits, lack of alternative income sources, and weak buffering capacity.
[c] For example, of ethnic minorities, women, lower class people or caste members.
[d] For example, shift from communal to private rights, tenure holdings and titles.

Other specific factors and processes bear low counts, and therefore tend to be of minor importance. This implies that, for example, the gradual erosion of market prices for primary products, the gradual deterioration of terms of trade, and the slow build-up of off-farm wage or employment opportunities has lower relevance than could have been expected. This holds also true for rapid changes in macro-economic and trade conditions leading to price changes.

*Gradual outside policy interventions* Syndromes associated with gradually implemented outside policy interventions are important (n=70, or 53%). They have differing significance though in various dryland regions of the world. In contrast, rapidly occurring external and politically motivated interventions – such as rapid policy change or war – are of considerably lower importance. Examples are economic development programmes (not mentioned to be a causative factor in Australian cases), insecurity of land tenure (not causative in European and Australian cases), perverse subsidies and fiscal incentives (not causative in Africa and North America), and frontier developments (not causative in Europe and North America). However, poor governance and corruption, which are generally thought to be causative to desertification, bear relatively low counts.

## Less Important Syndromes

Other syndromes are reported in less than half of the cases, i.e., the both rapid and gradual loss of adaptive capacity and increased vulnerability (n=61, or 46%), and changes in social organization, in resource access and in attitudes (n=60, or 45%), with slow changes outweighing rapid ones.

*Rapid and gradual loss of adaptive capacity, increased vulnerability* A syndrome related to the loss of adaptive capacity and increased vulnerability was found in less than half of the desertification cases. It appears that fast factors and rapid processes associated with the syndrome (n = 47, or 36%) are more important than slow factors or gradual processes (n=31, or 23%). Regional variations are considerable. Among the fast causative variables, by far the most widespread are risks associated with natural hazards such as droughts, bush fires, or invasions of rodents or insects leading to a crop failure or a loss of resources or productive capacity. Mostly affected are Australian rangelands, and to a lesser degree, drylands in Latin America and Europe (reportedly not important at all in North America). Among the slow causative variables, the most important ones seem to be features of impoverishment such as creeping household debts, no access to credits, lack of alternative income sources and weak buffering capacity. Again, mostly affected are Australian and Latin American cases (reportedly not important at all in Europe and Asia). Other factors or processes associated with the syndrome are found to be less important in the cases.

Table 5.6  Frequency of Syndromes Associated with Fast Land Change – Resource Scarcity, Changing Markets and Policies[a]

| | All Cases (N=132) | | Asia (n=51) | | Africa (n=42) | | Europe (n=13) | | Australia (n=6) | | N-America (n=6) | | L-America (n=14) | |
|---|---|---|---|---|---|---|---|---|---|---|---|---|---|---|
| | abs | rel | abs | rel | abs | rel | abs | rel | abs | rel | abs | rel | abs | rel |
| 1. Resource scarcity causing a production pressure | 37 | 28 | 30 | 59 | 5 | 12 | 0 | | 1 | 17 | 1 | 17 | 0 | |
| a. Spontaneous migration, increased mobility, forced displacement, refugees | 11 | 8 | 9 | 18 | 1 | 2 | 0 | | 1 | 17 | 0 | | 0 | |
| b. Decrease in land availability due to encroachment[b] | 25 | 19 | 21 | 41 | 3 | 7 | 0 | | 0 | | 1 | 17 | 0 | |
| 2. Changing opportunities created by markets | 65 | 49 | 35 | 69 | 12 | 29 | 2 | 15 | 6 | 100 | 3 | 50 | 7 | 50 |
| a. Capital investments | 62 | 47 | 34 | 67 | 10 | 24 | 2 | 15 | 6 | 100 | 3 | 50 | 7 | 50 |
| b. Changes in ntl./global macro-economic & trade conditions[c] | 3 | 2 | 0 | | 2 | 5 | 0 | | 0 | | 0 | | 1 | 14 |
| c. New technologies for land use intensification | 56 | 42 | 30 | 59 | 10 | 24 | 0 | | 6 | 100 | 3 | 50 | 7 | 50 |
| 3. Outside policy interventions | 27 | 20 | 7 | 14 | 6 | 14 | 3 | 23 | 5 | 83 | 0 | | 6 | 43 |
| a. Rapid policy change[d] | 20 | 15 | 7 | 14 | 0 | | 2 | 15 | 5 | 83 | 0 | | 6 | 43 |
| b. War, social uprising | 10 | 8 | 3 | 6 | 6 | 14 | 1 | 8 | 0 | | 0 | | 0 | |

[a] Multiple counts possible; abs = absolute number; rel = relative percentages; cum = cumulative percentages; percentages relate to the total of all cases for each category (column).
[b] Encroachment by other land uses or due to closing of frontier (e.g., urban sprawl, reservation of grazing areas), creating a tradedy of enclosure.
[c] Leading to change in prices such as a surge in energy prices or a global financial crisis.
[d] For example, application of market liberalization policies or European colonization.

Table 5.7  Frequency of Syndromes Associated with Fast Land Change – Adaptive Capacities and Social Organization[a]

| | All Cases (N=132) | | Asia (n=51) | | Africa (n=42) | | Europe (n=13) | | Australia (n=6) | | N-America (n=6) | | L-America (n=14) | |
|---|---|---|---|---|---|---|---|---|---|---|---|---|---|---|
| | abs | rel | abs | rel | abs | rel | abs | rel | abs | rel | abs | rel | abs | rel |
| 4. Loss of adaptive capacity, increased vulnerability | 47 | 36 | 13 | 25 | 18 | 43 | 1 | 8 | 6 | 100 | 1 | 17 | 8 | 57 |
| a. Internal conflicts[b] | 5 | 4 | 1 | 2 | 3 | 7 | 0 | 0 | 0 | 0 | 0 | 0 | 1 | 7 |
| b. Illness[c] & famine-related mortality | 6 | 5 | 3 | 6 | 2 | 5 | 0 | 0 | 0 | 0 | 1 | 17 | 0 | 0 |
| c. Risks associated with natural hazards[d] | 40 | 30 | 11 | 22 | 15 | 36 | 1 | 8 | 6 | 100 | 0 | 0 | 7 | 50 |
| 5. Changes in social organization, resource access and attitudes | 23 | 17 | 16 | 31 | 0 | 0 | 0 | 0 | 5 | 83 | 1 | 17 | 1 | 7 |
| a. Loss of entitlements to environmental resources[e] | 23 | 17 | 16 | 31 | 0 | 0 | 0 | 0 | 5 | 83 | 1 | 17 | 1 | 7 |
| Fast variables, total | 104 | 79 | 50 | 98 | 24 | 57 | 7 | 54 | 6 | 100 | 4 | 67 | 13 | 93 |

[a] Multiple counts possible; abs = absolute number; rel = relative percentages; cum = cumulative percentages; percentages relate to the total of all cases for each category (column).
[b] For example, about land between farmers and herders.
[c] For example, fluorine poisoning.
[d] For example, droughts, bush fires or rodent/insect invasion leading to a crop failure, loss of resource or productive capacity.
[e] For example, urban growth or expropriation for large-scale projects, leading to an ecological marginalization of poors.

*Changes in social organization, resource access and attitudes* Another syndrome, associated with changes in social organization, in resource access and in attitudes, was again found in less than half of the desertification cases. It appears that slow factors and gradual processes associated with changes in social organization, in resource access and in attitudes (n=51, or 39%) are more important than fast factors or rapid processes (n=23, or 17%). Regional variations are considerable, again. As for slow change, by far most widespread are combinations of the following variables: growth of urban aspirations of an industrializing society, growth of individualism and materialism, lack of training or education, and poor information flow on the environment. These features were mainly found in Asian and Latin American cases. Important, but to a far lesser degree, are changes in institutions governing the access to resources such as shifts from communal to private rights, tenure holdings and titles. These cases are solely reported from Asian, African and Australian drylands. Fast variables associated with the syndrome include the loss of entitlements to environmental resources leading to an ecological marginalization of poor people. Reported examples are urban growth and expropriation for large-scale projects, which predominantly stem from Asian and Australian cases.

## Process Rates

In 56 cases (42%), changes of the amount of dryland degradation per time period were specified in quantitative terms. Rates were translated into percentage change on a yearly basis, and a difference was made between slow and fast rates. They relate to meteorological changes (mainly, annual rainfall decrease), vegetation changes (decline and increase of undegraded *versus* degraded vegetation cover), wind and/or water erosion (sand encroachment and exposure of bare, rocky outcrop), water degradation (spread of salinity, changes in water surface and river discharge), and changes in pastoral and agricultural suitability.

### Meteorological Change

Meteorological changes (n=11) were most commonly described as a decrease in annual rainfall compared to the long-term average (n=5). The annual rate of decrease ranges between 0.6 and 1.5%, with all rates stemming from African cases and averaging 1.2% – see Table 5.8. Though data availability is limited, it appears that drylands in Southern Africa experienced more rapid decreases than other sites in Africa. In terms of time scale, there is slight indication that the decline in annual precipitation might be more distinct, or has even speeded up, during the last half or quarter of the $20^{th}$ century.

Other cases tend to confirm this trend on the basis of graphs and/or statements (rather than absolute numbers). A decrease in rainfall since the 1950s is reported from five sites in the Niger River Delta of central Mali, with extremely dry years in 1973 and 1984 (Benjaminsen, 1993). Rainfall data at St. Louis (1855-1993) and

Louga (1919-93) stations in the West African Sahel of NW-Senegal show a negative slope, in congruence with all Sahel data (Gonzalez, 2001). A reduction of mean rainfall of at least 30% relative to the 1960s is reported from two sites in the Sahel-Sudan Zone of Northern Nigeria (Mortimore; Harris and Turner, 1999). Similarly, rainfall has been reduced since 1955 at six sites in the Gorom Gorom district of N-Burkina Faso, with indication to be consolidated in the 1990s (Rasmussen, Fog and Madsen, 2001). And, the Alentejo Region in S-Portugal, showed a marked decline in precipitation in the 1960-90 period with measurements dating back to 1880 (Seixas, 2000).

**Table 5.8    Annual Decrease in Rainfall (%)***

| Rate | Time period | Site | Source |
|---|---|---|---|
| | | Slow change (<1.2%) | |
| 0.6% | 1920-84 | W&Central Sudan: Sahel-Sudan Zone, semi-arid zone | Ayoub, 1998 |
| 1.0% | 1960-91 | NE-Nigeria: Sahel-Sudan zone, Kano Close-Settled zone, Tumbau village area | Mortimore et al., 1999 |
| 1.0% | 1960-91 | NE-Nigeria: Sahel-Sudan zone, Kano Close-Settled zone, Tumbau village area | Mortimore et al., 1999 |
| | | Fast change (>1.2%) | |
| 1.5% | 1970-96 | NW-Botswana: Kalahari, Okavango Delta, Ngamiland district, one site | Dube and Pickup, 2001 |
| 1.5% | 1970-96 | NW-Botswana: Kalahari, Okavango Delta, Ngamiland district, another site | Dube and Pickup, 2001 |

\*   Mean value is 1.2%; n=5 sites.

The trend towards a marked precipitation decline in drylands during the second half of the 20$^{th}$ century, not only in African sites, seems confirmed by other measures of the rainfall variable, too. For example, in the Kordofan and Darfur Provinces of West and Central Sudan there had been one prolonged drought since the 1960s dating until 1985 (except for 1978), with 1984 being the lowest record in 100 years (Olsson, 1993). Similarly, Ayoub (1998) provides data from the semi-arid zone of Sudan according to which the length of the wet season period contracted by 1.5% annually in the 1960-91 period. Increasing aridity since the 1950s, apparently due to (global) climate change, including reductions in rainfall, is characteristic, too, for two sites in the Korqin (desert) region of the Inner Mongolian Plateau of NE-China (Sheehy, 1992). Another three cases are provided by Lin and Tang (2002) from the Central Asian desert and steppe region: the index of aridity increased at 0.03% per year in the 1910-60 period for the Great Plains of Northwest and North Central China, namely the Ordos and Inner Mongolian

Plateaus. For the upper reaches of the Yellow River source region at the Qinghai-Xizang (Tibet) Plateau, Wang, Qian, Cheng and Lai (2001) measured a mean rise of temperature at 0.05 degree C per year in the 1976-96 period, and a rise in ground temperature of the upper layer of frozen soil (0-20m) at 0.25 degree C per year over the last 30 years. Both increases are reported to run in parallel with a decline in annual summer precipitation at 0.3-0.5 mm/year for the same period. Mouat, Lancaster, Wade, Wickham, Fox, Kepner and Ball (1997) report that on-going drought conditions as measured by a drought severity index were characteristic for the Colorado Plateau of Utah State in the Southwestern United States for the 1986-91 period.

However, most of the rates of meteorological or climate change as given above lack comparability with other sites. They can hardly be transformed into annual percentage change rates, and/or have unique measures which were not used in other cases. An example of the latter originates from the Circum Aral Region of Central Asia (Saiko and Zonn, 2000): in the 1960-95 period, the annual precipitation of salt particles grew to 29 kg/ha, the annual transport of dust and salt occurred at 45 million tons per 400 km, and the yearly increase of dust storm days has been 2.5.

*Vegetation Change*

Among quantified vegetation changes (n=16), the most common rates were the decrease of undegraded vegetation cover (29 measures at n=13 sites), and, reciprocally, the increase of degraded vegetation cover (20 measures at n=5 sites). Rates of decrease or increase relate to both broad vegetation classes such as 'woody vegetation cover' or 'native grasses', and to specific denominations such as Creosote bush *(Larrea spp.)* or mesquite *(Prosopis spp).*

*Decrease of initial, undegraded vegetation* The most common measure of the decline of initial, undegraded vegetation in the cases was the (annual) decrease of vegetational cover of various floristic classes. The mean value of 3.4% is based upon 29 rates in total for various vegetation classes at 13 sites – see Tables 5.9 and 5.10.

Spanning from rather slow change as given in Table 5.9 – i.e., occurring at a rate of 0.1%, which was the gradual decline of woody vegetation cover in two village areas of the semi-arid Kano Close-Settled zone in NE-Nigeria in 1950-81 – to very fast change as given in Table 5.10 – i.e., occurring at a rate of 20.0%, which was the next to complete decline of blue gramme and Buffalo grasses at the semi-arid Colorado Plateau in southern Utah over six years only (1986-91), but under conditions of on-going droughts – the range of the value is high and indicates very high vegetation dynamics.

No difference exists between the various floristic classes when it comes to slow or fast change, i.e., herbaceous ground cover as well as native and/or desert grasses (eg. Buffalo grass, blue gramme grass), desert shrubs and other scrub or dense vegetation, woody vegetation (eg. Ulster wood, *Diversifolia schrenk*, or *Populus eupthatica*); hydrophytes, swamp vegetation, swamp and high-cold meadows are likewise affected by high or low decreases. Vegetation decreases at a given site can shift between slow and fast modes of change, at least for the time periods studied. There is hardly any indication that the decline in vegetational cover might be more distinct, or has increased even during the last half or quarter of the 20th century.

Table 5.9     Annual Decrease in the Areal Extent of Undegraded Vegetation Cover (%) – Slow Change (<3.4%)*

| Rate | Time period | Site | Source |
|---|---|---|---|
| 2.1% | 1957-87 | N-Tanzania: SW-Masai Steppe, Arusha district | Mwalyosi, 1992 |
| 0.1% | 1950-69 | NE-Nigeria: Sahel-Sudan zone, Kano Close-Settled zone, Dagaceri vge area | Mortimore et al., 1999 |
| 0.1% | 1969-81 | | |
| 0.1% | 1950-81 | | |
| 1.5% | 1982-95 | SW-USA: Chihuahuan Desert, southern New Mexico, Otero Mesa | Ludwig et al., 2000 |
| 0.8% | 1865-1995 | SW-USA: Sonoran Desert, Arizona, southern districts | Fredrickson et al., 1998 |
| 0.6% | 1915-28 | SW-USA: Chihuahuan Desert, southern New Mexcio, Jornada del Muerto Basin | Rango et al., 2000 |
| 0.9% | 1858-1928 | | |
| 1.0% | 1858-1963 | | |
| 1.1% | 1858-1915 | | |
| 2.1% | 1915-63 | | |
| 2.8% | 1928-63 | | |
| 2.0% | 1987-91 | S-Portugal: Alentejo Region | Seixas, 2000 |
| 0.7% | 1960-97 | N-China: Hexi Corridor, Hei River Basin, middle reaches | Genxu and Guodong, 1999 |
| 0.4% | 1976-86 | NW-China: Qinghai-Xizang (Tibet) Plateau, Yellow River source region | Wang et al., 2001 |
| 1.4% | 1976-96 | | |
| 2.4% | 1986-96 | | |
| 2.5% | 1976-86 | | |
| 3.0% | 1976-96 | | |
| 3.0% | 1980-90 | | |
| 2.1% | 1958-78 | NW-China: Taklimakan Desert, Tarim River Basin, various reaches | Feng et al., 2001 |

\* Mean value (all rates) is 3.4%; n=13 sites.

Not included in the generation of annual rates above is one case of annual vegetation cover increase at 0.6% (or 185 km$^2$ annually) for the 1982-94 period in the nomadic grazing area of Hail-Gassim in the northern Saudi Desert (Weiss, Marsh and Pfirman, 2001). Vegetation increases (or 'improvements') there, even on sandy and rocky surfaces, are reportedly due to more dryland being put under heavy utilization for irrigated wheat agriculture, and due to the increased mobility of nomadic herders using, for example, water trucks for pasture irrigation. Precipitation was found to be of no importance.

**Table 5.10  Annual Decrease in the Areal Extent of Undegraded Vegetation Cover (%) – Fast Change (>3.4%)***

| Rate | Time period | Site | Source |
| --- | --- | --- | --- |
| 5.6% | 1972-90 | N-Burkina Faso: Sahel-Sudan Zone, Gorom-Gorom District | Rasmussen et al., 2001 |
| 5.3% | 1982-95 | SW-USA: Chihuahuan Desert, southern New Mexico, Otero Mesa | Ludwig et al., 2000 |
| 5.8% | 1982-95 | SW-USA: Chihuahuan Desert, southern New Mexico, Otero Mesa | |
| 20.0% | 1986-91 | SW-USA: Colorado Plateau, southern Utah | Mouat et al., 1997 |
| 3.5% | 1986-96 | NW-China: Qinghai-Xizang (Tibet) Plateau, Yellow River source region | Wang et al., 2001 |
| 3.5% | 1958-78 | NW-China: Taklimakan Desert, Tarim River Basin, upper, middle, and lower reaches | Feng et al., 2001 |
| 3.8% | 1958-78 | | |

* Mean value (all rates) is 3.4%; n=13 sites

Some of the measures of vegetation decrease could not be transformed into annual percentage change rates. An example of this are the lower reaches of the Hei River Basin in the Hexi Corridor of N-China, where vegetation cover decreased in the 1958-80 period by ¾ (or less) of the original area, ie. Ulster wood, *Diversifolia schrenk*, and meadow by 26 km$^2$ per year, bush (with 70% canopy cover) by 87 km$^2$, and bush (with 30-70% canopy cover) by 36km$^2$ (Genxu and Guodong, 1999). Other measures of vegetation decrease have unique specifications which were not used in other cases. An example of this is the annual (southward) shift at 500-600 m of Sahel, Sudan, and Guinea vegetation zones into more humid locations in a Sahelian area of NW-Senegal, covering 2,600 villages over an area of 7,600 km$^2$ (Gonzalez, 2001). The shift is reportedly coupled with an annual decrease of tree densities (h>3m) from 10+-0.3 to 7.8+-0.3 trees/ha in the 1954-89 period (i.e., ca. 0.1/ha yearly), and an annual decrease of perennial plant species richness at 0.6% (from 64+-2 to 43+-2 species) in the 1945-93 period, i.e., by 33%.

The reduction in the production of biomass is another measure related to vegetation decrease, but not specified in terms of cover units. Gonzalez (2001), for example, reports for a large Sahelian study area in NW-Senegal that the standing biomass of trees decreased by 2.1 t/ha in the 1956-93 period. Finally, a single case was found only in which the annual decrease of vegetation cover is linked to related decreases of timber stock volume and biomass production – see Table 5.11.

On *Populus euphtatica*, mainly populating the Tarim River basin in NW-China, Feng, Endo and Cheng (2001) report that vegetation cover in the total river basin has declined by two thirds and biomass decreased by half in the 1958-78 period. They provide measures of change for various river reaches which show that, for arid zones only, annual vegetational cover decrease is immediately translated into related stock volume and biomass production measures (i.e., 2.1% - 2.8% - 2.8%, and 3.5% - 3.9% - 3.9%).

Table 5.11   Annual Degradation of *Populus euphtatica* (*Pe*) Vegetation in the Tarim River Basin of Taklimakan Desert in N-China, 1958-78

|  | Semi-arid upper reaches (200-800 mm/yr) | Arid middle reaches (ca. 105 mm/yr) | Arid lower reaches (50-100 mm/yr) |
|---|---|---|---|
| Annual decrease of Pe vegetation cover (rel.) | 3.8% | 2.1% | 3.5% |
| Annual decrease of Pe vegetation cover (in km$^2$) | 86 | 38 | 19 |
| Annual decrease of Pe timber stock volume (rel.) | 1.1% | 2.8% | 3.9% |
| Annual decrease of Pe timber stock volume (in m$^3$) | 12,750 | 91,350 | 10,400 |
| Annual decrease of mainly Pe biomass production (rel.) | 1.2% | 2.8% | 3.9% |
| Annual decrease of (mainly) Pe biomass production (in m$^3$) | 638 | 91,500 | 10,500 |

*Source*:   Adapted from Feng, Q., Endo, K.N. and Cheng, G.D. (2001), 'Towards sustainable development of the environmentally degraded arid rivers of China - A case study from Tarim River', *Environmental Geology*, vol. 41, pp. 229-38.

*Increase of degraded vegetation*   Most commonly, the increase of degraded vegetation in the cases was measured in terms of the (annual) area increase in vegetation cover of various floristic classes that were considered to be a factor, or an indication of degradation such as desert scrubs like mesquite (*Prosopis spp*) or Creosote bush (*Larrea spp*), desertified steppe, alpine steppe, and steppified meadow – see Table 5.12. The overall increase in the extension of degraded cover reportedly ranges between 0.2% (alpine steppe in the Qumarleb, Darlag, Maqen and Madoi counties of Qinghai Province in the Yellow river source region of NW-China, 1986-96) and 26.2% (desertified steppe, same area, same period). As with the decline in original vegetation cover, the broad range indicates high vegetational ecosystems dynamics. The mean value (4.3%) is based upon 20 measures for various vegetation classes which were done at five sites from Central Asian and US Southwest cases only.

**Table 5.12**     Annual Increase in the Areal Extent of Degraded Vegetation Cover (%)*

| Rate | Time period | Site | Source |
|---|---|---|---|
| | | Slow change (< 4.3%) | |
| 0.2% | 1976-86 | NW-China: Qinghai-Xizang (Tibet) | Wang |
| 0.5% | 1976-96 | Plateau, Yellow river source region | et al., 2001 |
| 0.7% | 1986-96 | | |
| 1.0% | 1980-90 | | |
| 1.7% | 1976-86 | | |
| 3.0% | 1976-96 | | |
| 4.0% | 1976-86 | | |
| 4.2% | 1986-96 | | |
| 1.2% | 1982-95 | SW-USA: Chihuahuan Desert, southern New Mexico, Otero Mesa, one site | Ludwig et al., 2000 |
| 4.0% | 1982-95 | SW-USA: Chihuahuan Desert, southern New Mexico, Otero Mesa, another site | |
| 0.8% | 1865-1995 | SW-USA: Sonoran Desert, Arizona, southern districts | Fredrickson et al., 1998 |
| 0.6% | 1915-28 | SW-USA: Chihuahuan Desert, southern New Mexcio, Jornada del Muerto Basin | Rango et al., 2000 |
| 1.0% | 1858-1963 | | |
| 1.2% | 1915-63 | | |
| 1.5% | 1928-63 | | |
| 1.5% | 1858-1928 | | |
| 1.8% | 1858-1915 | | |
| | | Fast change (> 4.3%) | |
| 26.2% | 1986-96 | NW-China: Qinghai-Xizang (Tibet) | Wang |
| 15.1% | 1976-96 | Plateau, Yellow river source region | et al., 2001 |
| 16.6% | 1977-95 | SW-USA: Chihuahuan Desert, southern Arizona, San Simon Valley, Portal area | Brown et al., 1997 |

* Mean value is 4.3%; n=5 sites.

First, no difference appears to exist between the various floristic classes when it comes to slow or fast change. Second, and different from vegetation decline, it appears that increases of degraded vegetation at a given site and per one time period can fall both under the category of either slow change (<4.3% annually) or fast change (>4.3% annually). Third, and different from original cover decline, there seems to be indication – as with meteorological change – that increases of degraded cover has accelerated during the last quarter of the 20$^{th}$ century. The cases of most rapid change are annual increases in total cover of desertified steppe (26%, 15%) at the Tibet Plateau, driven by the interplay of overgrazing and biophysical feedbacks from the soil and climate system (Wang, Qian, Cheng and Lai (2001), and, as a large and unexpected response from the regional climate system, increases in total cover of large, woody shrubs (17%) in the San Simon Valley of the Chihuahuan Desert in Arizona (Brown, Valone and Curtin, 1997). In the case of woody shrubs, both areal vegetation and shrub numbers increased.

Some of the measures of reported increases in degraded vegetation cover could not be transformed into annual percentage change rates of areal extent. For example, annual increases of degraded bush cover at lower than 30% canopy cover reportedly occurred at an annual rate of 123 km$^2$ in the lower reaches of the Hei River Basin, Hexi Corridor, N-China, in the 1958-80 period (Genxu and Guodong, 1999).

*Wind and Water Erosion*

Changes related to wind and/or water erosion (n=28) were most commonly specified in terms of increases in the areal extent of sand cover (18 measures at n=18 sites) and of eroded, bare, rocky ground cover (5 measures at n= 5 sites).

**Table 5.13**      **Annual Increase in the Areal Extent of Sand Cover (%) – Fast Change (> 4.7%)\***

| Rate | Time period | Site | Source |
|---|---|---|---|
| 18.1% | 1976-96 | NW-China: Qinghai-Xizang (Tibet) Plateau, Yellow river source region | Wang et al., 2001 |
| 34.7% | 1986-96 | | |
| 7.8% | 1950-95 | N-Central China: Great Plains, Ordos Plateau, Mu Us Region, Inner Mongolia, Ningxia Hui, Shaanxi Prov. | Lin and Tang, 2002 |
| 10% | 1983-89 | SW-Russia: Caspian (Sea Region) Plain and Yergenin Uplands, Kalmyk ASSR | Rozanov, 1991 |

\*   Mean value is 4.7%; n=19 sites (fast and slow change).

*Increase in sand cover* The most common measure of desertification related to wind/water erosion was the annual increase in the areal extent of sand cover, i.e., desert-like sandy conditions, or 'sandification' – see Tables 5.13 and 5.14. The mean value (4.7%) is based upon 28 rates from 19 sites. It ranges from as high as 34.7% in the Yellow river source region at the Tibet Plateau (Wang, Qian, Cheng and Lai, 2001) to as slow as 0.1% annually (or <1 km$^2$) in the case of the Gansu Province in the Hexi Corridor of N-China (Genxu and Guodong, 1999).

**Table 5.14 Annual Increase in the Areal Extent of Sand Cover (%) – Slow Change (< 4.7%)***

| Rate | Time period | Site | Source |
|---|---|---|---|
| 1.6% | 1950-89 | NE-China: Inner Mongolian Plateau, Keerqin or Korqin Desert ('Meadow') | Lin and Tang, 2002 |
| 1.5% | 1950-90 | NE-China: Songlia Basin, Songnen Plain, W-Jilin and W-Heilongjiang Provinces | |
| 0.5% | 1945-95 | N-Central China: Great Plains, Ordos Plateau, Mu Us Region (desert, 'meadow'), Jingbian county | Zhou et al., 2002 |
| 1.4% | 1976-86 | NW-China: Qinghai-Xizang (Tibet) Plateau, Yellow river source region | Wang et al., 2001 |
| 0.7% | 1958-87 | NW-China: Taklimakan Desert, Tarim River Basin, upper reaches | Feng et al., 2001 |
| 1.0% | 1958-87 | NW-China: Taklimakan Desert, Tarim River Basin, middle reaches | |
| 1.2% | 1958-87 | NW-China: Taklimakan Desert, Tarim River Basin, lower reaches | |
| 2.1% | 1949-97 | N-China: Hexi Corridor, Hei River Basin, Gansu Province, upper reaches of Shandan Region | Genxu and Guodong 1999 |
| 0.1% | 1949-97 | N-China: Hexi Corridor, Hei River Basin, Gansu Province, middle reaches of Jinta Region | |
| 0.1% | 1949-97 | N-China: Hexi Corridor, Hei River Basin, Gansu Province, middle reaches of Zhangye and Linze Regions | |
| 0.3% | 1949-97 | N-China: Hexi Corridor, Hei River Basin, Gansu Province, middle reaches of Gaotai Region | |
| 1.2% | 1949-97 | N-China: Hexi Corridor, Hei River Basin, Gansu Province, lower reaches of Ejina Region | |
| 2.9% | 1960-95 | Kazakstan, Uzbekistan, Turkmenistan: Circum Aral (Sea)Region | Saiko and Zonn, 2000 |
| 0.5% | 1982-94 | NE-Saudi Arabia: Arabian Peninsula, northeastern Saudi Desert | Weiss et al., 2001 |
| 3.3% | 1982-94 | W-Saudi Arabia: Arabian Peninsula, western Saudi Desert | |

\* Mean value is 4.7%; n=19 sites (slow and fast change).

It indicates high land cover dynamics for sand encroachment, higher even than cover conversion into bare, rocky, eroded ground. Slow and rapid sand encroachments can happen alternatively and abruptly at a given site within decades. Most strikingly, all cases of reported sand encroachment stem from Asia.

Some measures of sandification could not be transformed into annual percentage rates of cover change, and/or were specified in terms not used in other cases. For example, sand cover reportedly increased in two study areas of the west-central Sahel of Mali, Niger, and Burkina Faso by 2,720 km$^2$, and 23,200 km$^2$ per year, respectively, in the 1986-90 period (Ringrose and Matheson, 1992). In other cases, the activation of (semi)fixed sand dunes was measured in terms of velocity at which sand dunes move. High rates at 900-1200 m annually were reported for the Circum Aral (Sea)-Region in the 1983-95 period, an area being as large as 1,600,000.00 km$^2$ including parts of Kazakstan, Uzbekistan and Turkmenistan (Saiko and Zonn, 2000). At considerably lower rates, shifting sand dunes reportedly moved in North and Northwest China: for example, at 5-10 m annually in the western Kashi Plain of Tarim River Basin in the Xinjiang (Ulgur) Autonomous Region during the 1950-95 period (Lin, Tang and Han, 2001), and at 8-15 m per year in the middle reaches of Yarlung Zangbo, Lhasa and Nianchu Rivers of the mid-south Qinghai-Xizang (Tibet) Plateau, in the 1980-96 period (Liu and Zhao, 2001).

For the Autonomous Region of Tibet, frequent and periodic sand deposits were reported on both pastoral and cultivated land (including siltation of low-lying zones), amounting to an annual sand accumulation in the order of 7.5 cm, or 8 m$^3$ in the 1980-96 period. Similarly, Wijdenes, Poesen, Vandekerckhove and Ghesquiere (2000) report an annual rate of sedimentation of 4 m$^3$ (per gully head), or 820 m$^3$ (per river stretch), for the Guadalentin catchment area in SE-Spain during 1997/98. On annual sand transport (but not sedimentation), Wang, Qian, Cheng and Lai (2001) state that 88,140 tons of sand were moved each year in the Yellow River source region of NW-China in 1976-86.

*Increase in eroded, bare, rocky ground cover* The exposure of bare, rocky outcrop through erosion was mostly specified in terms of the increase in areal extent of eroded, bare rocky ground cover – see Table 5.15. The mean value (1.2%) is based upon 5 measures from 5 sites. It ranges from the slow exposure of bare, rocky parent material at 0.8% (or 2 km$^2$) to 1.7% annually (or 3 km$^2$), with both values originating from the same case, i.e., the SE-Californian Manix Basin (near Barstow) in the Mojave Desert of the Great Basin in the American Southwest (Okin, Murray and Schlesinger, 2001). The geomorphic setting there is responsible for the difference between rapid change (i.e., direct disturbance of the basin floor) and slower change (i.e., indirect disturbance of non-playa parts of the basin area). The mean value ranges between 0.8 and 1.7% only, which indicates that exposure of rocky outcrop is a rather gradual and slow process, especially if compared to other change rates such as sand encroachment and vegetation dynamics. There appears to be no indication that the process of exposing parent material has accelerated over time.

**Table 5.15  Annual Increase in the Areal Extent of Eroded, Bare, Rocky Ground Cover (%)***

| Rate | Time period | Site | Source |
|---|---|---|---|
| | | Slow change (<1.2%) | |
| 0.8% | 1979-97 | SW-USA: Mojave Desert (Great Basin), California, Manix Basin | Okin et al., 2001 |
| 0.9% | 1982-94 | W-Saudi Arabia: Arabian Peninsula, western Saudi Desert | Weiss et al., 2001 |
| 1.1% | 1957-87 | N-Tanzania: SW-Masai Steppe, Arusha district | Mwalyosi, 1992 |
| | | Fast change (>1.2%) | |
| 1.7% | 1979-97 | SW-USA: Mojave Desert (Great Basin), California, Manix Basin | Okin et al., 2001 |
| 1.3% | 1990-94 | S-Central Botswana: central Kalahari, Kweneng District, Malwele area | Ringrose et al., 1996 |

* Mean value is 1.2%; n=5 sites.

Some measures related to the exposure of ground cover could not be transformed into annual percentage, and/or were expressed in terms which were not used in other cases. Examples of the first are annual increases of rocky outcrop to the extent of 2,720 km$^2$, and 23,200 km$^2$, respectively, in two study areas of the west-central Sahel in the 1986-90 period (Ringrose and Matheson, 1992). Examples of the latter are various measures such as soil losses leading to the exposure of the matrix. Though limited in terms of comparability, Ludwig et al. (2000) report soil losses from Otero Mesa in the Chihuahuan Desert of New Mexico at 0.4 mm per year for the 1983-95 period, which is about half of what Rango, Chopping, Ritchie, Havstad, Kustas and Schmugge (2000) report from the nearby Jornada del Muerto Basin, i.e., soil losses on both grass and shrub (mesquite) covered soils at 0.9 mm/year in the 1934-80 period (however, if only mesquite coppice dune formation is taken, authors found soil gains at annually 0.4 mm for the same period). Annual soil losses have further been specified in terms of volume or amount eroded, or in terms of dry bulk density. For example, Wijdenes, Poesen, Vandekerckhove and Ghesquiere (2000) report 0.95 m$^3$/ha, or 1.2 t/ha (dry bulk density) to be eroded in the Rambla Salada River stretch in SE-Spain during 1997/98. This means a fraction only of the denudation rate which Mwalyosi (1992) cites for the catchment area of Kisongo Dam in the Ardai Plains of the Masai Steppe for 1959/69 to 1969/71, i.e., 6.6 t per hectar and year. In none of the cases, areal increases in eroded, bare, rocky ground cover are linked to related soil losses and/or other measures for the very same site and time period.

## Water Degradation

The most common measure of rates in water degradation (n=14) was the annual increase in the salinity of groundwater, the spread of salinity in cultivated land, and the annual decrease in water surface and of river discharge.

**Table 5.16  Annual Changes (in %) related to Water Degradation – Salinity Increase in Groundwater***

| Rate | Time period | Site | Source |
|---|---|---|---|
| | | Slow change (<6.8%) | |
| 4.0% | 1960-95 | NW-China: Taklimakan Desert, Tarim River Basin, middle reaches | Feng et al., 2001 |
| 3.2% | 1960-95 | NW-China: Taklimakan Desert, Tarim River Basin, lower reaches | |
| | | Fast change (>6.8%) | |
| 13.2% | 1960-95 | NW-China: Taklimakan Desert, Tarim River Basin, upper reaches | Feng et al., 2001 |

\*  Mean value is 6.8%; n=3 sites.

*Increase in groundwater salinity*  Groundwater salinity reportedly increased at a mean value of 6.8% annually, based upon 3 measures from 3 sites within a larger case study region – see Table 5.16. The rate ranges from 3.2% annually – as in the case of the lower reaches of Tarim River in NW-China in the 1960-95 period – to as high as 13.2%, as in the case of the upper reaches of the same river and over the same time period (Feng, Endo and Cheng, 2001). A comparable amount of change per time (but not available as %-increase) was reported for the 1960-95 period from the Circum Aral Sea-Region, where groundwater salinity increased annually at 0.2-0.3 g/l (Saiko and Zonn, 2000).

*Areal spread of salinized land*  The expansion of salinized land in all cultivated, irrigated land reportedly occurred at an annual mean value of 0.5%, based upon 5 sites and 5 measures – see Table 5.17. Thus, the spread of salinity occurs at a very low pace. It ranges between 0.01% annually (or, <1km$^2$) – as in the case of the Shandan Region in the upper reaches of Hei River of northern China in the 1930-97 period – and 1.5% (or, <1km$^2$), as in the Jinta Region of the middle reaches in the same area and over the same period (Genxu and Guodong, 1999). Again, there are measures of the amount of change per time, but no %-increases. As for the Jinta Region in the middle reaches of Hei River in N-China in the 1930-97 period, a more specific rate relates to the area having a specific salt content of 3-5 g/l, and which reportedly expanded at 6 km$^2$/year in the 1980-85 window (Genxu and

Guodong, 1999). Far lower expansion rates are reported from the terminal area of the Tarim River basin in the Taklimakan Desert, where salinity in irrigated land expanded at 1 km$^2$ annually in the 1950-98 period (Feng, Endo and Cheng, 2001). The same holds true for the Caspian Plain and Yergenin Uplands of the Caspian Sea Region, where salinized land expanded at a rate of 2 km$^2$ annually in the 1980-89 period (Rozanov, 1991).

Table 5.17    Annual Changes (in %) related to Water Degradation – Areal Spread of Salinity in Irrigation Schemes*

| Rate | Time period | Site | Source |
|---|---|---|---|
| | | Slow change (<0.5%) | |
| 0.4% | 1930-97 | N-China: Hexi Corridor, Hei River Basin, Gansu Province, middle reaches of Gaotai Region | Genxu and Guodong, 1999 |
| 0.2% | 1930-97 | N-China: Hexi Corridor, Hei River Basin, Gansu Province, middle reaches of Zhangye Region | |
| 0.3% | 1930-97 | N-China: Hexi Corridor, Hei River Basin, Gansu Province, middle reaches of Linze Region | |
| 0.01% | 1930-97 | N-China: Hexi Corridor, Hei River Basin, Gansu Province, middle reaches of Shandan Region | |
| | | Fast change (>0.5%) | |
| 1.5% | 1930-97 | N-China: Hexi Corridor, Hei River Basin, Gansu Province, middle reaches of Jinta Region | Genxu and Guodong, 1999 |

\* Mean value is 0.5%; n=5 sites.

*Decrease of water surface* Decreases of water surface were reported to occur at an annual average rate of 2.0%. The mean value stems from 3 cases and is based upon 9 measures – see Table 5.18. It ranges between 0.1% – i.e., lakes and swamps in the Yellow river source region in NW-China for the 1976-86 period (Wang, Qian, Cheng and Lai, 2001) – and 5.9%, as in the case of Lake Juyan in the Hei River Basin of N-China, in the 1980-97 period (Genxu and Guodong, 1999). The case of Lake Juyan covers a time interval of nearly 70 years and seems to suggest a high dynamics of water surface changes. There are more measures of rates related to water surface decreases, but no annual %-increases could be calculated. For example, the water surface of the Aral Sea decreased yearly by 997 km$^2$ in the 1960-95 period, the sea volume by 21.5 km$^3$ yearly, the length of shore line by 14 km per year, and the sea level dropped by 0.5 m annually, with all rates relating to the 1960-95 period (Saiko and Zonn, 2000).

**Table 5.18    Annual Changes (in %) related to Water Degradation – Decrease of Water Surface***

| Rate | Time period | Site | Source |
|---|---|---|---|
| | | Slow change (<2.0%) | |
| 0.7% | 1930-60 | N-China: Hexi Corridor, Hei River | Genxu and |
| 1.8% | 1930-80 | Basin, Gansu Province, lower reaches: | Guodong, 1999 |
| 1.5% | 1930-97 | Lake Juyan in Ejina Region | |
| 0.1% | 1976-86 | NW-China: Qinghai-Xizang (Tibet) | Wang |
| 0.5% | 1976-96 | Plateau, Yellow river source region: | et al., 2001 |
| 0.9% | 1986-96 | lakes & swamps | |
| | | Fast change (>2.0%) | |
| 4.3% | 1960-80 | N-China: Hexi Corridor, Hei River | Genxu and |
| 5.9% | 1980-97 | Basin, Gansu Province, lower reaches: Lake Juyan in Ejina Region | Guodong, 1999 |
| 2.0% | 1950-2000 | NW-China: Taklimakan Desert, Tarim River Basin, terminal area | Feng et al., 2001 |

\* Mean value is 2.0%; n=3 sites

*Decrease of river discharge* Decreases of river discharge reportedly occurred at an annual average rate of 2.1% in the cases. The mean value comes from 6 cases and 11 measures – see Table 5.19. It ranges between 0.3% (or $3 \times 10^6$ m$^3$) – i.e., Hunshui River area of the Hei River middle reaches in Jinta Region of Gansu Province in N-China in the 1930-60 period – and 5.0% (or $6 \times 10^6$ m$^3$), as in the same case, but for the 1960-80 period (Genxu and Guodong, 1999). Similarly, the discharge of Shandan River in the upper reaches of Hei River Basin decreased annually by 1.0% over 30 years before 1960, and in the following 20 years at a 5-fold higher rate (Genxu and Guodong, 1999). Thus, the Hei River Basin study suggests increasingly shrinking river discharges over time at two sites in a semi-arid to temperate region which has 300-500 mm rainfall yearly, and which covers mountain as well as basin units (more than 30 tributaries dried up in 1930-80). However, no comparative results from other river basins were available. As for water surface change, river discharge tends to be highly dynamic as the Hei and Tarim River Basis cases prove, and has comparable rates to those of water surface changes.

Other measures of rates than the ones given in the tables above were not widely used. For example, annual increases of surface (rather than ground) water salinity tend to be higher, but the number of cases for comparison is limited (n=2), and therefore not included in the table. Feng, Endo and Cheng (2001), for example, state that the middle reaches of Tarim River in northwestern China showed an annual increase of surface water salinity at 2.7% (or 0.1 g/l) in the 1960-95 period.

**Table 5.19    Annual Changes (in %) related to Water Degradation – Decrease of River Discharge***

| Rate | Time period | Site | Source |
|---|---|---|---|
| | | Slow change (<2.1%) | |
| 0.5% | 1960-95 | NW-China: Taklimakan Desert, Tarim River Basin, middle reaches | Feng et al., 2001 |
| 1.2% | 1976-96 | NW-China: Qinghai-Xizang (Tibet) Plateau, Yellow river source region | Wang et al., 2001 |
| 0.3% | 1930-60 | N-China: Hexi Corridor, Hei River | Genxu and |
| 2.0% | 1930-80 | Basin, Gansu Province, middle reaches of Jinta Region (Hunshui River) | Guodong, 1999 |
| 1.0% | 1930-60 | N-China: Hexi Corridor, Hei River Basin, Gansu Province, middle reaches of Jinta Region (Linshui River) | |
| 2.0% | 1930-80 | N-China: Hexi Corridor, Hei River | |
| 0.9% | 1950-95 | Basin, Gansu Province, upper reaches of Shandan Region (Shandan River) | |
| | | Fast change (>2.1%) | |
| 2.3% | 1950-98 | NW-China: Taklimakan Desert, Tarim River Basin, lower reaches | Feng et al., 2001 |
| 5.0% | 1960-80 | N-China: Hexi Corridor, Hei River Basin, Gansu Province, middle reaches of Jinta Region (Hunshui River) | Genxu and Guodong, 1999 |
| 3.3% | 1930-60 | N-China: Hexi Corridor, Hei River | |
| 5.0% | 1960-80 | Basin, Gansu Province, lower reaches of Ejina Region (terminal area) | |

* Mean value is 2.1%; n=6 sites

Similarly, Genxu and Guodong (1999) report an annual salinity increase of 4% (or 0.002 g/l) in the 1962-87 period for the Jinta Region of Hei River in N-China. Some measures related to water degradation show the absolute amount of change per time, but no annual %-increases were given. For example, Saiko and Zonn (2000) report about Aral sea surface water, the salinity of which increased by 0.6-0.7 g/l annually within 35 years (1960-95); and Rozanov (1991) gives the annual average input of salts into surface water in the Karakum Canal Zone of the Turkmenia Plain as high as 9.5 million tons in the 1978-89 period. Other measures which were not widely used are those on water pollution. There is just one rate given on the Circum-Aral Region stating that chemical pollution increased from 20 to 80% of all irrigated land in the 1960-95 period, ie. by 1.7% annually (Saiko and Zonn, 2000).

*Decline in Pastoral Suitability*

Among the reported decreases in pastoral suitability (n=15), the most common rates were the decrease in the areal extent of residual rangeland, and, reciprocally, the invasion or area increase of impalatable, undesirable species. However, the number of cases is limited and broad generalizations are difficult to make.

**Table 5.20    Annual Decline (in %) in Pastoral Suitability – Areal Decrease of Residual Rangeland***

| Rate | Time period | Site | Source |
|---|---|---|---|
| | | Slow change (<4.0%) | |
| 1.5% | 1986-90 | West-central Sahel, central Niger: Agadez-Bagzan-Abalak-Aderbissinat area | Ringrose and Mathseon, 1992 |
| 1.7% | 1949-64 | NE-China: Inner Mongolian Plateau, Keerqin Desert, Xilingole League | Sheehy, 1992 |
| | | Fast change (>4.0%) | |
| 5.0% | 1986-90 | West-central Sahel, Mali, Niger: Goundam-Niafounke-Gossi-Doro-Ansongo-Tahoua-Keita area | Ringrose and Mathseon, 1992 |
| 9.0% | 1986-90 | Sudan-Sahel, west-central Sahel, Mali, Niger, Burkina Faso: Dori-Ayorou-Tera-Abala-Filingue-Dogondoutchi-Dosso-Say Birni N'Konni-Madaqua area | |

* Mean value is 4.0%; n=4 sites

*Decrease of residual rangeland* The most common measure of the rate of decrease in remaining vegetation cover for pastoral uses was the (annual) decrease of the area of residual rangeland. The mean value (4.0%) is based upon four sites and four rates – see Table 5.20. With the rates ranging from 1.5% to 9.0%, there is slight indication that rangeland change – as it is with vegetation change – is very dynamic and occurs at rather high levels. Related measures in absolute terms per time (but no %-increases) were reported from northern China: residual rangeland in the lower reaches of Tarim River in the Taklimakan Desert decreased by 6 km$^2$

annually in the 1958-90 period (Feng, Endo and Cheng, 2001), and by 40 km$^2$ in the Zhaowuda League of the Inner Mongolia Autonomous Region, situated in the Korqin Desert, in the 1960-90 period (Sheehy, 1992).

**Table 5.21  Annual Decline (in %) in Pastoral Suitability – Areal Invasion of Undesirable (Impalatable) Species***

| Rate | Time period | Site | Source |
|---|---|---|---|
| | | Slow change (<1.5%) | |
| 0.2% | 1950-81 | NE-Nigeria: Sahel-Sudan zone, Kano | Mortimore |
| 0.3% | 1970-81 | Close-Settled zone, Tumbau & Dagaceri | et al., 1999 |
| 0.2% | 1950-69 | village areas | |
| 0.5% | 1969-81 | | |
| 0.4% | 1950-81 | | |
| | | Fast change (>1.5%) | |
| 7.5% | 1990-94 | S-Central Botswana: central Kalahari, Kweneng District, Malwele area | Ringrose et al., 1996 |

\* Mean value is 1.5%; n=3 sites

*Increase of undesirable species* The increase in the areal extent of undesirable species reportedly occurred at annually 1.5% on average, with the mean value derived from three sites and based upon six measures of change rates – see Table 5.21. The rates range between 0.2 and 7.5%. Not included in the table is one case of invading species which had been on a retreat during the period studied in Tumbau village area of Kano Close-Settled zone in NE-Nigeria, i.e., by –0.01% in the 1950-70 period (Mortimore, Harris and Turner, 1999). A measure related to the area expansion of undesirable species was the decline in productivity of palatable grass species. However, this measure was given only for the middle reach of Hei River in NW-China, where the yield of palatable grasses declined by 1.6% annually, or 24 kg/ha, in the Zhangye Region of Gansu Province during the 1960-97 period (Genxu and Guodong, 1999).

Other measures of pastoral suitability decline than the ones given in the tables above were not widely used, and/or lack comparability, such as changes in livestock carrying capacity. For example, Feng, Endo and Cheng (2001) report an annual decrease of livestock carrying capacity at 0.5% in the 1950-98 period for the middle reaches of Tarim River in NW-China, while Manzano, Návar, Pando-Moreno and Martinez (2000) specify pastoral degradation in terms of the annual growth of excess hectarage/animal unit (i.e., beyond carrying capacity) at 0.9 to 1.0% in the 1970-85 period for two sites (Sonora Desert in NE-Mexico, and Thamaulipan thornscrub area in northern coastal lowlands of Mexico). Mwalyosi (1992) provides a measure of pastoral suitability decline which states that livestock units per hectar increased annually by 3.3% in the 1957-87 period in the semiarid

Masai steppe of N-Tanzania, while the availability of cattle per pastoral household head decreased by 1.5% in the same period and area.

Other measures could not be transformed into in annual percentage values. For example, the yearly decline in livestock number has been specified in four cases in terms of million livestock units, ranging between annually 0.1 and 0.2, regardless of regional variations. For the arid and semi-arid zone of Australia, Pickup (1998) reports an annual decline (sheep) at 0.2 million livestock units in the 1890-1902 period for New South Wales, and a decline at 0.1 million units in the 1930-60 period for Queensland. At a similar rate (i.e., 0.1 million units of cattle per year), livestock declined in the Nacqu Prefecture of Tibet Autonomous Region in the 1990-98 period (Holzner and Kircher, 2001). Finally, Aagesen (2000) states that the annual decline of sheep in the Percey River watershed in the Chubut Province, situated in the Patagonian rangelands of southern Argentina, occurred at a similar order of 0.2 million livestock units in the 1952-88 period.

**Table 5.22**     Annual Decline (in %) in Agricultural Suitability – Decline in Crop Yields[a]

| Rate | Time period | Site | Source |
|---|---|---|---|
| | | Slow change (<30.3%) | |
| 2.2%[b] | 1945-91 | Southcentral Mexico: Tehuacán Valley, Río Zapotitlán, Metzontla village area | McAuliffe et al., 2001 |
| 8%[c] | 1977-88 | N-Tanzania: SW-Masai Steppe, Arusha district | Mwalyosi, 1992 |
| 10.0%[d] | 1985-95 | NW-India: Thar Desert, Rajasthan, Jodhpur District, Khabra Kalan vge area | Ram et al., 1999 |
| 1.4%[e] | 1951-95 | | |
| | | Fast change (>30.3%) | |
| 80.0%[f] | 1984-85 | Republic of Sudan: arid to hyperarid zone, Kordofan Province, El Obeid-Umm Ruwaba-Bara-Kagmar area | Olsson, 1993 |
| 80.0%[f] | 1984-85 | Republic of Sudan: arid to hyperarid zone, Darfur Province, El Fasher area | |

[a] Mean value is 30.3%; n=5 sites
[b] Maize
[c] Seed bean.
[d] Cluster bean
[e] Pearl millet
[f] Millet, sorghum

*Decline in Agricultural Suitability*

Among the cases in which a decrease in agricultural suitability was reported (n=9), the most common rate had been the decline in crop yields, and the loss of cultivated land. The reduction of agricultural potential due to salinization was not considered here, since dealt with earlier (see water degradation). The number of cases is limited, so that broad generalizations are difficult to make.

*Decline in crop yields* The most common measure of declining suitability for agricultural uses was the annual decrease of yields per various crops. The drastic mean value (30.3%) is based upon 5 sites and 6 measures – see Table 5.22. It ranges between as low as 1.4% (or 3 kg/ha) – i.e., annual average yield decrease of pearl millet in a Thar Desert village area of N-India in the period 1951-95 (Ram, Tsunekawa, Sahad and Miyazaki, 1999) – and as high as 80.0% - i.e., annual average yield decline of millet and sorghum in West and Central Sudan during two years of severe drought (Olsson, 1993). Different from the high and drastic declines in crop yield during drought conditions, slow and gradual crop yield declines without any oscillation were reported in the majority of cases. They could finally lead to full harvest losses and stop of production even, with cluster bean (Ram, Tsunekawa, Sahad and Miyazaki, 1999) and maize (McAuliffe, Sundt, Valiente-Banuet, Casas and Viveros, 2001) being crop-specific examples. Comparable to the collapse of agricultural suitability, Saiko and Zonn (2000) report a decline of fish catch at 1,000 t annually for the Aral Sea in the 1960-95 period, until full depletion of fish resources.

**Table 5.23 Annual Decline (in %) in Agricultural Suitability – Loss of Cultivated Land***

| Rate | Time period | Site | Source |
|---|---|---|---|
| | | Slow change (<2.4%) | |
| 1.6% | 1958-2000 | NW-China: Tarim River Basin, Taklimakan Desert, Xinjiang Region, Keriya River, Yutian county | Yang, 2001 |
| 2.0% | 1949-95 | N-China: Hexi Corridor, Hei River Basin, Gansu Province, lower (terminal) reaches of Ejina Region | Genxu and Guodong, 1999 |
| | | Fast change (>2.4%) | |
| 3.5% | 1985-95 | NW-China: Qinghai-Xizang (Tibet) Plateau, Yellow river source region | Wang et al., 2001 |

* Mean value is 2.4%; n=3 sites

*Loss of cultivated land* A measure for the annual loss of cultivated land in terms of areal extent was given for 3 sites – see Table 5.23. The mean value (2.4%) ranges between 1.6% – i.e., Keriya River artificial oases in the Tarim River Basin of the Taklimakan Desert in NW-China (Yang, 2001) – and 3.5%, or an equivalent of 55,650 km$^2$ annually – i.e., for the NW-Chinese Yellow river source region (Wang, Qian, Cheng and Lai, 2001). There had been more measures related to the loss of cultivated land. However, they lack comparability with other cases and are not included in the table. For example, Saiko and Zonn (2000) report that the annual increase in waterlogged soils on cultivated land amounts to 1.0%, and that the spread of soil modification through chemical pollution occurs at 1.1% of all cultivated land in the 1960-95 period.

*Other Socio-Economic Changes*

There are only few other socio-economic changes or declines in human welfare that were given a measure of rate, i.e., change in terms of amount per time, and/or annual %-values. Impaired or surpassed firewood carrying capacity – but not human population carrying capacity – is one example. Gonzalez (2001) reports an annual increase at 2.6 persons/km$^2$ in 1988-93 for an area of 2,600 villages in the West African Sahel of NW-Senegal in the St. Louis/Louga area, and Benjaminsen (1993) reports that wood collection distance increased by 90% (or 0.5 km) annually in the period 1980-90 in a dried up lake area of the alluvial inland plain of Niger River in the Gourma Region of Central Mali (Bambara-Maounde).

Concerning land-use related human disorders, Olsson (1993) specifies annual excess deaths due to starvation at 100,000 persons in 1984-85 for the Darfur Province of Western Sudan, and Saiko and Zonn (2000) provide annual health (or disease) measures from the Circum-Aral Region according to which hypertonia disease increased annually by 6/10,000 persons, heart attacks by 9/10,000, hepatitis by 20/10,000, and typhoid fever by 26/10,000 persons in the 1960-95 period.

Chapter 6

# Pathways

**Introduction**

Pathways or trajectories of dryland change are shaped by the land use and environmental histories of the respective sites under study – see chapter 3 on initial conditions – they are made up of the causative factors behind desertification, including system properties of the process – see chapter 4 – and they are featured by characteristic syndromes and process rates of dryland change – see chapter 5. Several tables summarize the basic and distinct features of these pathways by major geographic dryland regions, including ecological factors shaping the environmental history such as climate and topography (Table 6.1) and vegetation, soil and water ecologies (Table 6.2), a short history of changing institutional and socio-economic contexts including land use history (Table 6.3) and current land tenure (Table 6.4), a brief description of dryland formation and degradation in pre-historic, ancient and historical times (Table 6.5) as well as contemporarily (Table 6.6), the dominant proximate as well as underlying causes of desertification (Tables 6.7 and 6.8, respectively), and, finally, the major syndromes and mediating factors (Table 6.9) as well as the dominant system properties (Tables 6.10). However, pathways as found in the desertification cases are not necessarily regional pathways. Typical elements of regional pathways can obviously be shared by several geographical units either adjacent to each other (such as the grasslands of northern Mexico and the US Southwest), or far from each other but linked to similar cultural histories (such as the rangelands of Patagonia, Australia and southern Africa). In total, six dominant pathways of desertification or dryland degradation were identified.

**Two Pathways of Desertification in Central Asia**

Two pathways of desertification can be generalized from cases located in Central Asia. They are about are 'sandification' and the more or less uni-directional, double invasion, on the one hand, of grain farming onto steppe grazing land as well as, on the other hand, of hydraulic cultures into desert or subdesert ecosystems. Except for few dryland sites in Asia, the most important feature making up the paths of land change since millennia across wide parts of the region is strong directional change exerted by central state authorities in China, Russia (or former Soviet Union countries), and Mesopotamia in the past. Historically, the specific measures related to land consolidation and control over territories (and people) at the margin of civilizations or empires founded on settled agriculture rather than on nomadic pastoralism. Contemporary measures perpetuate the goal of nation-building and land consolidation (e.g., through forced population transfers of Han chinese into northwestern frontier provinces), but also relate to the need of attaining self-sufficiency in food and clothing for increasing human populations through massive agricultural production (e.g., of grain and cotton). These policy measures were always founded on fundamental national policy programmes and implemented through central command directives. They were meant to serve national goals, regardless of the specific cultural, socio-economic or ecological properties of marginal dryland areas.

In some cases, directional change as an overriding socio-economic driving force – i.e., policy-driven production growth and agricultural intensification, application of large-scale water technologies (of low efficiency), forced immigration, rising population densities, and urbanization as well as industrialization of frontier zones as most recent phenomena – already led to irreversible desertification. In most of these cases, outcomes are spectacular such as the increase of desert-like conditions ('sandification') over large areas, hardly reported from other dryland cases in the world. Clearly, the process of desertification is enhanced by a strong biophysical predisposement for desert formation, apparently valid for areas in northern China. The directionality of socio-economic change and the vastness of landscape units affected easily upscale land degradation to desertification. This is different, for example, from the multi-directional change and small-scale nested landscape units in southern European drylands. The process is occasionally irreversible, such as in the case of the Aral Sea basin, or it happens at high speed across several land change classes (vegetation degradation, sandification, water degradation, losses of cultivated land), such as in the Yellow river source region, Tarim River and Hei Basins of northern China, and Caspian Sea Region. The major syndromes of change are gradual as well as rapid, i.e., changing opportunities created by markets and outside policy interventions (economic development programmes, rapid capital investments and new technologies for intensification). They are associated with rapid process rates mainly in irrigation zones where water degradation is the most striking indication of degradation, with high degrees of severity not found in other dryland areas.

Initial conditions, however, mainly stemming from topography and geomorphological setting, divert the overriding impact of directional change into two major pathways, one for steppe lands on arid to semi-arid upland plateaus (or high plains), used for grazing and cropping, and one for irrigation zones in river, delta or lake basin ecosystems under arid to hyper-arid (sub)desert conditions. Along both pathways, inappropriate land uses invade fragile dryland ecosystems both in terms of areal extent and the degree of intensification, i.e., rainfed grain farming in dry grassland zones of steppe ecosystems (suitable for nomadic pastoralism), and excessive irrigation farming in combination with increasing non-agrarian uses in desert ecosystems (suitable for small-scale oasis agriculture). The pathways differ considerably in their degree of resilience to change.

*Grazingland > Sedentary Farming > Overcultivation/Grazing > Sandification*

On the semi-arid to arid upland steppe plateaus, which carry mainly dry grasslands, the dominant trajectory of land change, in particular at the ecotone of sandy and loessial soils, is the ancient, historical and contemporary transformation of millennia-old (semi)nomadic grazing into more intensified modes of extensive grazing. No cases, however, were found in which the transition itself has reportedly triggered desertification, because most steppe grazingland is best suited for extensively managed, but not sedentary grazing animals production.

Rather, the livestock transition has to be put in the context of another land use change which, in combination with livestock intensification, led to overstocking, i.e., expanding grain farming land onto what had been rangelands before. Rangelands, featured by previously high forage capacity, were reduced, and/or increasingly limited to sites with only marginal productivity. Sedentary rainfed agriculture, mainly for large-scale grain production, became located at the ecotone between steppe grazingland and sedentary agricultural areas, but ecosystem properties such as a long, cold, windy and dry winter/spring season, a short growing season and low, variable rainfall were hardly conducive to highly productive, crop-based sedentary agriculture. Often they reinforced further land reclamation. Actually, the short vegetation period and low temperatures on cold plateaus reportedly constitute control points in some cases beyond which land use is easily and irreversibly shifted onto a degrading path. And, grazing pressure on the reduced ranges was drastically increased through the addition of new livestock and inappropriate infrastructure investments that did not suit well to the fragile dryland ecosystem, such as mechanization of water delivery and built up of watering structures.

Geological sand formation, cold, dry, continental climates, the quaternary uplifting of the Tibet Plateau and, though to a lesser degree, contemporarily increased extreme weather events predispose the arid and semi-arid steppe grazingland biophysically to processes of desert formation. Consequently, desert margins oscillate since ancient times due to coupled human-biophysical impacts. This makes the process of desertification highly dynamics, as in the case of the Mu Us region (meadow, desert) located on the Ordos Plateau.

The pattern to be found since ancient times throughout of northern China, in particular, can be described as large-scale alternations of grain farming and livestock raising, with desertification happening intermittently: first, the cultivation of grazingland, second, the subsequent abandonment following misuse of water resources, overcultivation, overgrazing, or warfare, and, third, translocation of rural Han Chinese peasants to other grazinglands where this process was then repeated. The pattern very much follows an amplifying, self-perpetuating process of expanding cropland at the expense of rangeland, followed by overstocking and soil mining, leading to an increased demand for pasture and cropland, which again triggers expanding rangeland and/or cropland, etc., all in synergy with oscillating desert margins. No mitigating feedbacks were reported in the cases, and opinions are divided about success or failure of technological measures such as the Green Wall.

In contrast to river, delta or lake ecosystems in (sub)deserts, semiarid steppe land has the capacity to recover ecological conditions between periods of disturbance, thus making possible large-scale alternations between grain farming and extensive grazing. However, human efforts to mitigate the advancement of contemporary 'sandification' – such as the green wall project – very likely increased rather than controlled desertification by increasing the density of grazing animals on remaining grazing land further. While semi-arid steppe ecosystems do not necessarily follow a pathway towards irreversible degradation, cold mountain steppe zones, featured by alpine meadows and permafrost soils, seem especially fragile and extremely vulnerable. Different from steppe areas, vegetation degradation there occurs at high speed. Increases in desertified alpine steppe or steppified high-cold meadows are hardly reversible due to unidirectional changes in waterlogged soils and frozen underground – i.e., drying out of the waterlogged surface, decomposition of peat, soil losses due to erosion – so that high mountain steppe quickly turns into semi-desert vegetation, with desert, stone or gravel pavement being the final result.

*Traditional Oases > Large-Scale Irrigation > Booming Non-Agrian Uses > Water Degradation > Vegetation Degradation > Sandification*

At sites in dry and hot lowland plains, depressions or basins which carry river, delta or lake ecosystems under (sub)desert conditions, long-settled traditional land uses based on irrigation were supported for centuries, if not millennia's due to rich groundwater resources and constant river flows. Despite of geological sand formation, these dryland sites entered a pathway of contemporary desertification during the $20^{th}$ century (partly starting in late $19^{th}$ century) which is mainly featured by water degradation such as salinization, leading to vegetation degradation and sandification at the utmost. A key factor is the transition from small-scale irrigation farming into large-scale irrigation schemes, which expand even onto hitherto marginal or completely unsuitable sites for irrigation farming. Examples are widespread across all over the Central Asian desert region originating from northern China as well as the Turkmenia Plain and the Caspian and Aral Sea basin regions.

Advances in water technology, mainly large-scale hydro-technical installations, and strong directional policies which were motivated out of economic and demographic reasons led to the expansion and simultaneous intensification of irrigated farming land. This created additional demands on water resources. The intensification of farming involved increased artificial inputs such as fertilizer and improved seed varieties. It also meant changes in the composition of irrigated crops, in particular, a shift towards crops demanding more water than any previous crops. These have been cash crops such as cotton monocultures and high intensity grain and rice productions, occasionally aside with other water-demanding crops such as vegetables, fruits, and grapevines. In addition, pressures on water resources were amplified by the influx of booming industries (such as oil/gas, but also mining) and related infrastructures (such as power plants and factories) as well as by expanding settlements. The latter added to pressures coming from increasing water demands, generated here by the lifestyle of new worker and employee households and public services.

Thus, original sites of traditional oasis agriculture, where formerly productive land had mainly been used for small-scale food production, often became the primary sites of contemporary desertification in river and delta ecosystems. They got desertified due to the decay or destruction of traditional irrigation systems, due to soil salinization and the advancement of surrounding desert sands.

Different from the semi-arid steppe grazinglands, the capacity to restore land is low. There are several main reasons why the path of land change can easily lead to the irreversible reduction of land productivity. First, hydrotechnical installations had induced fundamental and mostly irreversible changes to natural hydrographic networks, altering the natural hydrological cycle and requiring even more and expensive technical solutions. Second, quaternary climate change towards arid conditions predisposes desert formation, leading to the development of mainly sandy soil properties, which seriously impede the replenishment of soil nutrients, thus putting limits to efficient large-scale water technologies. Finally, and since even driest ecosystems depend in their functioning upon ice and snowmelt waters from surrounding mountains, indicated climate change might worsen the predisposing conditions. Climate change, and reduced water provision from snow and ice-pack of mountains surrounding most of the basin sites, were reported to constitute thresholds of land use. The case of the Aral Sea region provides an indication of how ecological factors impeding ecosystem recovery operate in causal synergy with strong, directional socio-economic driving forces, inducing a feedback loop that shifts the system towards irreversible degradation.

## An African Pathway of Desertification

Different from Central Asia, various desertification cases across several dryland areas in Africa report a common pathway which includes the sedentarization of nomadic herders leading to overgrazing around infrastructure nuclei and related, increasingly intensified other uses (such as cropping and fuelwood collection), which, especially in combination with droughts, triggers vegetation degradation and loss of soil productivity.

The various African dryland regions share a common major syndrome which is the rapid loss of adaptive capacity and/or increased vulnerability. It relates to rising internal conflicts about land and risks associated with natural hazards such as droughts, leading to crop failure, loss of resources, or loss of productive capacity, especially during the second half of the 20$^{th}$ century. Lesser annual rainfall and rising population densities since about the 1950s clearly set the stage for this syndrome to gain wide importance. A typical sequence associated with this syndrome is the contemporary transformation (or sedentarization) of nomadic pastoralism which leads to extensive, village-based and open communal grazing systems, often grouped around infrastructure nuclei and made possible by advances in borehole technology and veterinary services. Increases in herd size and high grazing densities followed, as well as changes in livestock composition, i.e., introduction and addition of new to traditional species. Permanent grazing around new homes occurred aside with fuelwood collection and cropping, with the intensity patterns of these land uses radiating from villages or water points on permanent rangelands, which became the major sites of localized desertification in combination with the highly fluctuating resource bases of grass, water, and rainfall. The dynamics of vegetation, especially grasses and herbs, is reportedly considerable, but rainfall is not only variable in time and space, but has a considerable stochastic component, too. Thus, in most of the cases only the various impacts of droughts in combination with other factors – either as sole forces of land degradation, operating in synergy with land use, or concomitantly with other factors – led to various outcomes of dryland degradation. African dryland cases differ from other cases worldwide in that increased aridity is invariably linked with causative mechanisms or syndromes associated with desertification. Also, the sequence leading to desertification has a self-perpetuating mechanism which consists of amplifying feedbacks between land use change, albedo change, drought impact, and land degradation.

However, indicated outcomes of desertification are manifold, and not considered to be serious. For example, and quite unexpectedly, the highly weathered soils and widespread sandy deposits did not imply a particular predisposement for 'sandification', which is the most spectacular outcome of desertification (it is actually low in comparison with other dryland regions of the world). The reasons why desertification does not impact across wide areas of the region probably stem from both socio-economic and biophysical initial conditions. Rainfall is stochastic, but vegetation is highly dynamic and resilient because its development was driven over centuries or millennia by humans and fire in the Sudan-Sahel of West Africa, and by indigenous herbivores and fire in the East African grassland zone. Thus, the interplay of increased aridity with human activities, triggering, for various underlying reasons, overgrazing as well the extension of croplands (millet, sorghum, maize, wheat) at the expense of residual rangelands, occurred against a background which is rather immune against degradation, or largely dampens desertification. Also, national policies and development pressure were found to drive land change, but the directionality of change behind land use activities – either towards intensified, commercial livestock rearing or expansion of food and cash crop areas, or both – was found to be less striking than, for example, in the Asian cases. Rather, civil wars or political instability was often reported from African drylands. Also, there has been a tradition of growing traditional crops or herding animals under a system of seasonally differentiated, nested use rights which existed for agriculturalists and pastoralists alike (in West Africa, at least). This left investments of labour in land, manure, fallowing time, and fertilizers intrinsically short term but also highly flexible in situations when degradation occurs. The Sudan-Sahel of West Africa was the only region mentioned in the cases where integrated crop and livestock management could reach high levels of intensification, which were normally attributed to (semi)commercial ranching operations only.

Therefore, and typically, thresholds of land change or mitigating feedbacks upon land change were far less, or not at all reported (while the resilience of initial biophysical conditions was pointed out in almost all of the cases studied). However, amplifying feedbacks were plentiful and strong, especially a downward spiral which encapsulated land use change, albedo change, land degradation, drought, again land degradation, etc. Therefore, it comes not as a surprise that plentiful switch and choke points were mentioned which were all biophysical in character. These had namely been the impact of droughts and water stress, i.e., the incapacity of ecosystems to let soils regenerate naturally or vegetation develop back to dense growth. At exactly these control points of land change, a syndrome associated with the rapid loss of adaptive capacity and/or increased vulnerability came into play. It appears as if serious dryland degradation or desertification in Africa is not a phenomenon occurring across vast areas affected, but rather remains associated, in particular, with impoverished individuals, groups, or larger fragments of the society (actually, mediating, shaping or intervening factors of land change) which have low adaptive or coping capacity to deal with environmental change or fluctuating resource bases.

## Beefing up the New World's Drylands

Desertification in Australia, in the US Southwest, in Patagonia and in northern Mexico can be subsumed under the notion of 'beefing up' the drylands, taking into consideration though that pathways had been historically convergent, but are contemporarily divergent. Despite of their diverse ecological setting, cases of land degradation in the drylands of Australia, Patagonia, northern Mexico and the US Southwest share strikingly common features, until the mid-20$^{th}$ century at the latest. The commonalities are the rapid introduction, especially during the 19$^{th}$ century, of commercial pastoralism through European settlers into ecosystems that had not known such uses before. These uses were based on exotic animal species, they were often export-oriented from the very beginning, and they were unmatched by other land uses such as cropping or fuelwood harvesting, which were low or non-existent in these regions. There are distinct features, though, which allow to separate a pathway typically shared by Australia and the US Southwest from a pathway typically shared by Patagonia and northern Mexico. The most important one is the cost-price squeeze affecting agriculture since about the 1950s. It appears as if the Australian livestock industry and its related infrastructure establishments (roads, watering technology) developed a flexibility high enough to avoid localized and spectacular desertification outcomes (such as 'sandification'). Similarly, land uses in the US Southwest shifted away from cattle ranching as the principal land use, mainly due to socio-economic change and an emphasis on competing, protective or recreational uses. Consequently, desertification in both regions is more a historical phenomenon than is the case in Patagonia and northern Mexico. In the latter regions, there is often no alternative or diversification option for local farmers to continue with livestock raising. Advanced technologies to react with more flexibility to the vagaries of oscillating resources – such as in the Australian cases – are rare or non-existent, and – different from the North American cases – no socio-economic change towards a more urban population or more recreational activities occur. Rather, growing rural and partly impoverished populations add to the continuous pressure upon natural resources (an exception might be the Taumalipan matorral). Consequently, desertification is not a historical phenomenon, but is advancing contemporarily.

*Historical Desertification in Australia and the US Southwest*

The pathways of historical desertification in Australia and the US Southwest can be briefly summarized as follows. Favourable rainfall triggered livestock increase, which, in the case of recurring droughts, led to overstocking, and consequently vegetation degradation and land productivity declines.

*Australia* The Australian drylands had long been subject to major variations in climate throughout the Quaternary, but were exposed to desertification processes only recently with the advent of European colonization from about 1850 onwards. There is evidence that desertification since European colonization happens in an episodic manner, i.e., in a series of step changes linked to shifts in rainfall over (wet as well as dry) periods lasting several decades. Unusually long wet periods lead to drastic increases of livestock numbers and the expansion of cropping land into previously drier areas. Climatic variability operates as a triggering mechanism. Once drought periods return, there are immediate consequences for land use and land cover such as declining vegetation cover and shrinking land productivity, causing land degradation. The disappearance of vegetation is speeded up by the impact of grazing and/or changes in fire regimes. Valid only for the more arid parts of the drylands (central Australia), the huge and episodic variations follow a pattern of three long periods over the last 120 years – i.e., below-average growth of herbage cover and dry periods in the late 1890s, the late 1920s, and the early 1960s – and two exceptional pulses of vegetation growth and wet conditions – i.e., one during 1920-21 and the other during 1973-75.

Maximum land degradation had very likely occurred when grazing had been introduced during dry periods first time in ecosystems which did not know these uses before. In Australia, these had been the late 1890s, and probably the late 1920s. Degradation had shifted since then from very intense soil degradation around natural waters (early 20$^{th}$ century) to less intense, intermittent degradation across a wider area in the second half of the 20$^{th}$ century.

Australian dryland cases are featured by a bundle of various syndromes, but no quantification of the process at work can be provided. Neither amplifying or mitigating feedbacks, nor thresholds or control points are reportedly inherent to the alterations of ecosystem endowment, leading to land degradation through overgrazing in dry decade-long periods. It appears as if potentially resilient vegetation and contemporary road improvements mitigate the impacts of overstocking (roads and associated transport technology allow for large herd mobility). Features of impoverishment were mentioned that mediate the process of desertification, or mitigate it in combination with the cost-price squeeze affecting ranching, namely many indebted, economically not viable grazing enterprises with low capacity to cope with drought.

*US Southwest* Despite of some geological pre-disposement for desert formation, and comparable to the Australian cases, desertification in the US Southwest was mainly historical rangeland degradation due to overstocking during the late cattle boom as part of western frontier colonization (in the 1880s) until the early/mid $20^{th}$ century. Current desertification, for example in the form of coppice dune formation, appears to be low, not only because of the contemporarily light grazing pressure, but also because vegetation dynamics seems strongly driven by meteorological conditions others than droughts, and the natural germination of native, perennial vegetation set clear thresholds beyond which land use could hardly be moved without impairing the recovery potential of rangelands.

The effects of variable rainfall upon the livestock industry were similar in both the US Southwest and Australia, i.e., favourable rainfall triggered livestock increase, while droughts caused overstocking and, consequently, vegetation degradation and land productivity decline associated with desertification. However, drought sequences usually comprised several years only (no decade-long sequences such as in Australia), so that mitigating feedbacks such as livestock reduction in periods of droughts had immediate and limited, but not drastic consequences such as in Australia.

There is no unidirectional vegetation change such as grassland ecosystems being invaded by shrubs ('weedy weed', mesquite) due to newly introduced livestock uses. At least in southern Texas, woody species of the shrubby thornscrub ecosystem shows a degradation path just in the opposite direction of grasslands, i.e., from woody species to mainly annual grasses and herbs, leading to a patched plant community, simpler, and more prone to disturbances.

Different from the Australian cases, the syndrome of dryland change as it appeared from most of the US Southwest cases – i.e., resource scarcity due to loss of land productivity and failure to restore or maintain protective work – is rather simple. Therefore, it should not come as a surprise that citizen involvement from the 1950s onwards, together with the decay of large ranching interests, successfully evoked land use changes that finally contributed to a dampening of desertification.

## *Historical and Current Desertification in Patagonia and Mexico*

Different from the historical cases of 'beefing up' the drylands in Australia and the US Southwest, desertification in Patagonia and northern Mexico is both historical and advancing contemporarily. The path of land change in these two American drylands shows a commonality which can be briefly summarized as follows. The conversion of shrublands and modification of rangelands made possible intense, continuous grazing, which, in combination with increasing market opportunities and in periods of favourable rainfall, led to increasing livestock numbers, that finally and directly contributed to overgrazing and dryland degradation.

Thus, the pathway of desertification in the grassland and, partly, shrubland ecosystems of Patagonia and northern Mexico is a straightforward one, which consists of the following, few building blocks. First, be it the dry-cold climate of Patagonia or the dry-hot conditions in northern Mexico, large parts of thorny shrublands, which were the dominant life form for centuries in certain parts of the wider region, were cleared in the early 1990s to seed artificial pastures (but also for cropland and settlements). Second, and to a much larger extent, formerly extensive and open grasslands were transformed into rangelands especially from the last decades of the 19$^{th}$ century onwards. In an early phase, this occurred through the removal of woody individuals, triggering shrub encroachment and consequent clearing activities. Later on, rangelands were modified through fire, chemicals, and mechanical suppression. Rangeland modification was especially intense during and after the period of Green Revolution in the 1960s, when sown pastures (together with mechanized farming) started to expand dramatically. In both northern Mexico and Patagonia, new European land uses, livestock mainly, were introduced as early as in the 16$^{th}$ century. However, only from the late 19$^{th}$ century onwards initial frontier colonization for extensive private subsistence sheep farming in Patagonia (1880-1920) pushed into lands with last, richest vegetation and adequate water resources. Similarly, extensive areas of natural grasslands were increasingly used to freely graze numerous herds for private cattle ranching in northern Mexico during the same period.

As in the Australian and US Southwest cases, rainfall above long-term average leads to livestock increases, while drought spells reduce the number of cattle or ruminants, i.e., due to reduction of ecosystem productivity, namely soil erosion, degradation of plant communities, and henceforth reduction of rangeland carrying capacity. However, the overriding impact on livestock expansion, despite of clear signals of degradation, stems from economic and political incentives for livestock expansion such as tax releases, marketing assistance, and booming meat and wool markets. Market growth for livestock products went hand in hand with technological improvements in the transport and processing sector. Even when economic conditions worsened, overstocking was a rather rationale response of especially small, impoverished ranchers. This is in particular true for Patagonia where hardly any options for the diversification of sheep rearing exist, while the economic conditions in northern Mexico are less gloomy due to expanding (semi)urban settlements (and related industries, and their meat demand) and due to disintensification of rangeland uses (e.g., sports hunting). Therefore, it should not come as a surprise that the overriding syndrome of dryland change in these regions are changing market opportunities (through commercialization and capital investments) in combination with heavy surplus extraction away from the land managers, and a related loss of adaptive capacity due to impoverishment.

The frequency, intensity, extent and magnitude of grazing means a fundamental selection pressure in the ecosystems, which is highest when year-round livestock rearing is carried out such as in Patagonia. Pressure on vegetation had resulted in a reduction of densities and productivity of grasses and herbaceous plants, concomitant shrub invasion in grasslands and landscape patterning in thornshrub lands. These changes were considered to be undesirable because they had reduced carrying capacities for livestock, contributed to soil erosion and reductions in stream flows, altered wildlife habitat, and threatened pastoralists or ecosystem sustainability. Mainly due to overgrazing, desertification was a historical process in Patagonia (when sheep had been introduced on a large scale and livestock numbers reached peak values in 1885-1914), but it is a currently advancing process, too, as well as in northern Mexico (no data on historical desertification were provided in the cases). In Patagonia, advanced formation of desert-like conditions is reported from some localities where year-round grazing is carried out and overlaps exist with geological pre-disposement for 'badland' creation.

## Cropping (> Pasture > Fire) > Degradation – A Mediterranean Pathway

Despite 7,000 years of agro-pastoral land uses and widespread soil erosion in countries of southern Europe, which are fully or partly situated in dryland areas of the Mediterranean Basin, these areas still support a valuable agricultural base and have low, if at all outcomes of desertification if compared to other dryland areas of the world. Two pathways of dryland degradation (rather than desertification) appear, all of them starting with the impact of growing annual and/or perennial crops, and not with grazing impact. Cases indicate that the first trajectory is of low importance, and that more empirical studies are needed to support the second trajectory as a pathway ending up in desertification.

First, once the soil is eroded below a critical soil depth, mainly through cropping on slopes, and once extreme hot and dry meteorological conditions won't allow the regeneration of vegetation cover, irreversibly degraded lands, or 'badlands', are created. As a rule, hilly soils formed on consolidated parent materials such as limestone, sandstone, volcanic lava, etc. usually have a restricted effective rooting depth and will produce soils that are shallow and easily eroded. Under hot and dry climatic conditions, rainfed vegetation may not be supported in such areas, leading to desertification. However, 'badland' sites are few, and opinions are divided on whether they stem from highly erosive lithologies alone or had been due to ancient or historical land use practices, or both. 'Badland' sites are not only few, but rates of soil erosion are also low if compared to other dryland regions. Even where current and expanding cropping areas, such as for wheat and almonds groves, locate most active gully development, rates of change are comparatively low. Consequently, decreases in semi-natural vegetation cover also occur at relatively low rates. Also, the impact of oscillating rainfall conditions is low, increases in degraded, impalatable vegetation are next to nil, no serious water degradation appears to exist, and contemporary increases in sand or rocky cover are non-existent, or not reported in the cases, at least. In summary, the cropping-'badland' trajectory exists for some very localized areas, but does not dominate throughout the Mediterranean basin.

Second, once the soil is eroded and shallow, skeletal soils are formed, increasingly poor yields from the various agricultural crops trigger a change in land use which is usually a shift from cropping to pasture. Mechanization of the remaining, usually most fertile plots under cultivation usually leads to accelerated soil degradation, but the cropping-pasture transition is more typical due to large amounts of indirect subsidies for grazing, mainly provided by the European Common Agricultural Policy (CAP). Today, most pasture land in the Mediterranean basin is abandoned cropping land, and only a few areas, which have been fenced or strictly controlled by private and public land owners, remain ungrazed (besides grazing, hunting and fuelwood collection are traditional land uses also carried out on 'abandoned' land). As a matter of fact, and apart from few forest stands, almost all 'natural' phrygana vegetation in the Mediterranean basin is grazed to some extent, either by migrating or permanent flocks and herds. However, only narrative evidence exists that rangelands might be irreversibly desertified. If so, severe degradation seems to be limited to remote mountain ranges (covered by matorral, garrigue, maquis), where amplifying feedbacks from road extension were reported to lead to increasing human population and grazing stock numbers. Otherwise, moderate grazing on abandoned farming land is reportedly associated with land improvement such as increased plant species richness and improved soil quality. The last section of the Mediterranean trajectory, i.e., the fire-degradation path, is not empirically supported in the cases, but was assumed only. It had been mentioned that not grazing, but intense and uncontrolled fires could lead to the complete removal of especially woody vegetation, and thus drive accelerated vegetation degradation and soil erosion. However, no quantification had been provided.

Surprisingly, European dryland cases are strongly featured by syndromes such as the loss of land productivity following excessive or inappropriate use, and a failure to restore or maintain protective works, but these slow causative mechanisms are not matched by rapid process rates of change leading to severe desertification. Also, gradual socio-economic changes such as increasingly elderly, low density rural populations, and the breakdown of large, extended farming families favour the abandonment of cropping and might jeopardize the stability of formerly productive land. Again, the pattern cannot be associated with high rates of dryland formation. No large-scale and potentially irreversible desertification over wide areas of the region had reportedly occurred until now. It is assumed that this has been due to a combination of the following factors. They stem from initial, predisposing environmental factors, from contemporary land use history, and from system properties inherent to the process, especially easy-to-implement technological improvements. They stem also from the fact that no mediating factors (such as unequal land distribution or prevalence of poverty) or no control points of land change were reported in the cases.

First, massive degradation appears to be blocked by the seemingly high resilience of semi-natural vegetation such as matorral, garrigue, or phrygana, and more so through the particular nature of the Mediterranean relief, topography and lithology. The latter is a nested system of very different, small-scale ecosystem units that obviously prevents or limits any large-scale extension of self-perpetuating degradation processes. Thus, the major causative mechanism behind contemporary land degradation, which are the mechanization of cultivation and indirect subsidies for grazing under highly variable rainfall conditions, do not cause high rates of land degradation over wide areas.

Second, it further appears that socio-economic factors such as the abandonment of cropped land during the $20^{th}$ century – rather than rapid expansion of cultivated area for demographic and economic reasons as in other dryland regions – counterbalances massive land degradation. Land abandonment in remote, rural areas was reportedly due to soil productivity declines, but depopulation and reduced economic vitality of rural areas reinforce abandonment so that no or low investments are made to restore and maintain productivity. It was mentioned though that the effects of land abandonment on land quality may be positive or negative, depending on the interplay between various lithologies or parent material (governing soil development) and climatic conditions, both interacting with plant development, and the type of post-abandonment land use. Soils under favourable climatic conditions that sustain plant cover, for example, may improve with time, by accumulating organic materials, increasing floral and faunal activity, improving soil structure, increasing infiltration capacity, and therefore decreasing erosion potential. Both runoff and sediment loss decrease exponentially as the percentage of vegetation cover increases (on Lesvos, for example, the abandonment of cropping land has greatly increased soil organic matter content and soil aggregate stability, especially on soils formed on lava and schist marble, as compared to soils under cultivation). Also, moderate grazing pressure on abandoned agricultural land may lead to partial rejuvenation of vegetal communities, with high diversity even. Overgrazing triggers a decline in, but not the complete removal of the vegetation, including the loss of particular herbaceous families which help to maintain the soil structure. Fire impact is greatest only in areas with lowest fire frequencies, but there are only few dense woody vegetation stands to be potentially degraded.

Third, related to the quality of germination conditions are thresholds such as crucial minimum soil depth and critical vegetation cover, necessary for the regeneration of (semi)natural vegetation and mainly governed by the quality of soil lithology or parent material. Throughout the Mediterranean Basin, parent material was characterized as a soil-forming factor that not only affects soil properties, but also plant growth, soil erosion, and ecosystem resilience. The nature of the parent material becomes increasingly important in vegetation establishment and land protection once soil depth has been reduced due to erosion. In the Mediterranean, there is no vast and uniform occurrence of lithologies detrimental to plant and soil growth such as in the lowland plains and basins of the Central Asian desert and steppe region, but nested systems of small-scale and various landscape units dominate. Soils formed on pyroclastics had a reportedly lower capacity to regenerate natural vegetation, thus leading to higher erosion, than soils formed on shale, ignimbrite, schist-marble or volcanic lava, for example, which had a higher capacity for at least partial regeneration of natural vegetation.

Finally, and despite amplifying feedbacks, such as road extension creating more human and livestock pressure especially in mountain areas or increasing spatial heterogeneity of the landscape as a response to hydrological stress, minor technological changes were reported to dampen easily degradation. An example is the construction of small earth dams which limit river bank gully erosion, an application which is easy to implement, but has reportedly high impact.

Table 6.1  Basic and Distinct Features of Dryland Regions – Ecological Factors shaping the Environmental History (Climate, Topography and Vegetation)

| | Asia* | Africa* | Europe* | Australia* | N-America* | L-America* |
|---|---|---|---|---|---|---|
| Climate (including annual rainfall in mm) | Hyperarid > sub-humid: 30->500; monsoon regime; hot and cold; sparse, variable (in arid zones), seasonal rainfall (in less arid zones); more extreme events (storms). | Hyperarid > sub-humid: 100->500; mainly Inter-Tropical Convergence Zone; high stochastic component in temporal and spatial variability of rainfall. | Semi-arid > sub-humid: 150->500; hot, dry summer, temperate winter, seasonal rainfall with (more) intense events; high spatial and temporal rainfall variability. | Arid > semi-arid: 150-450; tropical cyclones and ENSO; sporadic, variable rainfall (in arid zones), seasonal rainfall (in less arid zones); rainfall shifts over decades. | Arid > semi-arid: 100-255; variable and seasonal rainfall; wet-cool, moist-warm condition; (more) droughts over several years; more dust/sand storms. | Arid > sub-humid: 120->500; hot (Mexico) and cold (Patagonia); sporadic rainfall (in arid zones), seasonal rainfall (in less arid zones). |
| Topography (including range of altitude in m) | Lowland plains, depressions and basins, upland steppe plateaus; high surrounding mountains; 0-6,000. | Floodplains, flat plateaus; low relief energy, broad undulations, except Morocco (Atlas); 0-3,400. | Nested system of small units (basins, slopes, lowlands, valley floors, hills, mountains); 120-2,000. | Alluvial, sandy plains with sand dunes (central Australia). | Plains, upland plateaus (mesas), mountains, some (ancient) inland basins, deltas; 540-1,550. | Lowland plains, uplands with intersected drainage basins, plateaus, hills, mountains; <150-2,000. |
| Vegetation | Dry grasslands, (sub)deserts; steppe (high forage capacity) versus cold mountain steppe (low capacity). | (Sub)desert, grass-/shrub- and woodlands; highly resilient, except trees; savannas altered by land use. | Semi-natural phrygana mosaics; relicts only of natural tree and woodland stands. | Shrubs and dry grasslands; associations of grasslands, tree-shrub mosaics and Acacia woodlands. | Dry grasslands and (thorny) shrubs; oscillating shifts between grass, bush, and woody weed. | Desert to steppe formations; grasslands; shrublands; tree-shrub mosaics; woodland and forest patches. |

* For a description of landscapes per major geographical area, see Table 6.2.

**Table 6.2** Basic and Distinct Features of Dryland Regions – Ecological Factors shaping the Environmental History (Soil, Water and Changes in Species Habitat)

| | Asia[a] | Africa[b] | Europe[c] | Australia[d] | N-America[e] | L-America[f] |
|---|---|---|---|---|---|---|
| Soil ecology | Sandy and loess/ial soils; permafrost soils with peat (Tibet). | Alluvial floodplain soils, highly weathered sandy deposits; widespread fossil dunes. | Shallow, skeletal soils; various associations; parent material governs erosivity. | Associations of alluvial, sandy, calcareous plain soils, sand dunes and stone-mantled soils. | Sandy deposits, clay soils; fine-grained, shallow, high erosivity. | Alluvial deposits, loams (basins), clay soils (lowlands); mixed upland/mountain soils. |
| Water ecology | Rich groundwater; constant surface water flow interrupted last century. | Fluctuating water resources; surface water overrides groundwater in relevance. | Ephemeral, torrential surface water flow; high sediment load. | Increased flooding. | Reduced stream flows; stream channel cutting; drop in water tables. | Surface water flow through canyons (barrancas), reduced stream flows. |
| Changes in species habitat | Wild animal habitats and swamps reduced; more halophytes; increase in rodents and insects (Tibet). | Wildlife habitats reduced, but no extinction of species; trees and woodland stands fragmented. | More plant species on abandoned cropland, but decline on grazed land; low regeneration of trees. | From about 1850 onwards, displacement of native herbivores by European pastoralists; | Decimation of wild grazing animals & prairie dogs; elimination of wolves & grizzly bears. | Mule deers reduced; more small mammals and breeding birds; no relics of Pre-European species (south). |

[a] Central Asian desert and steppe region (plateaus, basins, plains), East Mediterranean steppe, Arabian Peninsula, Thar Desert.
[b] Sudan-Sahel, North African mountains, East African grasslands (Turkana & Masai Steppe); Kalahari Sandveld.
[c] Mediterranean Basin of southern Europe (Portugal, Spain, Italy, Greece).
[d] New South Wales, Queensland, South Australia, Western Australia, and Northern Territory.
[e] US-Southwest (Great Basin): Chihuahuan, Sonoran and Mojave Desert, Colorado Plateau.
[f] Northern Mexico (Chihuahua, Sonora, Durango, Nuevo León, Coahuila), south-central Mexico, and Patagonia (S-Argentina).

Table 6.3  Basic and Distinct Features of Dryland Regions – Changing Institutional and Socio-Economic Contexts (Land Use History)

| Asia* | Africa* | Europe* | Australia* | N-America* | L-America* |
|---|---|---|---|---|---|
| (Semi) nomadic grazing on plateaus, ancient oasis farming in basins; large-scale alternations of grain farming and livestock (steppe); gradual transformation of oasis agriculture into large-scale irrigation; rapid expansion of water technology; contemporary urbanization & industrialization; outside Central Asia: gradual intensification of mixed crop-animal farming. | Sedentarization of nomads; shifting and rotational fallow farming change to permanent modes; low mechanization; low relevance of irrigation farming; partial land conservation; intensive collection of fuelwood; integral role of transhumance in sustaining agricultural livelihoods; shifts to integrated mixed animal-crop husbandry and intensive ranching; frequent land zoning and redistribution. | Millennia-old tradition of agro-pastoralism, i.e., intense (slope) cultivation and excessive, extensive grazing by sheep and goats, including use of fires; long tradition of growing annuals (wheat) and perennials (almonds, olives); abandonments of farmland, especially in remote areas, during second half of the 20th century. | From about 1850 onwards, displacement of native hunter-gatherers (Aboriginals) by European pastoralists; concentration of newly introduced cattle and sheep around watering points; commercial pastoralism for export markets. | Extensive grazing; cattle boom in late 19th century due to colonization of the frontier; decline of livestock industry until present; shift to urban-driven land uses such as recreation, conservation and housing development; shift of dryland farming (in 1800s) to irrigation farming in 20th century (central-pivot agriculture; few suitable farming land only. | Predominantly intensive and extensive livestock production; some urban and cropland extensions since early 1900s (in northern Mexico); initial colonization by extensive sheep farming (1880-1920), with shifts to intense grazing, woodcutting and mineral extraction (in Patagonia). |

\* For a description of landscapes per major geographical area, see Table 6.4.

**Table 6.4** Basic and Distinct Features of Dryland Regions – Changing Institutional and Socio-Economic Contexts (Current Land Tenure)

| | Asia[a] | Africa[b] | Europe[c] | Australia[d] | N-America[e] | L-America[f] |
|---|---|---|---|---|---|---|
| Current land tenure | Government (development) land aside with collectively and some privately held land (Central Asia); communal and private land-holdings (in the Thar and Saudi Deserts as well as in the East Mediterranean steppe zone). | Seasonally differentiated, nested use rights by farmers and herders (West Africa); private small holdings (East Africa); village-based, open communal (small) holdings aside with individually owned (large) holdings (in southern Africa). | Individual, privately held properties. | Individual properties, mainly restricted purposes leaseholds. | Privately held land (ranches, housing, mines) aside with public land (conservation zones, military areas). | Semi-privatized communal holdings (ejido) aside with private ranches (in northern Mexico); private ranches (in Patagonia). |

[a] Central Asian desert and steppe region (plateaus, basins, plains), East Mediterranean steppe, Arabian Peninsula, Thar Desert.
[b] Sudan-Sahel, North African mountains, East African grasslands (Turkana & Masai Steppe); Kalahari Sandveld.
[c] Mediterranean Basin of southern Europe (Portugal, Spain, Italy, Greece).
[d] New South Wales, Queensland, South Australia, Western Australia, and Northern Territory.
[e] US-Southwest (Great Basin): Chihuahuan, Sonoran and Mojave Desert, Colorado Plateau.
[f] Northern Mexico (Chihuahua, Sonora, Durango, Nuevo León, Coahuila), south-central Mexico, and Patagonia (S-Argentina).

**Table 6.5  Basic and Distinct Features of Dryland Regions – Pre-historic, Ancient and Historical Dryland Formation and Degradation**

| | Asia* | Africa* | Europe* | Australia* | N-America* | L-America* |
|---|---|---|---|---|---|---|
| Pre-historic | Predisposition of desert formation through both arid conditions of dry interval and dry-cold plateau climates. | Shift from unusually wet conditions (9,500-4,500 BP) to more arid conditions (4,500-3,000). | None | None | Wind erosion as geological process leads to frequent desertification events. | Soil erosion due to climatic transition (south central Mexico); localized geological pre-disposement (Patagonia). |
| Ancient | Oscillating desert margins due to coupled climatic variations (11/13th, 17th century) and destructive land uses under several dynasties (in NW-China). | Cycles of degradation and restoration of fossil dunes due to century-old cultivation under fluctuating rainfall. | (Assumed) overuse of vegetation due to deforestation and forest/woodland degradation; formation of skeletal soils and gullies due to intense (hillside) cultivation. | | | Soil erosion through ancient and historical land use (south-central Mexico). |
| Historical | Drier and warmer (more droughts) since 19th century (in NW-China). | | | Episodic, with maximum impact due to grazing in dry periods (late 19th century). | Degradation of rangeland due to overstocking by drastic livestock increases (in 1880s onwards). | |

\* For a description of landscapes per major geographical area, see Table 6.6.

Table 6.6   Basic and Distinct Features of Dryland Regions – Contemporary Dryland Formation and Degradation

| | Asia[a] | Africa[b] | Europe[c] | Australia[d] | N-America[e] | L-America[f] |
|---|---|---|---|---|---|---|
| Contemporary | Sandification of basin and plateau sites (settlements, cropland, grazing land); frequent sand and salt storms; excessive water degradation at basin and low plain sites in ca. 1950 onwards (not uplands, except Tibet); shrinkage of water volume, salinization, chemical pollution; nutrient decline in steppe. | Droughts and marked precipitation decline since 1950 onwards; woodland and soil degradation; some reactivation, followed by stabilization of fossil dunes; soil and vegetation degradation around watering points; very localized water degradation (salinization). | Decreased rainfall during 2nd half of 20th century, with more seasonal extremes triggering more soil erosion; high spatial variability of localized soil productivity decline; no large-scale events of dryland degradation; low desertification outcomes. | Episodic desertification; shift from very intense soil degradation localized around natural waters (early 20th century) to less intense, intermittent degradation across a wider area in 2nd half of 20th century. | Formation of coppice dunes (but only light grazing pressure); drastic reduction of prairie wildlife and biodiversity; localized, mainly urban growth-driven degradation of land and water resources; more sustained winds triggering sand and dust storms. | Incipient and advancing rangeland degradation since the 1980s (in northern Mexico); soil erosion through agricultural practices; severe grazing pressure due to rangeland degradation from ca. 1920 onwards (in Patagonia). |

[a]   Central Asian desert and steppe region (plateaus, basins, plains), East Mediterranean steppe, Arabian Peninsula, Thar Desert.
[b]   Sudan-Sahel, North African mountains, East African grasslands (Turkana & Masai Steppe); Kalahari Sandveld.
[c]   Mediterranean Basin of southern Europe (Portugal, Spain, Italy, Greece).
[d]   New South Wales, Queensland, South Australia, Western Australia, and Northern Territory.
[e]   US-Southwest (Great Basin): Chihuahuan, Sonoran and Mojave Desert, Colorado Plateau.
[f]   Northern Mexico (Chihuahua, Sonora, Durango, Nuevo León, Coahuila), south-central Mexico, and Patagonia (S-Argentina).

Table 6.7  Basic and Distinct Features of Dryland Regions – Proximate Causes of Desertification

| | Asia* | Africa* | Europe* | Australia* | N-America* | L-America* |
|---|---|---|---|---|---|---|
| Proximate causes | Expansion of cropland (rainfed, irrigation); nomadic and extensive grazing on reduced rangeland; wood and related extractional activities (medicinal herbs); infrastructure extension (hydro-technology, settlements, industries); increased aridity. | Extensive and some nomadic grazing on reduced rangeland; expansion of rainfed cropland (some overuse of irrigation land in West Africa); harvesting of fuelwood; increased aridity. | Extensive grazing, especially on mountains; annual/perennial cropping on hill slopes; some wood extractional activities; increased aridity. | Sheep and cattle grazing; watering technologies, industry extension and road extension ('industrial-road-water complex'); episodic, decade-long droughts. | Extensive (historical) grazing; irrigation farming and water extraction for modern urban uses; increased aridity. | Extensive and intensive cattle and sheep grazing (northern Mexico, Patagonia), and subsistence maize farming (in south-central Mexico); wood and related extractional activities; droughts. |

* For a description of landscapes per major geographical area, see Table 6.8.

**Table 6.8** Basic and Distinct Features of Dryland Regions – Underlying Driving Forces of Desertification

| | Asia[a] | Africa[b] | Europe[c] | Australia[d] | N-America[e] | L-America[f] |
|---|---|---|---|---|---|---|
| Underlying driving forces | Strong, directional, national policy-driven production growth and intensification; newly introduced livestock species; large-scale water technologies of low efficiency; (forced) immigration of farmers and rising population densities; urbanization and industrialization; warmer, drier. | Agricultural growth policies; newly introduced technologies for mechanization (in East Africa); export orientation of beef market (southern Africa); rising population densities; malfunct traditional land tenure; wars; lesser annual rainfall due to increased climatic variability. | Mechanization of arable agriculture; indirect subsidies for grazing; reduced economic vitality & low training of farming population; increased climatic variability; | Export market orientation of livestock sector, newly introduced watering technologies; individual responses to unusually wet climate periods and market signals; indebtedness; poor management of livestock industry; direct impact of droughts on surface cover. | Newly introduced animal species and watering technologies (for irrigation farming); frontier colonization based on cattle industry; malfunct land laws; direct impact of droughts on surface cover. | Introduction of grazing animal species; improved transport and processing technologies; export market orientation; human population growth and poverty (south-central Mexico); direct impact of climate on surface cover. |

[a] Central Asian desert and steppe region (plateaus, basins, plains), East Mediterranean steppe, Arabian Peninsula, Thar Desert.
[b] Sudan-Sahel, North African mountains, East African grasslands (Turkana & Masai Steppe); Kalahari Sandveld.
[c] Mediterranean Basin of southern Europe (Portugal, Spain, Italy, Greece).
[d] New South Wales, Queensland, South Australia, Western Australia, and Northern Territory.
[e] US-Southwest (Great Basin): Chihuahuan, Sonoran and Mojave Desert, Colorado Plateau.
[f] Northern Mexico (Chihuahua, Sonora, Durango, Nuevo León, Coahuila), south-central Mexico, and Patagonia (S-Argentina).

Table 6.9  Basic and Distinct Features of Dryland Regions – Major Syndromes and Mediating Factors

| | Asia[a] | Africa[b] | Europe[c] | Australia[d] | N-America[e] | L-America[f] |
|---|---|---|---|---|---|---|
| Syndromes | Changing opportunities created by markets; outside policy interventions: economic development programmes, capital investments, new technologies for intensification; decrease in land availability. | Loss of adaptive capacity, increased vulnerability; internal conflicts about land and risks associated with natural hazards (such as, droughts) leading to crop failure, loss of resources, or loss of productive capacity. | Loss of land productivity following excessive or inappropriate use; failure to restore or maintain protective works of resources. | Resource scarcity and production pressure; changing market opportunities; outside policy interventions; loss of adaptive capacities and increased vulnerability; changes in institutions governing resource access. | Resource scarcity due to loss of land productivity and failure to restore or maintain protective works. | Changing market opportunities (commercialization, capital investment); heavy surplus extraction away from land manager; loss of adaptive capacity due to impoverishment; population growth-driven shrinking land availability. |
| Mediating factors | | Impoverished individuals or groups are most vulnerable to desertification. | | Indebted, economically not viable grazing enterprises with low capacity to cope with drought. | Historical 'grazing out' smallholder squatters by overgrazing land claimed by large livestock and mining interests. | Unequal land distribution and impoverishment of small ranchers and farmers. |

a   Central Asian desert and steppe region (plateaus, basins, plains), East Mediterranean steppe, Arabian Peninsula, Thar Desert.
b   Sudan-Sahel, North African mountains, East African grasslands (Turkana & Masai Steppe); Kalahari Sandveld.
c   Mediterranean Basin of southern Europe (Portugal, Spain, Italy, Greece).
d   New South Wales, Queensland, South Australia, Western Australia, and Northern Territory.
e   US-Southwest (Great Basin): Chihuahuan, Sonoran and Mojave Desert, Colorado Plateau.
f   Northern Mexico (Chihuahua, Sonora, Durango, Nuevo León, Coahuila), south-central Mexico, and Patagonia (S-Argentina).

**Table 6.10** Basic and Distinct Features of Dryland Regions – System Properties of Desertification

| | Asia[a] | Africa[b] | Europe[c] | Australia[d] | N-America[e] | L-America[f] |
|---|---|---|---|---|---|---|
| Amplifying feedbacks | Self-perpetuating, current process of expanding cropland (at the expense of rangeland). | Land use-albedo-drought-degradation (downward) spiral. | Increasing spatial heterogeneity of landscapes; policy-driven access roads. | | Exotic plants increase fuel load and fire frequency. | Favourable rainfall speeds up livestock expansion. |
| Mitigating feedbacks | | | Small technological measures such as earth dams. | | Citizen involvement in evoking land use change; severe droughts. | Severe droughts dampen livestock expansion. |
| Thresholds | Climate and snow/icepack-dependent water provision. | | Crucial minimum soil depth and vegetation cover. | | Natural germination of native perennial vegetation. | Recovery of adequate water-storage capacity of soils. |
| Control points | Short vegetation period and low temperature (plateaus). | Incapacity to let soils regenerate naturally or vegetation develop to dense growth. | | | | |

[a] Central Asian desert and steppe region (plateaus, basins, plains), East Mediterranean steppe, Arabian Peninsula, Thar Desert.
[b] Sudan-Sahel, North African mountains, East African grasslands (Turkana & Masai Steppe); Kalahari Sandveld.
[c] Mediterranean Basin of southern Europe (Portugal, Spain, Italy, Greece).
[d] New South Wales, Queensland, South Australia, Western Australia, and Northern Territory.
[e] US-Southwest (Great Basin): Chihuahuan, Sonoran and Mojave Desert, Colorado Plateau.
[f] Northern Mexico (Chihuahua, Sonora, Durango, Nuevo León, Coahuila), south-central Mexico, and Patagonia (S-Argentina).

Chapter 7

# Indicators

**Introduction**

Indicators used in the case studies were grouped into three broad clusters, showing the ecological, meteorological, and socio-economic dimensions of desertification. Each broad cluster was further subdivided into more specific categories. Only direct indicators were considered here, and consequently coded. Indicators may overlap, but since most of them were given different quantitative measures, they were coded individually. They were clustered into four degrees of dryland degradation as a first step for the further development of desertification indicators for monitoring. Since environmental indicators are a phenomenon or statistic so strictly associated with a particular environmental condition that its presence can be taken as indicative of that condition, typically indicated degradation pattern per major geographic region were identified finally.

**Broad Clusters**

Most of the desertification cases are described by ecological indicators (n = 121, or 92%), be they, for example, vegetation change, water degradation or wind/water erosion. Considerably less of the cases use socio-economic (n = 74, or 56%) and meteorological indicators (n = 51, or 39%). As a characteristic feature it appears that only up to one third of the cases is identified by indicators related to a single type of variables, i.e., exclusively ecological indicators (n=41, or 31%), or meteorological indicators (n=4, or 3%). However, no socio-economic indicators alone were found to be used. This implies that combinations of indicators are the most common form to describe desertification. They are applied in two third of the cases, namely, the combination of ecological and socio-economic indicators (n=40, or 30%), the combination of ecological and meteorological indicators (n=14, or 11%), the combination of meteorological and socio-economic indicators (n=7, or 5%), and the combination of all three clusters (n=26, or 20%). This suggests that a reasonably good multidisciplinary approach to measuring desertification is often applied, but most case studies fail to adopt a truly integrative approach, especially with regard to an integration of biophysical and human indicators. The frequency of specific indicators as found in 25% (and more) of the cases is given in Table 7.1.

**Table 7.1**  **Frequency of Desertification Indicators[a]**

|  | Abs | Rel |
|---|---|---|
| Ecological indicators | 121 | 92% |
|  |  |  |
| Vegetation[b] | 105 | 80% |
|     Reduced vegetation cover | 79 | 60% |
|     ... scrub, grass, herbaceous cover | 50 | 38% |
|     ... forest, woodland | 40 | 30% |
|     ... both scrub etc. and forest etc. | 33 | 25% |
|     Encroachment of less palatable vegetation (grass, bush, 'woody weed') | 44 | 33% |
|  |  |  |
| Soils[c] | 98 | 74% |
|     Erosion | 66 | 50% |
|     ... wind erosion (sheet) | 9 | 7% |
|     ... water erosion (gully) | 7 | 5% |
|     ... wind and water erosion | 22 | 17% |
|     ... unspecified | 28 | 21% |
|     Physical degradation | 33 | 25% |
|     ... compaction, soil crusting (pan formation, surface soil sealing) | 22 | 17% |
|     ... increase in loosened soil, soil crust destruction | 8 | 6% |
|     ... reduced ability to absorb/store moisture, increase in waterlogging | 19 | 14% |
|  |  |  |
| Water[d] | 38 | 29% |
|  |  |  |
| Wildlife[e] | 9 | 7% |
|  |  |  |
| Socio-economic indicators | 24 | 56% |
|  |  |  |
|     Changes in livestock parameters[f] | 37 | 28% |
|     (Negative) changes in crop harvest and/or fish catch[g] | 36 | 27% |
|     Demographic, human welfare and conflict indicators[h] | 26 | 20% |
|  |  |  |
| Meteorological indicators[i] | 51 | 39% |
|  |  |  |
|     Increased aridity: increasingly drier, warmer, more droughts, less rain, and/or shift of rainfall to winter season | 42 | 32% |

<sup>a</sup> Multiple counts possible; percentages relate to the total of N = 132 cases; frequency only of indicators in 25% or more cases.

<sup>b</sup> Other vegetation-related indicators (<25% of all cases) include losses in natural plant/biomass production (n=30 cases, or 23%), change (i.e., reduction, elimination) of plant species richness/diversity (n=28 cases, or 21%), and reduced plant density (n=17 cases, or 13%). Not considered here are unspecified vegetation changes.

<sup>c</sup> Other soil-related indicators (<25% of all cases) include chemical degradation, mainly soil nutrient losses, (n=21, or 16%).

<sup>d</sup> Hydrological indicators are, for example, salinization (n=25, or 19%), or less (potable) water in rivers, boreholes, etc., mainly in combination with a drop in water tables, decreased discharge of rivers and/or shrinkage, shortening or disappearance of rivers, lakes, shorelines, etc. (n=21, or 16%).

<sup>e</sup> Wildlife indicators are mainly reduction or disappearance of species (n=6, or 5%).

<sup>c</sup> For example, reduced, impaired or surpassed livestock carrying capacity (n=19, or 14%), declining quality/productivity of palatable plant biomass (n=14, or 11%), increased cattle mortality, and drastic reduction of livestock numbers (n=7, or 5% each).

<sup>g</sup> For example, (i) lost crop/catch, falling or lower yields/productivity, and constantly low yields/productivity despite of major, continued improvements (n=23, or 17%), or (ii) abandonment and/or shift of cultivated land (n=16, or 12%).

<sup>h</sup> For example, shrinking or reduced human carrying capacity (n=7, or 5%), forced outmigration from rural to urban areas (n=5, or 4%), high and/or increasing poverty, excess economic (production) losses in monetary terms for public, and/or private entities, e.g., due to sandstorms (n=7, or 5%), reduced life expectancy and/or excess mortality rates (e.g. due to famine) (n=4, or 3%), decreasing or constantly low economic vitality, and increasing social conflicts about land use and access (e.g. between farmers and herders) (n=4, or 3% each).

<sup>i</sup> Other meteorological indicators (<25% of all cases) are (i) more sustained winds triggering more (dust/sand/snow) storms and emissions (n=14, or 11%), and (ii) less snow cover due to withdrawing glaciers (n=3, or 2%).

## Specific Indicators

*Ecological Indicators*

Ecological indicators as found in the cases can broadly be grouped into vegetation change, wind/water erosion, water degradation, and other degradation of terrestrial ecosystem functions. Vegetation-related variables (n=105, or 80%) and soil-related variables (n=98, or 74%) are the most important and used to about equal shares – see Table 7.2. The indicators most prevalent in the cases are: (i) the reduction of vegetative cover (60%), with decreases of scrub, grass and/or herbaceous cover slightly outweighing decreases in forest, woodland and/or tree cover; (ii) soil erosion (50%), mainly as combined wind and water erosion in one fifth of all cases, though the share of unspecified cases is rather high; (iii) encroachment of undesirable, less palatable vegetation species (grass, bush, 'woody weed') (33%); (iv) physical soil degradation (25%), mainly in the form of soil moisture changes, and with compaction and soil crusting clearly dominating over soil crust destruction (or loosening of the soil); (v) increases in bare, rocky, eroded ground cover (27%), mainly linked to soil erosion; and (vi) increases in (desert-like) sand cover (25%), mainly through the revitalization of (semi-)fixed fossil dunes in the form of shifting sands and/or coppice dunes. If ecological indicators are categorized according to varying degrees of degradation, a range from slight to extreme degradation appears, with indicative quantitative data attached to each category and type of indicator – see Tables 7.2 to 7.5.

Table 7.2 Vegetation Change – Ecological Degrees of Dryland Degradation

| Slight | Moderate | Strong | Extreme |
|---|---|---|---|
| Canopy of undegraded perennial plant cover extends over 50-70% of surface, or more. | Canopy of undegraded perennial plant cover reduced to 20-50% of surface. | Canopy of undegraded perennial plant cover reduced to 5-20% of surface. | Canopy of undegraded perennial plant cover reduced to <5% of surface. |
| Decrease of (undegraded) vegetation by ¼ (or less) of initial area. | Decrease of (undegraded) vegetation by ½ (or less) of initial area. | Decrease of (undegraded) vegetation by ¾ (or less) of initial area. | Decrease of (undegraded) vegetation by > ¾ of initial area. |
| Increase of (degr.) vegetation by ¼ (or less) of initial area. | Increase of (degr.) vegetation by ½ (or less) of initial area. | Increase of (degr.) vegetation by ¾ (or less) of initial area. | Increase of (degr.) vegetation by >¾ of initial area. |
| Decrease of timber stock by ¼ (or less) of initial volume. | Decrease of timber stock by ½ (or less) of initial volume. | Decrease of timber stock by ¾ (or less) of initial volume. | Decrease of timber stock by >¾ of initial volume. |
| Hardly any shifts in dryland vegetation. | Dryland vegetation zones shift to more humid locations. | | Remaining tree cover is lost. |
| Tree density decreased by ¼ (or less) of initial density. | Tree density decreased by ½ (or less) of initial density. | Tree density decreased by ¾ (or less) of initial density. | Tree density decreased by >¾ of initial density. |

## Table 7.3    Erosion – Ecological Degrees of Dryland Degradation

| Slight | Moderate | Strong | Extreme |
|---|---|---|---|
| Part of top soil removed (deep soils: 10-40% of area); shallow rills with a spacing of 20-50 cm at least. | All or large parts of top soil removed (deep soils: 40-70% of area, thin soils: 10-40% of area); rills of <20 cm apart; active gully development at a spacing of 20-50 cm. | Thin soils: all top soil removed, exposing bedrock; deep soils: all top soil and parts of subsoil removed moderately at >70% of area;. deep gullies < 20 cm apart. | Increase of eroded land on >¾ of area; land unreclaimable and impossible to restore. Increase of sand cover by >¾ of initial sand cover; shifting sands lead to the burial of pastoral/ agricultural terrain and human settlements. |
| Increase of eroded land on ¼ (or less) of area. | Increase of eroded land on ½ (or less) of area. | Increase of eroded land on ¾ (or less) of area. | |
| Increase of bare, rocky ground cover on ¼ (or less) of area. | Increase of bare, rocky ground cover on ½ (or less) of area. | Increase of bare, rocky ground cover ¾ (or less) of area. | |
| Increase of sand cover by ¼ (or less) of initial sand cover. | Increase of sand cover by ½ (or less) of initial sand cover. | Increase of sand cover by ¾ (or less) of initial sand cover. | |
| No activation of sand dunes. | Activation of <50% (semi)fixed sand dunes. | Activation of >50% (semi)fixed sand dunes. | |
| Hardly any sand deposits. | Frequent, periodic sand deposits of 5-10 cm (or more) on pastoral and cultivated land, and/or siltation of lowlying zones. | | |

**Table 7.4** **Water Degradation – Ecological Degrees of Dryland Degradation**

| Slight | Moderate | Strong | Extreme |
|---|---|---|---|
| Increase of groundwater salinity by ¼ (or less) of initial value. | Increase of grd.water salinity by ½ (or less) of initial value. | Increase of grd.water salinity by ¾ (or less) of initial value. | Increase of grd.water salinity by >¾ of initial value. |
| Salinized area is ¼ (or less) of all cultivated (irrigated) land. | Salinized area is ½ (or less) of all cultivated (irrigated) land. | Salinized area is ¾ (or less) of all cultivated (irrigated) land. | Salinized area is >¾ of all cultivated (irrigated) land. |
| Decrease of water surface, volume, shorelines, and/or drop in water tables or river discharge by ¼ (or less) of initial value. | Decrease of water surface, volume, shorelines, and/or drop in water tables or river discharge by ½ (or less) of initial value. | Decrease of water surface, volume, shorelines, and/or drop in water tables or river discharge by ¾ (or less) of initial value. | Decrease of water surface, volume, shorelines, and/or drop in water tables or river discharge by >¾ of initial value. |
| No or hardly any drying up of (natural) rivers, lakes, etc. | Occasional drying up of (natural) rivers, lakes, etc. | Frequent drying up of (natural) rivers, lakes, etc. | Unreclaimable drying up of (natural) rivers, lakes, etc. |
| Increase of chemical water pollution by ¼ (or less) of initial value. | Increase of chemical water pollution by ½ (or less) of initial value. | Increase of chemical water pollution by ¾ (or less) of initial value. | Increase of chemical water pollution by >¾ of initial value. |
| Chemically polluted area is ¼ (or less) of all cultivated, irrigated land. | Chemically polluted area is ½ (or less) of all cultivated, irrigated land. | Chemically polluted area is ¾ (or less) of all cultivated, irrigated land. | Chemically polluted area Is >¾ of all cultivated, irrigated land. |

**Table 7.5**  **Other Deterioration of Terrestrial Ecosystem Functions – Ecological Degrees of Dryland Degradation**

| Slight | Moderate | Strong | Extreme |
|---|---|---|---|
| Original biotic functions are still largely intact. | Original biotic functions are degraded. | Original biotic functions are largely degraded. | Original biotic functions are Fully degraded. |
| Biomass production (yield) reduced by ¼ (or less) of initial, or reduced by ¼ (or less) on used sites if compared to unused sites. | Biomass production (yield) reduced by ½ (or less) of initial, or reduced by ½ (or less) on used sites if compared to unused sites. | Biomass production (yield) reduced by ¾ (or less) of initial, or reduced by ¾ (or less) on used sites if compared to unused sites. | Biomass production (yield) reduced by >¾ of initial, or reduced by >¾ on used sites if compared to unused sites. |
| Perennial plant species richness reduced by ¼ (or less) of long-term average. | Perennial plant species richness reduced by ½ (or less) of long-term average. | Perennial plant species richness reduced by ¾ (or less) of long-term average. | Perennial plant species richness reduced by >¾ of long-term average. |
| Some Wildlife species reduced in numbers. | Some wildlife species disappeared, while others invaded and/or hazardous species increased greatly. | Some wildlife species disappeared, but no others invaded. | Large parts of Plant and wildlife species disappeared. |

*Socio-Economic Indicators*

In more than half of the cases, socio-economic variables are used to indicate desertification. The most important ones relate to three broad groups: (i) changes in livestock parameters (28%); (ii) (negative) changes or downward trends in crop harvest and/or fish catch (27%); and (iii) demographic, human welfare and social conflict indicators (20%). Some of the human variables are linked to, or even overlap with, ecological key parameters. If one is to categorize socio-economic indicators according to varying degrees of degradation, a range from slight to extreme degradation appears with indicative quantitative data attached to each category – see Tables 7.6 to 7.8.

**Table 7.6** **Pastoral Suitability – Socio-economic Degrees of Dryland Degradation**

| Slight | Moderate | Strong | Extreme |
|---|---|---|---|
| Total cover of forage species is >30%. Remaining vegetation cover decreased by ¼ (or less) of initial cover. Invasion of undesirable species on ¼ (or less) of area. | Total cover of forage species is 10-30%. Remaining vegetation cover decreased by ½ (or less) of initial cover. Invasion of undesirable species on ½ (or less) of area. | Total cover of forage species is 5-10%. Remaining vegetation cover decreased by ¾ (or less) of initial cover. Invasion of undesirable species on ¾ (or less) of area. | Total cover of forage species is <5%. Remaining vegetation cover decreased by >¾ of initial cover. Invasion of undesirable species on >¾ of area. |
| Productivity of palatable grasses reduced by ¼ (or less) of initial. | Productivity of palatable grasses reduced by ½ (or less) of initial. | Productivity of palatable grasses reduced by ¾ (or less) of initial. | Productivity of palatable grasses reduced by >¾ of initial productivity. |
| Impaired livestock carrying capacity by ¼ or 125% (or less) of initial number. | Surpassed livestock carr. capacity by ½ or 150% (or less) of initial number. | Surpassed livestock carr. capacity by ¾ or 175% (or less) of initial number. | Surpassed livestock carr. capacity by >¾ or >175% of initial number. |
| Decline in livestock number by ¼ (or less) of initial. | Decline in livestock number by ½ (or less) of initial. | Decline in livestock number by ¾ (or less) of initial. | Decline in livestock number by >¾ of initial. |
| Terrain with somewhat reduced pastoral suitability, but suitable for use in local grazing system; restoration to full productivity possible by modification of the management system. | Terrain with greatly reduced pastoral suitability, may be no more reclaimable at the household level; increasing economic losses occur and/or out-migration from rural to urban areas; degraded rangelands suffer from insect/rodent damages and/or are increasingly abandoned; increased conflicts may arise with cultivators about resources; major improvements and/or engineering works are required to restore productivity. | | Terrain with lost suitability for pastoral (and wild animal) uses; terrain is unreclaimable and beyond restoration. |

**Table 7.7** **Agricultural Suitability – Socio-economic Degrees of Dryland Degradation**

| Slight | Moderate | Strong | Extreme |
|---|---|---|---|
| Fluctuating, but stagnating crop yields despite of continued technological improvements (seed, fertilizer, etc.). | | Drastic decline of crop yields without any fluctuation, up to full harvest losses (and production stop). | |
| Decline of crop yields by ¼ (or less) of long-term avge. | Decline of crop yields by ½ (or less) of long-term avge. | Decline of crop yields by ¾ (or less) of long-term avge. | Decline of crop yields by >¾ of long-term avge. |
| Cultivated land is reduced by ¼ (or less) of initial area. | Cultivated land is reduced by ½ (or less) of initial area. | Cultivated land is reduced by ¾ (or less) of initial area. | Cultivated land is reduced by >¾ of initial area. |
| Terrain w. somewhat reduced agricult. suitability; suitable for use in local cropping; restoration to full productivity possible by modification of mgt. | Terrain w. greatly reduced agricult. suitability, may be no more reclaimable at farm level; increasing econ. losses occur and/or out-migration; fields and, partly, settlements are abandoned; increased conflicts may arise with herders about resources; major improvements are required to restore productivity. | | Terrain with lost suitability for agricultural uses; terrain is unreclaimable and beyond restoration. |
| Soil nutr. decline by ¼ (or less) of initial value. | Soil nutr. decline by ½ (or less) of initial value. | Soil nutr. decline by ¾ (or less) of initial value. | Soil nutr. decline by >¾ of initial value. |
| Chem. soil pollution on ¼ (or less) of cultivated area. | Chem. soil poll. on ½ (or less) of cultivated area. | Chem. soil poll. on ¾ (or less) of cultivated area. | Chemical soil pollution on >¾ of cultivated area. |
| Increase of waterlogged soils on ¼ (or less) of cultivated area. | Increase of waterlogged soils on ½ (or less) of cultivated area. | Increase of waterlogged soils on ¾ (or less) of cultivated area. | Increase of waterlogged soils on >¾ of cultivated area. |
| Impaired agrarian carrying capacity by ¼ (or less) of original (theoretical) number. | Surpassed agrarian (human) carrying capacity by ½ (or less) of original (theoretical) number. | Surpassed agrarian carrying capacity by ¾ (or less) of original (theoretical) number. | Surpassed agrarian carrying capacity by >¾ of original (theoretical) number. |

**Table 7.8  Other Human Welfare Degradation – Socio-economic Degrees of Dryland Degradation**

| Slight | Moderate | Strong | Extreme |
|---|---|---|---|
| Economic production losses of up to 25% of initial production. | Economic production losses of up to 50% of initial prod. | Economic production losses of up to 75% of initial prod. | Economic production losses of more than 75% of initial prod. |
| Impaired firewood carrying capacity by ¼ (or less) of initial number. | Surpassed firewood carrying capacity by ½ (or less) of initial no. | Surpassed firewood carrying capacity by ¾ (or less) of initial no. | Surpassed firewood carrying capacity by >¾ of initial no. |
| Firewood collection distance <1km. | Firewood collection distance 1-5 km. | Firewood collection distance >5 km. | Wood brought in from distant places. |
| Decline of fish catch by ¼ (or less) of long-term avge. | Decl. of fish catch by ½ (or less) of long-term avge. | Decl. of fish catch by ¾ (or less) of long-term avge. | Decl. of fish catch by >¾ of long-term avge. |
| No to minor land-use related societal disturbances (e.g., hunger, famine, violent clashes about land betw. farmers and herders). | Land-use related societal disturbances in small portions of local populations. | Land-use related societal disturbances, incl. famine, prevail in <50% of local populations. | Large parts of populations (>50%) suffer from land-use related human disorders or societal disturbances. |
| Only few land-use related health problems. | Some land-use related health problems. | Widespread land-use related health problems. | Dramatic increase of land-use related health problems. |

*Meteorological Indicators*

Meteorological indicators are least widely applied. The ones mostly reported in the cases are increased aridity, measured in terms of temperature and/or precipitation (amount, intensity, duration, variability) (32%), and changed wind patterns (triggering more storms and emissions) (11%). If one is to categorize meteorological indicators according to varying degrees of degradation, a range from slight to extreme degradation appears with indicative quantitative data attached to each category, as shown in Table 7.9.

**Table 7.9  Meteorological Degrees of Dryland Degradation**

| Slight | Moderate | Strong | Extreme |
|---|---|---|---|
| Decrease in annual rainfall by ¼ (or less) of long-term avge. | Decrease in annual rainfall by ½ (or less) of long-term avge. | Decrease in annual rainfall by ¾ (or less) of long-term avge. | Decrease in annual rainfall by ¾ (or more) of long-term avge. |
| Some winds triggering storms and emissions (dust, sand, snow, salt), with minor precipitation and low spatial reach of particle transport. | More sustained winds triggering more storms and emissions, with precipitation of <0.5 t/ha (salt), and/or annual transport of (salt, dust) particles at <10,000,000 t up to 100 km, and/or dust storms up to 25 days/yr. | More sustained winds triggering more storms and emissions, with precipitation of 0.5-1t/ha (salt), and/or annual transport of (salt, dust) particles at 10-40,000,000 t up to 250 km, and/or dust storms up to 50 days/yr. | More sustained winds triggering more storms and emissions, with precipitation of 1t/ha (salt), and more, and/or annual transport of (salt, dust) particles at >40,000,000 t up to 400km, and/or dust storms up to 100 days/yr. |
| Contraction of wet season by ¼ (or less) of initial length, and/or shift of rainfall to winter season. | Contraction of wet season by ½ (or less) of initial length, and/or shift of rainfall to winter season. | Contraction of wet season by ¾ (or less) of initial length, and/or shift of rainfall to winter season. | Contraction of wet season by >¾ of initial length, and/or shift of rainfall to winter season. |
| Desiccation over 1-2 years. | Desiccation over 3-4 years. | Desiccation over 4-5 years. | Desiccation over >5 years. |
| Index of aridity increases by ¼ (or less) of long-term avge. | Index of aridity increases by ½ (or less) of long-term avge. | Index of aridity increases by ¾ (or less) of long-term avge. | Index of aridity increases by >¾ of long-term avge. |
| Mean rise of annual (soil, air) temperature by ¼ (or less) of long-term avge. | Mean rise of annual (soil, air) temperature by ½ (or less) of long-term avge. | Mean rise of annual (soil, air) temperature by ¾ (or less) of long-term avge. | Mean rise of annual (soil, air) temperature by >¾ of long-term avge. |

## Indicated Types of Regional Dryland Degradation

Assuming that the value of a set of indicators alone reflects a particular human-environment condition, it appears that the broad category 'reduction of vegetation cover' indicates desertification in a robust manner, with regional variations of desertification emerging only if specific vegetation subclasses and more indicators are considered – see Table 7.10. Most strikingly, a difference between cases from the 'New World' (Australia, North America, Latin America) and from other regions, notably Asia, can be observed in terms of their human-environment condition as indicated by the following variables.

There are cases which are solely characterized by the removal of original scrub, grass and/or herbaceous cover, and not by decreases of trees, forests, or woodlands. These cases are settled in the dry grass- and shrubland ecosystems of Australia's arid and semi-arid zones, covering parts of the states of New South Wales, Queensland, South Australia, Western Australia and Northern Territory, and in the US American Southwest, namely, the Chihuahuan and Mojave Deserts ecosystems, and the Colorado Plateau. In these cases, natural plant cover reduction is interlinked with the encroachment of sparse and undesirable, because less palatable, grass and bush vegetation ('woody weed'). Again, this encroachment is most pronounced in the Australian and US American cases.

Together with cases from Mexico and Patagonia, the 'New World' cases as described above are also the ones in which changes in livestock parameters are most commonly reported as indicative of desertification. In contrast, cropland-related indicators are non-existent or far less important than in the cases from Asia, Africa, and Europe.

Most strikingly, it seems that soil erosion features more of the 'New World' cases than any other regional cases (except for Europe), and that meteorological indicators such as increased aridity are less frequently, if at all, reported in 'New World' cases. In the US American Southwest, reduced plant cover is reported as linked to soil erosion, not leading, however, to an increased exposure of rocky outcrops. Australian cases are different in that reduced plant cover is reportedly coupled with both soil erosion and increases in eroded, bare, and rocky ground cover. It is further noteworthy that Australian cases of desertification are indicated by both one of the highest regional shares of soil erosion and the highest share of exposed parent material. Finally, the most evident difference between 'New World' and other cases in terms of indicative desertification are increases in desert-like sand cover, mainly shifting sands due to the re-activation of (semi-)fixed fossil dunes. These increases are almost exclusively reported in cases from the arid Great Plains of the Central Asian desert and steppe region, the Arabian Peninsula, and the old-settled parts of the Sudano-Sahelian region in Africa. There is a single exception to this rule, which is a site showing coppice dune formation in a formerly heavily grazed area in the Jornada del Muerto Basin, located in the Chihuahuan Desert of the US Southwest in New Mexico.

### Table 7.10  Indicated Types of Regional Dryland Degradation*

| Indicator | All cases (N=132) | Asia (n=51) | Africa (n=42) | Europe (n=13) | Australia (n=6) | North America (n=6) | Latin America (n=14) |
|---|---|---|---|---|---|---|---|
| **Reduced vegetation cover, total** | 79  60% | 25  49% | 27  64% | 9  69% | 6  100% | 4  67% | 8  57% |
| Scrubs, grass, herbs (subtotal) | 50  38% | 17  33% | 16  38% | 6  46% | 1  17% | 3  50% | 7  50% |
| Forest, woodland, trees (subtotal) | 40  30% | 10  20% | 18  43% | 5  39% | | | 7  50% |
| Scrubs etc. and forest etc. (subtotal) | 33  25% | 10  20% | 10  24% | 6  46% | | | 7  50% |
| **Soil erosion** | 66  50% | 16  31% | 15  36% | 12  92% | 5  83% | 4  67% | 14  100% |
| **Grass, bush, woody weed encroachment** | 44  33% | 13  26% | 13  31% | 3  23% | 5  83% | 5  83% | 5  36% |
| **Increased aridity** | 42  33% | 10  20% | 25  60% | 5  39% | | 2  33% | |
| **Water-related indicators** | 38  29% | 22  43% | 5  12% | 3  23% | 1  17% | 2  33% | 5  36% |
| **Changes in livestock parameters** | 37  28% | 4  8% | 12  29% | 1  8% | 5  83% | 3  50% | 12  86% |
| **Increase in bare, rocky, eroded ground cover** | 37  28% | 11  22% | 10  24% | 6  46% | 5  83% | 2  33% | 3  21% |
| **Changes in crop harvest and/or fish catch** | 36  27% | 15  29% | 14  33% | 4  31% | | 1  17% | 2  14% |
| **Physical soil degradation** | 33  25% | 6  18% | 2  6% | 6  18% | 5  15% | 1  3% | 13  39% |
| **Increase in desert-like sand cover** | 33  25% | 25  49% | 7  17% | | | 1  17% | |

\* Absolute figures and relative percentages; multiple counts possible; percentages relate to the totals of broad geographical regions.

# Chapter 8

# Discussion

**Introduction**

In the following, outcomes of the study are discussed in the context of what had been laid down in chapter 1 on the issue of desertification and its identification. In summary, the meta-analysis of desertification cases does not provide support for either one of the two large groups of explanations (Watts, 1985; Helldén, 1991, Thomas, 1997), whether it is irreducible complexity revealing no distinct pattern (Dregne, 2002; Warren, 2002), or some primary cause held responsible for an irreversible extension of desert landforms and landscapes (Le Houérou, 2002; Breckle, Veste and Wucherer, 2002), be they irrational land management by nomadic pastoralists, growing populations, macro-economic forces, unjust class and power relations, or climatic factors alone. Rather, 'multiplicity' is identified as the most common theme reported in the empirically supported narratives of the impact on drylands of actors and activities (Rindfuss, Walsh, Mishra, Fox and Dolcemascolo, 2003). Case studies reveal multiple uses of the land with multiple environmental histories (chapter 3), multiple agents of change with multiple connections in social and geographical space, with multiple ties between people and land, and with multiple responses to social, climatic and ecological changes (chapter 4), multiple syndromes or associations of change rates with underlying causes (chapter 5), multiple spatial and temporal scales in the causes of and responses to desertification leading to multiple pathways of dryland change (chapter 6), and multiple combinations of indicators to best describe and monitor desertification (chapter 7). The theoretical framework that best accounts for this complexity is system dynamics, with particular emphasis on the history of the system (initial conditions), heterogeneity between actors, hierarchical levels of organization, non-linear dynamics caused by feedback mechanism, and system adaptation (Lambin, Geist and Lepers, 2003).

The most important observation is that multiplicity and complexity are not irreducible, but associated with a limited number of recurrent causative variable combinations and pathways of desertification, which makes the problem tractable (Stafford Smith and Reynolds, 2002). In the following, desertification as described in the cases is first put into the context of land use change. By demonstrating that desertification is strongly associated with, but not necessarily driven by everincreasing land use intensity, a hitherto neglected dimension of the definition of desertification is revealed. Second, mode, direction and pathways or trajectories of dryland change are discussed, as well as what are the key causative factors, which indicators are best, and whether it is a fast or slow process.

## Land Use Transitions Towards Increasing Intensity

In the discussion of what actually is desertification and, in particular, how the linked biophysical and socio-economic processes affect human welfare, common reference is made to land uses such as cropping and livestock raising when it comes to the so-called human dimensions (Reynolds, 2001; Reynolds and Stafford Smith, 2002), or 'human activities' (UNEP, 1977, 1994). The notion might arise that all is about static land uses. The meta-analysis of desertification cases, however, suggests that a focus needs to be adopted that deals with highly dynamic transitions towards ever-increasing land use intensity. This is especially true for the latter half of the $20^{th}$ century and applies to both the cropping and livestock sectors.

The majority of proximate factors involve greater input commitments per unit of land area compared to traditional land uses (Margaris, Koutsidou and Giourga, 1996; Niamir-Fuller, 1999). From a large part of the dryland studies, it appears as if for about two centuries, namely 1700 to circa 1900, land use change has been gradual in a sense that both in terms of areal extent and intensity, cropping of drylands rarely exceeded the natural boundaries set by fragile ecosystems which have limited production potential only for large-scale farming. An exception was, and still is, the steppe zone of northern China where, tied to biophysical changes, large-scale alternations of grain farming and pastoralism have rather been the rule. Pastoral uses had been, both in areal extent and intensity, well adapted to the fluctuating resource bases of grass and water provision, mainly in the form of flexible, highly mobile nomadic grazing.

Significant changes started in the 1900 to 2000 period. They could be characterized as a simultaneous or coupled land use expansion and gradual land use intensification. These had been both large scale extensions of cropland and managed rangelands onto fragile sites of variable natural ecosystems, and intensification of these land uses, be they livestock rearing or farming. One could be tempted to view this coupled effect of both expansion and intensification (under the premise of fluctuating natural resource bases) as the final link in the causal chain tying socio-economic and biophysical factors leading to desertification, i.e., greater cropping and stocking intensity per unit area of land that had been expanded onto previously marginal sites, used extensively for centuries or millennia. Where land was competed for by both livestock and crops, the comparative advantage for crops had in most cases been superior to those of livestock. Accruing investments in especially irrigated farming land for food crops (such as grain and rice to feed growing populations) as well as cash crops (such as cotton to provide the base product for clothing) had raised the level of land exploitation to new heights since about the late $19^{th}$ century, and again during the latter half of the $20^{th}$ century. An exception appears to be the Sudano-Sahelian zone in West Africa where a system of nested land use rights so far had prevented the directional, intended 'valorization' of fragile dryland sites. This included the absence of the development of land markets (land sales, etc.) and of land enclosures (with some closing of land frontiers in heavily populated zones, only), which had been characteristic features of other dryland areas.

Finally, it appears as if the low versus high level of integration between crop and livestock husbandry had been critical in defining the level of farming intensity. Cropping intensity, in turn, determined the extent of the replacement of natural dryland ecosystems by managed ecosystems. Where the level of integration had been low, such as in northern China or East Africa – with competing rather than integrative modes of the two broad land uses – the extent of ecosystem degradation appears to be high. And, on reverse, where the level of integration had been high, such as in the Sudano-Sahelian zone of West Africa or in the East Mediterranean steppe zone, the extent of ecosystem degradation seems to be low.

*Livestock Transitions*

Livestock and grazing activities have been extended in terms of areal extent, and also became more intensified, namely during the $20^{th}$ century. In most of the cases, the extension of rangelands had to compete with extending croplands, but had lower comparative advantage, with the consequence of encroaching croplands. As a consequence, intensified animal husbandry happened mostly on less available land for animals, both wild and domesticated, if compared to the pre-intensification period. Through intensified land use on increasingly limited land, a twofold pressure had been exerted on fragile dryland ecosystems. In the cases, hardly any traditional or undisturbed nomadic pastoralism was found, despite of the fact that its flexible and highly mobile mode has been well suited to fluctuating dryland ecosystem properties (Niamir-Fuller, 1999). In most of the few cases in which nomadic grazing was reported, changes in rights and traditions occurred aside with more fundamental transformations of the nomadic to semi-nomadic and further on to extensive modes of pastoralism (Rozanov, 1991; Runnström, 2000; Feng, Endo and Cheng, 2001; Lin, Tang and Han, 2001). A previously highly mobile land use practice, suited to fluctuations in the biophysical resource base, was transformed into the various and today most widespread modes of extensive grazing, either in the form of sedentary livestock rearing or as transhumance. Cases show that extensive land use system could quickly turn into intensive modes of livestock keeping (Manzano, Návar, Pando-Moreno and Martinez, 2000; Dube and Pickup, 2001), though this trend had limited geographical distribution only.

Tables 8.1 and 8.2 summarize the characteristic sequence of land use transformation in the livestock sector. They give the type of land use in the time period selected by the case study authors (see also Table 2.2). It appears that in only few cases (n=29, or 22%) the livestock land use system remained stable, while in more cases (n=48, or 36%) transitions have occurred, partly across distinct livestock management systems and always towards increasing intensification. In about one third of the cases (n=40, or 30%), which had all been cases in which extensive grazing was reported, it was not made explicit whether transhumant and sedentary modes existed concomitantly, or whether the first land use type was about to be transformed in the second one. From these data, it appears that the history of land degradation in dryland areas worldwide is linked to the transformation of pastoral economies to ever increasing intensity, mostly on less available grazing land.

**Table 8.1**  Livestock Transition Towards Increasing Intensity in Asia and Africa

| (Semi)nomadic pastoralism | | Extensive grazing | | Intensive livestock rearing[a] |
|---|---|---|---|---|
| Traditional nomadic pastoralism | Modified (semi)nomadic pastoralism | Transhumance | Permanent, sedentary | |
| **Asia** | | | | |
| Brown et al., 1997-98 | Holzner et al., 1950-98; Weiss et al., 1982-94 | | Ram et al., 1951-95 | |
| | | Ho, 1949-96; Khresat et al., 1985-95; Rozanov, 1980-89; Saiko et al., 1890-1995; Genxu et al., 1700-1997[b]; Yang, 1700-2000[b]; Zhou et al., 1700-1995[b] | | |
| Liu et al., 1980-96; Runnström 1982-93; Sheehy, 1700-1990[b], 1945-78; Wang et al. 1976-96 | | | | |
| | Feng et al., 1950-98 | | | |
| Lin et al., 1700-2000[b]; Runnström, 1982-93; Rozanov, 1978-89 | | | | |
| **Africa** | | | | |
| Ayoub, 1960-91; Benjaminsen, 1955-90; Keya, 1988-1990 | Ayoub, 1960-91; Benjaminsen, 1955-90 | | Dube et al., 1983-96; Palmer et al., 1989-94 | Dube et al.,1983-96; Mortimore et al., 1950-96 |
| | | Ayoub, 1960-91; Gauquelin et al., 1700-1995[b]; Keya, 1988-90; Mwalyosi, 1957-87; Palmer et al., 1989-94; Ringrose et al., 1986-90; Ringrose et al., 1990-94; Turner 1999a, 1987-89; Turner 1999b, 1993-95 | | |
| Keya, 1988-1990 | | | | |
| | Gonzalez, 1945-93; Rasmussen et al., 1955-95 | | | |
| Olsson, 1984-1985; Ringrose et al., 1986-90; Venema et al., 1904-90 | | | | |

[a] Ranching as well as integrated animal-crop production systems.
[b] Time period might date back later than 1700.

**Table 8.2    Livestock Transition Towards Increasing Intensity in Europe, Australia, North and Latin America**

| (Semi)nomadic pastoralism | | Extensive grazing | | Intensive livestock rearing[a] |
|---|---|---|---|---|
| Traditional nomadic pastoralism | Modified (semi)nomadic pastoralism | Transhumance | Permanent, sedentary | |
| Europe | | | | |
| | | | Seixas, 1985-94; Vandekerckhove et al., 1996-97 | |
| | | Gauquelin et al., 1700-1995[b]; Hill et al., 1700-1996[b]; Kosmas et al., 1700-1998[b]; Wijdenes et al., 1997-98 | | |
| Australia | | | | |
| | | | Pickup, 1850-1995, 1900-95 | |
| | | Bastin et al., 1988-89 | | |
| North America | | | | |
| | | | Brown et al., 1977-95 | |
| | | Fredrickson et al., 1700-1995[b]; Ludwig et al., 1982-95; Mouat et al., 1986-91; Rango et al., 1880-1999 | | |
| Latin America | | | | |
| | | Aagesen, 1876-1995 | | Valle et al., 1986-92 |
| | | Manzano et al., 1950-85 | | |

[a]  Ranching as well as integrated animal-crop production systems.
[b]  Time period might date back later than 1700.

In the Asian cases, the main transformations happened both within the broad group of (semi)nomadic pastoralism – i.e., from traditional to modified nomadism – and within the broad group of extensive grazing – i.e., from transhumance to sedentary grazing. In some of these cases, transitions could stretch over a time period of 300 years, and even more. Transitions also occurred from the first broad group (traditional and/or modified nomadism) to the second broad group (extensive grazing), although to a lesser degree. Interestingly enough, no intensive livestock activities and no intensive, integrated mixed cropping-animal production activities were reported from Asian drylands – see Table 8.1.

In the African cases, transitions appear to be similar to those in the Asian dryland cases, with two important exceptions though. First, intensified modes of livestock keeping had developed such as commercial cattle ranching in southern Africa. Second, highly integrated livestock and crop production systems had developed, including tree management even as a third component of the land use system, for example, in the Sudan-Sahel of West Africa. Different from the Asian cases, long time periods of transformation were reported to a far lesser degree, i.e., turnover rates are higher than in most of the other cases – see Table 8.1.

In the European, Australian, North and Latin American cases, no mention was made of (semi)nomadic pastoralist activities and integrated production systems. Cases of intensive livestock rearing stem exclusively from northern Mexico and Patagonia. The characteristic sequence in the transformation of livestock uses was the transition from transhumant to sedentary grazing. Only in the European drylands of the Mediterranean basin, livestock uses and transitions stretch over 300 years, and more, while they were tied to European colonization within the last 150 years, or so, in cases from Australia and the Americas – see Table 8.2.

*Cropping Transitions*

During the period under study in this meta-analysis, i.e., 1700 to 2000, crop production activities have also increased, both in terms of areal extent and intensity – except for drylands with only few suitable farming land such as those in Australia and the US Southwest.

Shifting cultivation, dryland farming and rotational fallow farming were mentioned in some cases to suit well to the fluctuating, scarce resource base of dryland ecosystems. They allow time for ecosystem recovery between periods of localized but more intensive use, namely, a long enough period to allow for the recovery of soil fertility, and would have usually been sustainable in most areas (except the Aral Sea Basin). However, their frequency of occurrence in the cases was reportedly low. This is especially true for shifting cultivation with cases stemming from Africa only. Cases of dryland farming were reported to a far lesser degree than expected, and traditional farming – be it shifting cultivation, rotational fallow farming, or traditional oasis agriculture – was always reported to be under transformation to other uses (usually, more intensive uses practiced over large areas). Shifting cultivation, similar to the transformation of nomadic pastoralism, turned into more or less sedentary agriculture with an increasing tendency towards continuous cropping during the $2^{nd}$ half of the $20^{th}$ century.

## Table 8.3  Cropping Transition Towards Increasing Intensity in Asia

| [a] | [b] | Non-irrigated cropping | | | [c] | Irrigated cropping | |
|---|---|---|---|---|---|---|---|
| | | Rotational (rainfed) farming, not mechanized | Continuous (rainfed) farming | Mechanized continuous farming | | Traditional oasis agriculture | Irrigation farming (large-scale) |
| | | Ho, 1949-96; Khresat et al., 1985-95; Lin et al., 1700-2000[d]; Sheehy, 1945-78; Zhou et al., 1700-1995[d] | | | | Feng et al., 1950-98; Genxu et al., 1700-1997[d]; Lin et al., 1950-95; Runnström, 1982-93; Yang, 1700-2000[d] | |
| | | Liu et al., 1980-96; Sheehy, 1700-1990[d] | | Khresat et al., 1985-95 | | Sheehy, 1700-1990[d] | Khresat et al., 1985-95; Rozanov, 1978-89; Saiko et al., 1890-1995; Weiss et al., 1982-94; Zhou et al., 1700-1995[d] |
| | | Runnström, 1982-93; Rozanov, 1980-89 | | | | | |
| | | | Ram et al., 1951-95 | | | | |

[a]  Shifting cultivation.
[b]  Dryland farming.
[c]  Wetland cropping.
[d]  Time period might date back later than 1700.

Similarly, wetland cropping (plots cultivated by using natural floods) and the century- and millennia-old oasis agriculture reportedly triggered hardly any ecosystem alterations, until about 1900. From then on, traditional oasis farming turned into small-scale irrigation agriculture, but only later it shifted to large-scale irrigation schemes. With the application in the second half of the 20th century of newly available water technology (canal systems, in particular) and hydrotechnical installations (e.g., dams, reservoirs, groundwater pumping schemes, etc.), there had been fundamental transformation of land use, and – partly irreversible – changes of the hydrological cycle. Rice and cotton were highly water demanding key crops in this crucial phase of land use transition, and massive irrigated grain production was meant to support increasing populations (the introduction of cotton served to fulfill the needs for clothing of increasing human populations, too). Some new irrigation areas at dryland sites in basins, plains, and along (ancient) rivers developed into nationally leading production zones of grain and cotton, and oasis sites with long traditions of farming practices became one of the first desertification hot spots, especially in Central Asia. At the end of the 20th century, modern irrigation techniques such as central pivot irrigation of diameters became directly introduced at arid to hyper-arid (sub)desert sites, which had no tradition of oasis or irrigation practices, and which were concomitantly used by indigenous nomadic herders. In the early cycle of these introductions, increases in biomass production occurred, and only later in the agricultural cycle, when farmland was abandoned and exposed to wind and water erosion, land degradation happened.

On the steppe soils of semi-arid climates, land intensification occurred in the form of mechanized rainfed agriculture, especially through the use of tractors and ploughs. Shrinking land availability for various reasons triggered further intensification such as biotechnological improvements of seed and new varieties, or the integration of livestock and crop husbandry (i.e., a shift to agro-pastoral land use). However, in most of these cases soil fertility decline and erosion quickly set thresholds to land use intensification.

**Table 8.4**     **Cropping Transition Towards Increasing Intensity in Africa**

| Non-irrigated cropping | | | | | Irrigated cropping | | |
|---|---|---|---|---|---|---|---|
| [a] | Dryland farming | Rotational (rainfed) farming, not mechanized | [b] | [c] | Wetland cropping | [d] | Irrigation farming (large-scale) |
| Ayoub, 1960-91; Mwalyosi, 1957-87 | | | | | Ayoub, 1960-91 | | |
| | Benjaminsen, 1955-90; Palmer et al., 1989-94; Ringrose et al., 1986-90 | | | | Benjamin-sen, 1955-90; Dube et al., 1983-96 | | Venema et al., 1904-90 |
| Dube et al., 1983-96; Ringrose et al., 1990-94 | | | | | | | |
| | Turner, 1993-95 | Ayoub, 1960-91; Gonzalez, 1945-93; Helldén, 1962-84; Mortimore et al., 1950-96; Olsson, 1984-85; Ringrose et al., 1986-90; Turner, 1987-89 | | | | | |
| | | Rasmussen et al., 1955-95 | | | | | |

a   Shifting cultivation.
b   Continuous rainfed farming.
c   Continuous mechanized farming
d   Traditional oasis agriculture.

## Table 8.5  Cropping Transition Towards Increasing Intensity in Europe, Australia, North and Latin America

| | Non-irrigated cropping | | | | Irrigated cropping | |
|---|---|---|---|---|---|---|
| [a] Dryland farming | Rotational (rainfed) farming, not mechanized | [b] | Mechanized continuous farming | [c] | [d] | Irrigation farming (large-scale) |
| Europe | | | | | | |
| | Basso et al., 1950-96; Gauquelin et al., 1700-1995[e]; Kosmas et al., 1700-1998[e] | | | | | |
| | Seixas, 1985-94 | | Vandekerck-hove et al., 1996-97 | | | |
| | | | Wijdenes et al., 1997-98 | | | |
| Australia | | | | | | |
| Pickup, 1850-1995 | | | | | | |
| North America | | | | | | |
| Mouat et al., 1986-91 | | | | | | |
| Okin et al., 1979-97 | | | | | | Okin et al., 1979-97 |
| Latin America | | | | | | |
| Aagesen, 1876-1995 | McAuliffe et al., 1700-1991[e] | | | | | McAuliffe et al., 1700-1991[e] |
| | Manzano et al., 1950-85 | | | | | |

a   Shifting cultivation.
b   Continuous rainfed farming.
c   Wetland cropping.
d   Traditional oasis agriculture.
e   Time period might date back later than 1700.

Tables 8.3 to 8.5 summarize the characteristic sequence of land use transformation in the cropping sector. They give the type of land use in the time period selected by the case study authors to study desertification. In only few cases (n=26, or 20%), the cropping system remained stable over the time period considered, while in more cases (n=74, or 56%) transitions have occurred, partly across very distinct cropping systems and always towards increasing intensification. Thus, cropland degradation in dryland areas worldwide is linked to the transformation of cropping systems to ever increasing intensity.

In Asia, two main transitions dominate, i.e., the transition from rainfed, rotational cropping (with low degrees of mechanization) to continuous, mechanized farming, and the transformation of traditional oasis agriculture into large-scale irrigation schemes. Shifting cultivation, dryland farming and wetland cropping, all of them with no or low degrees of intensification, were not mentioned in the cases. The transition towards highest intensity levels is already very advanced – see Table 8.3

The major transition in the African cases is from dryland farming, occasionally shifting cultivation, to rotational, rainfed (fallow) farming. In only few cases, transitions included higher levels of intensification such as continuous, mechanized farming in East Africa. Irrigated farming is reportedly traditional wetland cropping (such as the cultivation of paddy rice on naturally flooded plots) aside with large-scale irrigation schemes, mainly for rice (such as along the Senegal River) – see Table 8.4.

The major cropping transition in the European cases is from rotational fallow farming to continuous, mechanized farming, taking into consideration that cases of abandoned farming are not considered here. None of the selected case studies reported that irrigation farming was associated with desertification. Cropping was of relatively low importance in the Latin American, but especially Australian and North American cases – see Table 8.5.

Could it be that the empirically supported narrative of increasing land use intensity in dryland ecosystems, the resource basis and productivity of which fluctuates in terms of water provision and vegetation, is the final link in the causal chain tying social to environmental change? This would imply that putting desertification in the context of land use change is about greater intensity of rural labour investments and other inputs to production. All of these activities involve greater commitments per unit of land area than traditional nomadic pastoralism or shifting cultivation require, such as the addition of new and more livestock species (Keya, 1998), year-round grazing (Aagesen, 2000), increased soil preparation through ploughing and continuous cropping (Mwalyosi, 1992; Kosmas, Gerontidis and Marathianou, 2000), increased diversion of artificially gained water onto marginal land (Genxu and Guodong, 1999). Is it really as simple as this?

## Directionality and Adaptation

Most of the major syndromes of dryland change indeed confirm the notion of higher land use intensity being associated with desertification processes, and most of the case studies of desertification report variants of general syndromes that relate to resource scarcity causing a gradual pressure of production on resources (e.g., inappropriate or excessive uses ending up in losses of soil productivity) or that relate to changing opportunities created by markets (e.g., new technologies for the intensification of land uses) (Mooney, Cropper and Reid, 2003; Lambin, Geist and Lepers, 2003). Could it be that, in global drylands, intensified land uses simply expanded on to spatial scales which were not known before? Could it be that desertification is land use intensification into dryland ecosystems that were immune from such land uses before, therefore increasing the ecosystem vulnerability to dry episodes? Large parts of the land change literature would indeed go conform with this (Kharin, 2003), i.e., be it the notion of demographic pressure inducing higher intensity (Boserup, 1965; Bilsborrow, 1987; Turner, Hyden and Kates, 1993; Mortimore, 1993), the exclusion of impoverished people from the most fertile land (Blaikie and Brookfield, 1987; Leonard, 1989; Kates and Haarmann, 1992), common property institutional failure (Ostrom, 1990), the heavy imposition of taxes by the state, or the 'simple reproductive squeeze' associated with commodity production by subsistence producers (Watts, 1985, 1987).

However, case studies reveal that intensive land uses do not necessarily end up with increased land-use pressures associated with desertification (Rasmussen, Fog and Madsen, 2001). There are examples of integrated animal husbandry/cropping systems in the West African Sudan-Sahel zone (Mortimore, Harris and Turner, 1999), and disinvestments in labour or abandonments even of agricultural production in remote drylands of the European Mediterranean basin (Kosmas, Gerontidis and Marathianou, 2000). The critical point is thus that pressures derive not just from changes in intensity and magnitude of resource extraction, but also from how resources are extracted (Watts, 1985; Turner, 1999). With a view upon the design of future case studies, this implies a better understanding of the material reality of land users, which is shaped not only by an oscillating resource base (rainfall, biomass) (Zeng, Neelin, Lau and Tucker, 1999; Seixas, 2000; Nicholson, 2002), but also by labour processes (Turner, 1999; Fernandez, Archer, Ash, Dowlatabadi, Hiernaux, Reynolds, Vogel, Walker and Wiegand, 2002), and the social relations of livestock production or farming operations, or both, over time (Rasmussen, Fog and Madsen, 2001; Zhou, Dodson, Head, Li, Hou, Lu, Donahue and Jull, 2002). With few exceptions, the cases, indeed, provided little profound information about changes in the mode of resource extraction, but were focussing more on the magnitude of extraction, more or less 'labelling' causative factors rather than exploring how they were embedded and transformed in a net of social ties and ecosystem properties. These 'labels' – e.g., 'malfunct customary land tenure', 'agricultural development policy', or 'population pressure' – were used to code the information, but hardly any further and more detailed information was gained this way on how resources were extracted. Most strikingly, the variables of what exactly accounts for demographic pressure had been very often not clearly named. Therefore, and at least, a better nuanced population analysis is required (Lambin, Geist and Lepers, 2003), among other improvements in the future design of case studies on desertification.

Another recurrent theme in desertification case studies is that socio-cultural changes have modified the adaptive strategies of dryland societies in the face of natural variability (which is inherent to dryland ecosystems) and have therefore reduced the resilience of socio-ecological systems, hence creating instability (Vogel and Smith, 2002; Fernandez, Archer, Ash, Dowlatabadi, Hiernaux, Reynolds, Vogel, Walker and Wiegand, 2002). In traditional rural societies, land productivity had often been multiply constrained and cultural organizations – be they nomadic pastoralists using a fluctuating grass resource base, or traditional oasis farmers with complex social arrangements for water use on scarce farming land – were adapted to episodic, but recurrent changes. Cultural change occurred in a cyclical manner. In the West African Sudan-Sahel, for example, land productivity is linked to oscillating rainfall and constrained by a nested system of seasonally differentiated use rights to a piece of land by various sedentary farming as well as pastoral groups (households, lineages, villages, groups of settlements, pastoral clans (Turner, 1999; Hiernaux and Turner, 2002). During periods of drought, modifications of land management were negotiated between all of these groups. Adaptation required high degrees of flexibility and cooperation as paramount in the highly mobile grazing systems practiced by nomads (Niamir-Fuller, 1999).

In modern societies, cultural change becomes directional rather than cyclical, because land users as well as investors and consumers of agricultural produce, embrace the promise of progress, particularly in technology and material well-being, as noted by Jiang (2002, p. 182).

> This directionality is manifested in intensified land use and the increasing tendency to view the environment as a means to material or economic gain. This type of cultural change brings about a directional component of change in the environment, thus accelerating dynamic changes in the ecosystem and landscape.

In many cases, strong directionality of dryland change was found that is manifested in land use intensification, and could be interpreted as an increasing tendency to view the environment as a medium for rapid material or economic gains. Such cases tend to prevail in the Central Asian desert and steppe region, be these livestock improvements on semi-arid rangelands, water-technology driven irrigation farming at arid basin sites, or mechanized farming of steppe soils responding to demands of remote populations. With the shift from traditional to modern rural societies, cultural change brings about a directional transformation of the environment, too. Typically, rates of dryland degradation had been nil in a few cases from the Sudan-Sahel where land productivity has been multiply constrained, and in cases from the European Mediterranean Basin. In some instances, highly intensive agro-silvo-pastoral production systems have developed. Data from some densely populated village areas in the Kano Close-Settled zone of northeastern Nigeria, for example, suggest that the transformation of woodland or shrubland into farmland not only added economic value to farming, but also led to an increase of plant biomass on farmed parkland as well as high livestock densities on crop residues (Mortimore, Harris and Turner, 1999).

## What are the Key Causative Factors?

This study found that the relative weight and particular sequences of the causes of desertification vary from region to region. The results run contrary to undue simplifications by proponents of single-factor causation who suggest various primary causes such as irrational or unwise land management by nomadic pastoralists or growing farming populations. Rather, outcomes of the study go conform with arguments such as that causes of desertification vary regionally, mainly in terms of localized intensity, and that, as one of the consequences, 'large areas of these drylands (in Asia, the Mediterranean, Africa, Oceania, and the Americas) are considered to be experiencing differing degrees of desertification' (Reynolds and Stafford Smith, 2002). Detailing these previous, broad observations, results are presented from cross-case and cross-regional analysis that allow to assess even the relative weight of both broad and specific proximate as well as underlying causes, be they biophysical or socio-economic in nature. The meta-analysis also revealed complex dynamical patterns of dryland degradation, given the importance of mediating factors, feedback loops, and initial conditions. This moves the understanding of causative mechanisms of desertification far beyond previous compilations of causes, which were often static and made no difference between proximate and fundamental causes (Geist and Lambin, 2004).

This study suggests that claims that desertification is either a human-made or purely natural (i.e., climate-driven) process should be nuanced. Rather, case studies indicate that desertification is a coupled biophysical and socio-economic process (Puigdefábregas, 1998; Reynolds and Stafford Smith, 2002), the details of which nonetheless need to be specified for the various regions affected by desertification. Therefore, the focus of desertification research has to move away from natural versus human-made causes or working hypotheses on causes, and better deal, in conceptual terms, with a coupled process. On the basis of causal synergies which were identified to lie behind desertification, an improved understanding of the process has to move far beyond a very often unstructured summarization or 'shopping list' (Watts, 2000) of causative factors. For example, UNEP still continues to consider 'deforestation, overgrazing, overcultivation, [and] other' as the key proximate factors (Middleton and Thomas, 1997), and a pluralist grab-bag of factors still continues to be fed into the IPCC process such as 'overcultivation, overstocking, fuelwood collection, salinization, urbanization, [and] other' (Le Houérou, 1995, 1996).

Rather, evidence was found to support notions as put forward, for example, by Reynolds (2001) such as that 'land degradation resulting from human settlement in the rangelands of the United States, Argentina, and Australia share many of the same ecological consequences and social issues'. Likewise, Pickup (1998) claimed that '[t]here are parallels between what is happening in Australia and events in parts of southern Africa, South America and the southwestern USA'. Indeed, these areas have commonalities in both causative mechanisms and initial conditions which make up important building blocks of historically convergent pathways. As laid down in chapter 7, comparative case studies reveal, for example, that the arid and semi-arid lands of northern Mexico share similar commonalities in their trajectories of land change (with more uncertainties, however, surrounding cases from southern Africa to be included in this group). Moving beyond it, the study also found that pathways of dryland change in these regions are historically convergent, but divert contemporarily.

In contrast to what is circulated in the desertification literature, this study cannot support the notion that the principal causative mechanism behind dryland degradation is over-working of the land by rural poor. Although poverty was found to be inherent to cases associated with desertification, these cases were few in number and originated from other areas (e.g., Patagonia) than indicated in the mainstream literature (Africa). The particular focus on Africa, triggered by the great Sahelian drought/famine (1968-73) and evident also in the full wording of the UNCCD, needs to be revisited and very likely revised. Neither in terms of poverty as the key causative mechanism nor in the magnitude of change, African drylands experience 'serious desertification' (UNEP, 1994). Claims such as that half of the farmland in Subsaharan Africa is affected by soil degradation and erosion, and that up to 80% of its pasture and range areas show signs of degradation due to a downward spiral of population pressure, agricultural decline and land degradation (Cleaver and Schreiber, 1994) must be treated cautiously.

Similarly, cases which were analysed here suggest that the empirical basis is weak to relate desertification to traditional nomadic pastoralism (Eckholm and Brown 1977; Cleaver and Schreiber, 1994; Mannion, 2002). First, hardly any nomadic activities were described in the cases which had been traditional or unaffected by modernization. Second, and if so, the mobility and flexibility of the system was esteemed to suit best to, or be sustainable with, the fluctuating resource bases of water and grass in drylands. Third, not nomadic pastoralism and not even semi-nomadic transhumance was cited as a causative factor, but rather processes which led to the transformation of these highly flexible and well adapted land use systems such as policy-driven sedentarization of nomads and the transfer of livestock wealth to agriculturalists or urban investors, following a rationale of land use which is completely different from the nomadic mode of livestock raising. Like in tropical deforestation – i.e., with impoverished rural poor, mainly shifting cultivators being allegedly the key agents of deforestation – the very similar notion regarding desertification should better be treated as a 'myth' (Lambin, Turner, Geist, Agbola, Angelsen, Bruce, Coomes, Dirzo, Fischer, Folke, George, Homewood, Imbernon, Leemans, Li, Moran, Mortimore, Ramakrishnan, Richards, Skånes, Steffen, Stone, Svedin, Veldkamp, Vogel and Xu, 2001).

Another controversial issue in desertification research is the question of whether there 'is ... a relationship between how land is used (accountability) and ownership (which tends to be low in poor and high in rich ones)' (Reynolds and Stafford Smith, 2002). Taking the various examples of reported conflicts about land between farmers and herders, it was found that they occur under various forms of land ownership, and thus seem independent from the socio-economic system, thus supporting a view expressed by Rozanov (1991), for example. Land conflicts were found to happen in the West African Sudan-Sahel as well as in Asia under communal ownership, in the US with private land ownership, and in the former Soviet Union with a planned economy and public property. Also, it was found that

Discussion 215

private ownership has indeed been high in 'rich' countries such as Australia, but that the major trend there had been economically not viable, indebted farms which lost the adaptive capacity to successfully deal with droughts. In contrast, private ownership might have been 'low' in 'poor' countries such as Burkina Faso or Nigeria, but increased vulnerability or losses of adaptive capacity were rather low or nil. Thus, it could be assumed that the land ownership linkage with desertification should better not be taken for granted.

**Indicators – Which ones are Best?**

The comparative study of desertification cases in chapter 6 shows that there are no authoritative, key and easy to apply indices available for the monitoring and evaluation of desertification processes. Given the diversity of social and biophysical processes found in the cases, a key indicator of land cover change – such as the annual rate of deforestation in the study of forest cover change – won't be achievable. Could there be a small set of relevant indicators, though?

What could be learnt from the cases is that combinations of indicators are mostly applied, because combinations of environmental (ecological, meteorological) and human (socio-economic) variables are important. A reasonably good multidisciplinary approach to measuring desertification is often applied, but in reality most case studies fail to adopt a truly integrative approach, especially with regard to an integration of biophysical and human indicators. What looks like a simple statement, is not trivial. Considerable importance has been attached to dryland degradation as a state of criticality which affects human use values (Blaikie and Brookfield, 1987; Kasperson, Kasperson and Turner, 1995; Reynolds and Stafford-Smith, 2002; Vogel and Smith, 2002). In particular, the application of socio-economic variables indicating critically affected human use values so far has been poor: in the cases, only changes in terms of reduction or elimination of species are indicated which critically affect the diversity of wild animal or plant species habitats. Also, indicators of desertification need to better capture the most common type of land change in semi-arid regions which is land-cover modification rather than land cover conversion (Diouf and Lambin, 2001). Indeed, findings of the study reveal that indicators, for example, of chemical soil degradation – i.e., describing subtle changes in land quality which affect the character of land cover without changing its overall classification – are far less used than others, mainly cover-related vegetation indicators. Also, no species or biomass type appears to be a good indicator to be generalized for all dryland areas, and only few species are good indicators of drought conditions, even.

This is not to propose a design of indicators using the concept of state, driving force, impact, and response – i.e., have in separate 'state indicators', 'driver indicators', 'pressure indicators', etc. – but rather to take up the notions of causal synergies involved as major causative 'mechanisms' and link them to criticality towards human use values. Indicators could be, for example, syndrome-oriented, following the insight that not all causes of land-use change and not all levels of organization are equally important, as laid down in chapters 4 and 5. For any given human-environment system, a limited number of causes are essential to predict the general trend in land use (Lambin, Geist and Lepers, 2003). In this study, a matrix of specific ecological, socio-economic and meteorological indicators was extracted from a wide array of case studies. In a next step, it could further be related to a proposed limited set of syndromes of dryland change as distilled again from the cases. There are other syndrome approaches and taxonomies, such as describing archetypical, dynamic, co-evolutionary patterns of human-environment conditions, reflecting expert opinions based on local case examples (Petschel-Held, Lüdeke and Reusswig, 1999; Petschel-Held, 2004), or developing qualitative trajectories in terms of the development of the wealth of local populations and the state of the environment (Kasperson, 1993; Kasperson, Kasperson and Turner, 1999).

In summary, there are some good arguments that, despite of the large diversity of causes and situations leading to desertification, there are some generalizable patterns of dryland change that result from recurrent interactions between causative factors, following specific sequences of events. Even though, at the detailed level, these sequences play out differently in specific situations, their identification may confer some predictive power by analogy with similar pathways in comparable regional and historical contexts (Lambin, Geist and Lepers, 2003). Thus, the identification of typical pathways leading to desertification, as done in chapter 7, can set a stage for indicator development, coupled with the insight from the cases that dryland degradation is associated with a sequence of events that relates to greater intensity of land use in fluctuating environments.

As for monitoring purposes, indicators should be sensitive to capture what was called environmental patterning (Schlesinger, Reynolds, Cunnigham, Huenneke, Jarrell, Virginia and Whitford, 1990; Keya, 1998; Seixas, 2000), indicating the enormous and increasing spatial variability of land change classes. It appears that extrapolating or upscaling site-specific information to relatively large areas is not a solution for monitoring, if compared to addressing right away the complexity of desertification which directly stems from typical process changes over space and time, i.e., monitoring increasing environmental heterogeneity at the landscape scale.

**Rates – Is it a Slow or Fast Process?**

Results from the meta-analytical study, as laid down in chapter 5, can contribute only marginally to an improvement in the estimation of the rates of desertification, indicating the speed at which dryland degradation occurs. In chapter 1, the point was made that current estimates of the rates are based on limited data and generally rough at best, while in some instances gross exaggerations were produced. As for the latter, an often cited example is the study of Lamprey (1988) who reported that the Sahara was expanding southward at a rate of 5.5 km per year in the Republic of Sudan. Case studies, however, which originate from the area and were included in the meta-analysis, did neither confirm the magnitude of claimed sandification in the area nor the expansion of desert-like conditions at all. In cases where increases in sand cover were reported, with most of them stemming from the Central Asian desert and steppe region, sandification happened at a maximum of 900 to 1,200 m annually in the 1983-95 period (Circum Aral Sea Region), which is one fifth of the spectacular value cited above. It was found that strikingly more cases of 'creeping sands', the most spectacular outcome of desertification, occur in Central Asia than in Sudano-Sahelian Africa, where the northern limit of rainfed cultivation, or the southern fringe of the Sahara, turn out to be more or less stable, what has been largely confirmed by other analyses, too (Tucker, Newcomb and Dregne, 1991; Schulz, 1994; Schlesinger and Gramenopoulos, 1996; Tucker and Nicholson, 1999).

In more general terms, insights from case studies suggest that there will be no quick answer to the question whether desertification is a slow or fast process, unless the indicators to be used have been exactly defined. Results here show a variety of syndromes unfolding at various, but mainly slow speed, and a variety of ecological, meteorological and even socio-economic indicators to which some empirical data had been put. Except for very few dryland sites at which rapid change at a large magnitude was reported from, most of the process rates were not spectacular. This is true for fast variables which are highly variable, such as the annual decrease in rainfall at 1.2% annually on average, the annual decrease in the areal extent of undegraded vegetation cover at 3.4% yearly on average, or the annual increase in the areal extent of degraded vegetation cover at 4.3% annually on average. The same is true for slow, gradually operating variables such as the increase in eroded, bare, rocky ground cover (1.2%) or the areal spread of salinity in irrigation schemes (0.5%). These values are hardly compatible with rates of desertification as circulated in the literature, such as that the world's drylands have an average rate of current desertification progress of 3.5% per year (Dregne, Kassas and Rozanov, 1991; Dregne 2002). Therefore, there is some indication to reasonably assume that many estimates of desertification might indeed be overestimated. Thus, verification of some key process rates of major indicators as proposed earlier in this chapter is urgently needed. In addition, evidence was found from the cases (especially from northern China) that process rates need to be studied over long time periods, dating back to historical and even ancient times. For example, a working hypothesis for future research, stemming from a wide array of cases analysed here, could be as follows. Desertification speeded up in the last 19$^{th}$ century, while, in some localities, it declined during the 20$^{th}$ century (e.g., Mediterranean Europe, US Southwest), but occurs at even faster speed in other localities such as Central Asia and southern Argentina currently, compared to the past and, in particular, from the 1990s onwards.

A more solid empirical basis for discussing rates of desertification needs to be established very soon. The study of Lepers, Lambin, Janetos, DeFries, Achard, Ramankutty and Scholes (2003) could be seen as a first step into this direction. The suggestion to treat the process of desertification as a gradual or sudden decline in 'slow' variables (Reynolds and Stafford-Smith, 2002) suffers from the fact that these slow variables have not yet been supported by quantifiable, reliable data until now. Slow variables with long turnover times, though they determine the boundaries of sustainability and collectively govern the land use trajectory, were hardly given empirical data in the cases studied here. This relates to both socio-economic data such as natural fertility increases or mortality decreases of human populations as well as to biophysical data such as soil nutrient decline or the areal spread of salinity in irrigation schemes (the latter was quantified in a single case study only).

**Pathways – All Necessarily Headed for Irreversibility?**

Concerning the debate on whether desertification necessarily implies a unidirectional or irreversible trajectory of land change, as put forward, for example, by Eckholm and Brown (1977) and Mainguet and Chemin (1991) case study evidence suggests that there are regional as well as temporal variants of dryland change with various outcomes. For example, desertification happens in an episodic manner in the Australian drylands, it is advancing in the drylands of northern Mexico and southern Argentina, it is on a reverse in the US Southwest, it is fluctuating with short oscillations in African drylands, and progression is low in the European part of the Mediterranean Basin. Clearly, a better focus on thresholds and control points of land use and ecosystem change is needed to better address these issues. Indications were found, however, that most rapid and very likely irreversible changes occur in (palaeo) basin and (palaeo) lake or sea sites under semi- or subdesert conditions. These sites are mainly used for large-scale irrigation farming, mainly in Central Asia. High vulnerability, both in social and ecological terms, was also reported from high-cold mountain ecosystems, again mainly in Central Asia. These sites showed a comparatively low resilience against coupled human and climatic impacts, chiefly because the functioning of these ecosystems depends upon water provision from ice and snowpack of adjacent high mountain zones so that thresholds of land-use and ecosystem change are easily surpassed. In contrast, steppe soil sites showed a reportedly much higher resilience, though lying roughly in the same zone. Paleao basin, sea or lake as well as high-cold mountain ecosystem sites are in particular prone to worsening climatic conditions, so that especially climatic factors need to be better framed in dealing with the issue of an assumed irreversibility of dryland degradation. 'Badlands' were cited as experiencing irreversible degradation due to geological pre-disposement for the formation of desert-like conditions, mainly through highly erosive lithologies. These areas, mostly lacking any discernible human impact in their causation, were cited not to spread across vast spatial scales. These localized 'badlands', however, have to be held apart from the above mentioned paleao-sites which usually have a century- or millennia-old tradition of land uses.

The issue of pathways relates further to insights from the theory of non-equilibrium range ecology (Behnke, Scoones and Kerven, 1993; Walker, 1993). From cases located in arid sites, which explicitly addressed the issue, biotic factors such as animal grazing were unanimously cited to be of much higher importance than rainfall variability governing the dynamics of ecosystem development. This runs contrary to non-equilibrium assumptions. Keya (1998), for example, claims that in many areas of the arid Turkana Steppe in northern Kenya domestic livestock have effects that override not only those of indigenous species but have also more weight than abiotic factors. Likewise, Ho (2001) found no evidence from several sites in the Ningxia Hui Autonomous Region of northern China for the proposition that in a non-equilibrium system animal herd growth and decline are primarily determined by the level of precipitation. He concluded that other factors may be at play such as soil properties and the socio-economic or institutional context, mainly. Indeed, the historically similar, but contemporarily divergent pathways which were identified for the predominantly arid parts of Patagonia and the arid US Southwest in chapter 7, for example, point to the outstanding importance of socio-economic context variables. However, this meta-analytical study was not designed to explicitly test a working hypothesis such as that semi-arid ecosystems behave as non-equilibrium systems. As a recommendation for future research on non-equilibrium systems, a wide array of cases from various dryland areas in the world probably needs to be studied in a comparative perspective.

# Chapter 9

# Conclusions

**Introduction**

Following the discussion of results from the meta-analysis of 132 subnational cases of desertification in the previous chapter, conclusions are drawn here in a threefold manner. First, implications are outlined with a view on future research on desertification, including modelling. Second, further conclusions are drawn with a view on policies designed to control desertification. Finally, it is concluded with a short outlook on the future design of case studies, namely case study comparisons to explore the dynamical causal patterns of desertification.

**Implications for Future Research on Desertification**

Some progress in the quantification and understanding of land-use/cover changes has been achieved since the 1970s when desertification reached the world stage, but much remains to be learned before one can fully assess and project the future role of land-use/cover change in drylands in the functioning of the earth system and identify conditions for sustainable land use in these zones (Lambin, Geist and Lepers, 2003; Mooney, Cropper and Reid, 2003; Steffen, Sanderson, Tyson, Jäger, Matson, Moore, Oldfield, Richardson, Schellnhuber, Turner and Wasson, 2004).

While new estimates of areas and rates of major land-use/cover conversions have narrowed down some uncertainties in various land change classes, a number of more subtle land changes still need to be better quantified at a global scale. This is particularly the case for desertification or dryland degradation, i.e., anthropogenic changes that strongly interact with natural environmental variability and therefore require longitudinal data over a long time period for a reliable assessment. Prominent among these changes are soil degradation in arid or semi-arid croplands, changes in the extent and productive capacity of pastoral lands in drylands, and degradation as well as deforestation of tropical dry forests. These changes are thought to be widespread, because local- to national-scale studies demonstrate their importance for livelihoods as well as their ecological significance, but they remain poorly documented at the global scale. For example, as a result of a most recent effort to overcome limitations of the GLASOD database by identifying areas of most rapid and most recent land cover changes in global drylands, a quantification of contemporary 'hot spots' of desertification still suffers from large uncertainties (Lepers, Lambin, Janetos, DeFries, Achard, Ramankutty and Scholes, 2003).

Analyses of the causes of land-use change have moved from simplistic single-cause explanations to an understanding that integrates multiple causes and their complex interactions. A few general pathways leading to land-use change in drylands have been identified from a wealth of local case studies in this study, and a systematic analysis of local scale land-use change studies, conducted over a range of time scales, clearly helped to uncover general principles. This inductive process of generalization paves the way for the development of more realistic models of land-use change. Lambin, Geist and Lepers (2003) note that an 'improved understanding of the complex dynamic processes underlying land-use change will allow more reliable projections and more realistic scenarios of future changes'. An understanding of factors that control positive and negative feedback in land-use change will be crucial, especially, how the 'relative strength of amplifying and attenuating feedback can be influenced by policies that control switches between land-use/cover change regimes dominated by positive or negative feedback'. Nevertheless, different perspectives of understanding still tend to follow different lines of explanation of the causes of land-use change because each focuses on specific organizational levels and temporal scales of the human-environment systems, as Lambin, Geist and Lepers (2003, p. 231) pointed out.

> Whereas a system perspective tends to focus on gradual and progressive processes of change at the scale of large entities, the agent-based perspective deals with people's own foreseeable futures at the individual level, and the narrative perspective adopts a much longer time horizon and focuses on critical events and abrupt transitions. Different assumptions about temporality lead to varying explanations and interpretations of the causes and significance of environmental changes. These assumptions should be made explicit to facilitate the development of an integrative theory of human-environment relationships.

The analysis, however, of interaction, coherence, or conflict between social and biophysical responses to changes in both ecosystem services and earth system processes caused by land changes is still a largely unresearched area. It will be a central focus of the new Global Land Project (GLP) of the International Geosphere-Biosphere Programme (IGBP) and the International Human Dimensions Programme (IHDP) of Global Environmental Change. Lambin, Geist and Lepers (2003, p. 232) described the path to GLP as follows.

> [I]nitial concerns about global land-use/cover change in drylands arose from the realization that transformation of the land surface there influences climate change and reduces biotic diversity, hence the interest in deforestation, desertification and other changes in natural vegetation. The more recent focus on issues related to ecosystem goods and services, sustainability, and vulnerability has led to a greater emphasis on the dynamic coupling between human societies and their ecosystems at a local scale

Despite the dooming prospects which are commonly associated with contemporary desertification, especially in the aftermath of the great Sahelian drought/famine (1968-73), an improved understanding of processes of land-use change in drylands has already led to a shift from a view condemning human impact on dryland environments as leading mostly to a deterioration of earth system processes to an emphasis on the potential for ecological restoration through land management. This change, paramount, for example, in the focus on highly intensive agro-sivo-pastoral land use systems in drylands, reflects an evolution of the research questions, methods, and scientific paradigm such as the Dahlem Desertification Paradigm (Stafford Smith and Reynolds, 2002), which was used to guide the framework of this study.

## Implications for Controlling Desertification

Despite claims to the contrary, the amount of suitable land remaining for crops is very limited in most developing countries, where most of the drylands are located and where most of the growing food demand originates. Where there is a large surplus of cultivable land, land is often under rain forest or in marginal areas such as drylands (Young, 1999; Döös, 2002). The need to manage drylands in a sustainable manner is obvious, but given the large variety of ecosystems and land use histories involved, universal assessments and policies to guide the design of future land use patterns will necessarily fail. Likewise, case study based evidence reveals that no global set of indicators as part of a global-scale assessment of the desertification status could reveal the complexity of human-environment systems inherent to dryland change (Geist and Lambin, 2004).

To achieve sustainable agricultural management, any policy intervention has to be region-specific, and sometimes even adapted to local particularities of the 'real world' pathways of dryland change, involving trade-offs between economic gains and conservation. This is why a thorough understanding of the main driving forces, key actors and processes of agricultural change and land use patterns in drylands, as done in this study, is vital to better assess the long-term change occurring in rural lands at the global agricultural dryland frontier. If land use patterns in drylands are to be sustainable, they need to balance the legitimate interests of development and equally legitimate global concerns over the environmental consequences of land cover change in drylands. This implies that trade-offs need to be considered between what is to be sustained, and what is to be developed. From the viewpoint of managing highly dynamic land use transitions in drylands, as laid down in the previous chapter, there must be an incentive structure introduced for various actors operating at different scales influencing negotiations about outcomes that suit the various interests involved. 'Multiplicity' as the most common theme identified in the desertification studies must be matched by equally multiple intervention measures.

A matrix, developed through the Alternatives to Slash-and-Burn (ASB) Programme, is put forward here as a method for assessing trade-offs and to draw implications for land use policies (Geist, Lambin, Palm and Tomich, 2005). The ASB matrix, originally designed for application in tropical humid forest ecosystems, provides an approach to assess the degree of trade-offs (and complementarities) between global environmental objectives served, for example, by mitigating desertification and national and local objectives, often involving conversion or modification of drylands to other uses, and to identify innovative policies and institutions needed to reconcile ecosystems and human well-being at the local level. It is also a powerful tool for looking at specific trade-offs between provisioning and regulating services in various dryland ecosystems under human uses, i.e., losses of certain ecosystem functions of global importance such as biodiversity or albedo, affecting central functions of the climate system, *versus* provision of food, fibre and feed services for local livelihoods as well as national economic development. The matrix also provides a basis for policymakers and stakeholders to assess trade-offs across land use systems regarding development options and ecosystem services (Tomich, Cattaneo, Chater, Geist, Gockowski, Lambin, Lewis, Palm, Stolle, Valentim, Noordwijk and Vosti, 2004).

In principle, the ASB matrix approach could be applied to ecoregions and land use systems outside the humid tropics. It could be applied to and modified for dryland areas, thus helping to reveal the mechanisms to create incentives for the conservation of dryland ecosystems that are sufficient to offset the powerful incentives for dryland modification or conversion (Geist, Lambin, Palm and Tomich, 2005). Details of an application in drylands still need to be developed and integrated, for example, in LADA– and DDP–driven work. As part of this exercise, technologies, and to a much larger extent, institutional capacities and policies will be key instruments to affect the rate and pattern of dryland degradation. Location-specific studies or benchmark sites need to be located in active zones of desertification, and a careful identification of the factors at work in a given location will be a prerequisite for getting the mix of interventions right, while minimizing the cost to local peoples' livelihood opportunities and other legitimate development objectives. Policy makers need accurate, objective information regarding the private and social costs and benefits of alternative land use systems on which to base their inevitably controversial decisions. To help them weigh up the difficult choices they must make, the matrix may help them as a tool (Tomich, van Noordwijk, Vosti and Witcover, 1998; Tomich, Cattaneo, Chater, Geist, Gockowski, Lambin, Lewis, Palm, Stolle, Valentim, Noordwijk and Vosti, 2004).

In the matrix, dryland ecosystems and the land use systems that replace them will have to be scored against different environmental, socio-economic and institutional criteria reflecting the objectives of different interest groups. To enable results to be compared across sites, the systems specific to each site are grouped according to broad categories such as (dry) forests and agroforests to grasslands and pastures. The criteria may be adjusted to specific locations, but the matrix always comprises indicators for global environmental concerns such as biodiversity and water cycle, agronomic sustainability, employment opportunities and economic growth, among many others. As with all the indicators used in the matrix, agronomic sustainability is a plot level indicator. It refers specifically to yield levels over time as a result of continuation of that particular land use. If yields under continued land use would be stable or increasing, then the land use is considered to be agronomically sustainable. If yields would be decreasing, it is considered unsustainable. The reference point is the farmer's ability to manage the resources, and indicators are based on an expert panel assessment of each land use regarding a range of soil characteristics, including trends in nutrients and organic matter over time (Tomich, Cattaneo, Chater, Geist, Gockowski, Lambin, Lewis, Palm, Stolle, Valentim, Noordwijk and Vosti, 2004).

Over a decade or so, researchers will have to fill in the matrix for representative benchmark sites across the world's drylands. The social, political and economic factors at work at these sites vary greatly, as also do their current resource endowment and initial conditions (land use and environmental histories). Typical recurrent causative factor combinations and typical pathways of land change, originating from this meta-analysis, may help to give a structure to the selection of benchmark sites globally. At each site, researchers will need to evaluate land use systems both as they are currently practised and in the alternative forms that could be possible through policy, institutional and technological innovations. As in the humid forest zones, a key question could be whether the intensification of land use through technological innovation could reduce both poverty and desertification (Geist, Lambin, Palm and Tomich, 2005).

The matrix allows researchers, policymakers, environmentalists and others to identify and discuss trade-offs among the various objectives of different interest groups, and/or to discuss ways of promoting land use systems that could provide a better balance among trade-offs without making any group worse off, but that still were not broadly adopted. If available, attractive 'middle paths' could be revealed between environmental benefits and equitable economic growth. However, for both economic and ecological reasons, no single land use system should predominate at the expense of all others. Mixes of land uses usually increase biodiversity at a landscape level, if not within individual systems, and also can enhance economic and ecological resilience. Where the productivity of the natural resource base has already sunk to low levels, concentrating development efforts on the simultaneous environmental and economic restoration of degraded landscapes is an option well worth exploring. The precise mix of interventions needed – hence the benefits and costs of restoration – will vary from place to place (Geist, Lambin, Palm and Tomich, 2005).

A critical point which has arisen from this study is that pressures upon dryland resources derive not only from changes in intensity and magnitude of resource extraction (grass, water), but also in how resources are extracted. This would leave some prospect, at least, to increase, for example, (re)investments in herding labour, mediating the often disturbed social relations between herding and farming populations, and safeguarding land use practices in dryland ecosystems which are based upon multiply constrained land productivity.

**Implications for Future Case Study Comparisons**

While the Dahlem Desertification Paradigm has been conceived to drive several networks of new, original case studies of desertification (Reynolds, Stafford Smith and Lambin, 2003), insights from this meta-analytical study can help to inform the design of future case studies. They can also contribute to, or better inform programmatic efforts such as the Land Degradation Assessment in Drylands

(LADA) project, including case study activities, with both initiatives briefly described in the introductory chapter.

In summary, and as discussed in the previous chapter, evidence presented in this study from empirical case studies that identify both proximate causes and underlying driving forces of desertification show that dryland degradation – with desertification as a potential but not necessary outcome – is determined by different combinations of various proximate causes and underyling driving forces in varying geographical contexts. Next to all of these combinations include coupled socio-economic and biophysical factors. Some of these combinations are robust geographically – such as the spread of watering and related infrastructures driven by growth-oriented policies and economic demands stirred by demographic changes – whereas most of them are region and time specific. The observed causal factor synergies and pathways of dryland change challenge single-factor explanations that put most of the blame of desertification upon the overworking of land by increasing numbers of rural poors and by nomadic populations. Rather, this analysis reveals that, at the underlying level, public and individual decisions largely respond to mostly national-scale policies promoting advanced land use technologies and creating new economic opportunities. These responses are mediated by land tenure systems and other local-scale institutions. They need to be incorporated into any policy-oriented consideration to assess trade-offs between what needs to be sustained and what needs to be developed. The analysis further reveals that, at the proximate level, regional distinct modes of increased aridity, expansion of cropping/grazing activities, infrastructure extension, and, to a lesser degree, wood extraction prevail in causing desertification. Also, land uses were found to be not static, but highly dynamic and, in next to all cases, showing a distinct directionality towards increasing land use intensity. How could these insights guide the design of future case studies of desertification, in particular, using a comparative perspective? Nine major points have been extracted from the previous chapters.

First, future case studies of desertification need to address, at the proximate level, highly dynamic cropping and livestock industry transitions, and shifts between these two options, rather than static land uses. Thresholds and switch and choke points need to be addressed, which are inherent to these dynamic transition processes. A listing of causative factors in a 'shopping list' approach should be avoided, as well as feeding these factors, for example, into the IPCC process in a theoretically rather untidy manner. Improved insights gained on the nature of land use transitions will also help to fill in a matrix of trade-offs between legitimate concerns coming from the global environmental change community and the equally legitimate development objectives.

Second, and related to the implication above, future case studies could be built around one major working hypothesis which stems from the insight that mainly variants of general syndromes were reported, including a mix of slow and fast variables, such as resource scarcity causing a gradual pressure of production on resources (e.g., inappropriate or excessive uses ending up in losses of soil productivity), or changing opportunities created by markets (e.g., new technologies for the intensification of land uses). The final link to be tested could be whether desertification is about land use intensification into dryland ecosystems that were immune from such land uses before, therefore increasing the ecosystem vulnerability to dry episodes. The test should be done empirically, on the basis of household survey data, secondary land use statistics and solid biophysical information such as on changes in soil nutrient content, etc. The quantitative approach could very well address and test the slow *versus* fast variables of change – see Tables 2.2 and 2.3. However, insights from case studies suggest that there will be no quick answer to the question whether desertification is a slow or fast process, unless the measure of quantification, or the indicator to be used, have been exactly defined.

Third, and on the other hand, future case studies need to consider that pressures derive not just from changes in intensity and magnitude of resource extraction, but also from how resources are extracted. This implies a better understanding of the material reality of land users, which is shaped not only by an oscillating resource base (rainfall, biomass), but also by labour processes, and the social relations of livestock production or farming operations, or both, over time. Clearly, this would imply a narrative, descriptive, qualitative and within-case perspective, as opposed to the quantitative, empirical and cross-case approach described before.

Fourth, future case studies need to better address, among other causes, the blurred notion of 'demographic pressure' often associated with desertification, while an assumed 'ownership linkage' (Reynolds and Stafford Smith, 2002) should better not be taken for granted. Clearly, a better nuanced population analysis is required (Lambin, Geist and Lepers, 2003). Detailing the often broad observations about impacts of causative factors such as population, cross-case and cross-regional analyses are vital because they allow to assess the relative weights of proximate as well as underlying causes, be they biophysical or socio-economic in nature. Mediating factors, feedback loops, and initial conditions are important in the analysis of pathways. Thus, the understanding of causative mechanisms of desertification is moved far beyond previous compilations of causes, which were often static and made no difference between proximate and fundamental causes.

Fifth, the details of desertification as a coupled biophysical and socio-economic process need to be specified for the various regions affected by desertification. The nine assertions of the Dahlem Desertification Paradigm may provide a framework (Stafford Smith and Reynolds, 2002).

Sixth, a particular focus on Africa is misleading and should be avoided. Rather, African cases need to be compared with other dryland zones, and the working hypothesis could very well be that neither in terms of poverty nor in the magnitude of change, African drylands experience 'serious desertification' (UNEP, 1994), but show higher resilience than other dryland areas.

Seventh, no investments should be done to arrive at a key indicator of desertification such as the annual rate of deforestation in the study of forest cover change, given the diversity of social and biophysical processes found in the cases. Indicators of desertification need to better capture the most common type of land change in semi-arid regions which is land-cover modification rather than land cover conversion (Diouf and Lambin, 2001). Indicators could be, for example, syndrome-oriented, following the insight that not all causes of land-use change and not all levels of organization are equally important In this study, a matrix of specific ecological, socio-economic and meteorological indicators was extracted from a wide array of case studies – see Tables 7.1 to 7.9. In a next step, this information could be related to a proposed limited set of syndromes of dryland change as distilled again from the cases, or developed elsewhere (Kasperson, 1993; Kasperson, Kasperson and Turner, 1999; Petschel-Held, Lüdeke and Reusswig, 1999; Petschel-Held, 2004).

Eighth, a more solid empirical basis for discussing rates of desertification needs to be established. The study of Lepers, Lambin, Janetos, DeFries, Achard, Ramankutty and Scholes (2003) could be seen as a first, still investigating step into this direction. The suggestion to treat the process of desertification as a gradual or sudden decline in 'slow' variables (Reynolds and Stafford-Smith, 2002) suffers from the fact that these slow variables have not yet been supported by quantifiable, reliable data until now. Slow variables with long turnover times, though they determine the boundaries of sustainability and collectively govern the land use trajectory, were hardly given empirical data in the cases studied here. This relates to both socio-economic data such as natural fertility increases or mortality decreases of human populations as well as to biophysical data such as soil nutrient decline or the areal spread of salinity in irrigation schemes (the latter was quantified in a single case study only).

Finally, future case studies should be designed in such a way that they can be a direct input to fill in lacking information for a trade-off matrix of land-use transitions in drylands (Geist, Lambin, Palm and Tomich, 2005).

# Bibliography

Adams, C.R. and Eswaran, H. (2000), 'Global land resources in the context of food and environmental security', in S.P. Gawande (ed), *Advances in land resources: Management for the 20th century*, Soil Conservation Society of India, New Delhi, pp. 35-50.

Adger, W.N., Benjaminsen, T.A., Brown, K., Svarstad, H. (2001), 'Advancing a political ecology of global environmental discourses', *Development and Change*, Vol. 32, pp. 681-715.

Aubreville, A. (1949), *Climats, forêts, et désertification de l'Afrique Tropicale*, Societé d'Editions Géomorphiques, Maritimes et Coloniales, Paris.

Barbier, E.B. (2000a), 'Links between economic liberalization and rural resource degradation in the developing regions', *Agricultural Economics*, Vol. 23, pp. 299-310.

Barbier, E.B. (2000b), 'The economic linkages between rural poverty and land degradation: Some evidence from Africa', *Agriculture, Ecosystems and Environment*, Vol. 82, pp. 355-70.

Bassett, T.J. (1993), 'Land use conflicts in pastoral development in northern Côte d'Ivoire', in T.J. Bassett and D. Crummey (eds), *Land in African agrarian systems*, University of Wisconsin Press, Madison, pp. 131-54.

Bassett, T.J., Zuéli, K.B. (2000), 'Environmental discourses and the Ivorian savanna', *Annals of the Association of American Geographers*, Vol. 90, pp. 67-95.

Behnke, R., Scoones, I., Kerven, C. (1993), *Range ecology at disequilibrium: New models of natural variability and pastoral adaptation in African savannas*, Overseas Development Institute, London.

Behrenfeld, M.J., Randerson, J.T., McClain, C.R., Feldman, G.C., Los, S.Q. Tucker, C.J., Falkowski, P.G., Field, C.B., Frouin, R., Esaias, E., Kolber, D.D., Pollack, N.H. (2001), 'Biospheric primary production during an ENSO transition', *Science*, Vol. 291, pp. 2594-7.

Benjaminsen, T.A. (2001), 'The population-agriculture-environment nexus in the Malian cotton zone', *Global Environmental Change*, Vol. 11, pp. 283-95.

Bilsborrow, R.E. (1987), 'Population pressures and agricultural development in developing countries: A conceptual framework and recent evidence', *World Development*, Vol. 15, pp. 183-203.

Biswas, A.K., Masakhalia, Y.F.O., Odego-Ogwal, L.A., Palnagyo, E.P. (1987), 'Land use and farming systems in the Horn of Africa', *Land Use Policy*, Vol. 4, pp. 419-43.

Blaikie, P. and Brookfield, H. (1987), *Land degradation and society*, Methuen, London.

Bohle, H.-G., Downing, T.E., Field, J.O., Ibrahim, F.N. (eds) (1993), *Coping with vulnerability and criticality: Case studies on food-insecure people and places*, Freiburg Studies in Development Geography No. 1, Breitenbach, Saarbrücken, Ft. Lauderdale.

Boserup, E. (1965), *The conditions of agricultural growth: The economics of agrarian change under population pressure*, Aldine: Chicago.

Breckle, S.W., Veste, M., Wucherer, W. (2002), 'Deserts, land use and desertification', in S.W. Breckle, M. Veste and W. Wucherer (eds), *Sustainable land use in deserts*, Springer Verlag, Heidelberg, New York.

Briceño, S. (1998), 'Desertification, biodiversity and climate change: The path to sustainable development', *The Courier*, Vol. 172, pp. 69-71.

Brooks, E., Emel, J. (1995): 'The Llano Estacado of the American southern high plain', in J.X. Kasperson, R.E. Kapserson and B.L. Turner II (eds), *Regions at risk: Comparisons of threatened environments*, UNU Studies on Critical Regions, United Nations University Press, Tokyo, New York, Paris, pp. 255-303.

Bunney, S. (1990), 'Prehistoric farming caused devastating soil erosion', *New Scientist*, Vol. 125, p. 20.

Charney, J. and Stone, P.H. (1975), 'Drought in the Sahara: A biogeophysical feedback mechanism', *Science*, Vol. 187, pp. 434-5.

Chasek, P.S., Corell, E. (2002), 'Addressing desertification at the international level: The institutional system', in J.F. Reynolds and D.M. Stafford Smith (eds), *Global desertification: Do humans cause deserts?*, Dahlem Workshop Report No. 88, Dahlem University Press, Berlin, pp. 275-94.

Claussen, M., Brovkin, V., Ganopolski, A. (2002), 'Africa: Greening of the Sahara', in W. Steffen, J. Jäger, D.J. Carson and C. Bradshaw (eds), *Challenges of a changing earth*, Proceedings of the Global Change Open Science Conference, Amsterdam, The Netherlands, 10-13 July 2001, Springer, Berlin, Heidelberg, New York, pp. 125-28.

Cleaver, K., Schreiber, G. (1994), *Reversing the spiral: The population, agriculture, and environment nexus in sub-Saharan Africa*, World Bank, Washington, D.C.

Darkoh, M.B.K. (1998), 'The nature, causes and consequences of desertification in the drylands of Africa', *Land Degradation & Development*, Vol. 9, pp. 1-20.

Desert Encroachment Control and Rehabilitation Programme (1976), General Administration for Natural Resources, Ministry of Agriculture, Food and Natural Resources and the Agricultural Research Council, National Council for Research (Republic of Sudan) in collaboration with UNEP, UNDP and FAO.

Diamond, J. (1999), *Guns, germs, and steel: The fates of human societies*. Norton, New York.

Diouf, A., Lambin, E.F. (2001), 'Monitoring land-cover changes in semi-arid regions: Remote sensing data and field observations in the Ferlo, Senegal', *Journal of Arid Environments*, Vol. 48, pp. 129-48.

Dirzo, R., Raven, P.H. (2003), 'Global state of biodiversity and loss', *Annual Review of Environment and Resources*, Vol. 28, pp. 137-67.

Downing, T.E., Lüdeke, M. (2002), 'International desertification: Social geographies of vulnerability and adaptation', in J.F. Reynolds and D.M. Stafford Smith (eds), *Global desertification: Do humans cause deserts?*, Dahlem Workshop Report No. 88, Dahlem University Press, Berlin, pp. 233-52.

Döös, B.R. (2002), 'Population growth and loss of arable land', *Global Environmental Change*, Vol. 12, pp. 303-11.

Dregne, H.E. (1977), *Generalized map of the status of desertification of arid lands*, A/CONF 74/31, United Nations Conference on Desertification, Nairobi, Kenya.

Dregne, H.E. (1983), *Desertification of arid lands*, Harwood Academic Publishers, New York.

Dregne, H.E. (1996), 'Desertification: Challenges ahead', *Annals of Arid Zone*, Vol. 35, pp. 305-11.

Dregne, H. (2002), 'Land degradation in the drylands', *Arid Land Research and Management*, Vol. 16, pp. 99-132.

Dregne, H.E., Chou, N. (1992), 'Global desertification and cost', in H.E. Dregne (ed), *Degradation and restoration of arid lands*, Texas Tech University, Lubbock, pp. 249-82.

Dregne, H., Kassas, M., Rozanov, B. (1991), 'A new assessment of the world status of desertification', *Desertification Control Bulletin*, Vol. 20, pp. 7-18.

Dumanski, J. and Pieri, C. (2000), 'Land quality indicators: Research plan', *Agriculture, Ecosystems and Environment*, Vol. 81, pp. 92-102.

Dwyer, E., Pereira, J.M.C., Grégoire, J.M., De Camara, C.C. (2000), 'Characterization of the spatio-temporal patterns of fire activity using satellite imagery for the period April 1992 to March 1993', *Journal of Biogeography*, Vol. 27, pp. 57-69.

Eckholm, E., Brown, L.R. (1977), *Spreading deserts: The hand of man*, Worldwatch Paper, No. 13, Washington, D.C.

Eltahir, E.A.B., Bras, R.L. (1996), 'Precipitation recycling', *Review of Geophysics*, Vol. 34(3), pp. 367-78.

Ellis, J., Galvin, K. (1994), 'Climate patterns and land use practices in the dry zones of Africa', BioScience, Vol. 44, pp. 340-49.

European Commission (2000), *Addressing desertification and land degradation*, Office of the Official Publications of the European Commission, Luxembourg.

Fairhead, J., Leach, M. (1996), *Misreading the African landscape: Society and ecology in a forest-savanna mosaic*, Cambridge University Press, Cambridge.

Fernandez, R.J., Archer, E.R.M., Ash, A.J., Dowlatabadi, H., Hiernaux, P.H.Y., Reynolds, J.F., Vogel, C.H., Walker, B.H., Wiegand, T. (2002), 'Degradation and recovery in socio-ecological systems: A view from the household/farm level', in J.F. Reynolds and D.M. Stafford Smith (eds), *Global desertification: Do humans cause deserts?* Dahlem Workshop Report No. 88, Dahlem University Press, Berlin, pp. 297-323.

Forse, B. (1989), 'The myth of the marching desert', *New Scientist*, Vol. 4, p. 31.

Geist, H.J., Lambin, E.F. (2002), 'Proximate causes and underlying driving forces of tropical deforestation', *BioScience*, Vol. 52, pp. 143-50.

Geist, H.J., Lambin, E.F. (2004), 'Dynamical causal patterns of desertification', *BioScience*, accepted for print.

Geist, H., Lambin, E., Palm, C., Tomich, T. (2005), 'Agricultural transitions at dryland and tropical forest margins: Actors, scales, and trade-offs', in F. Brouwer and B. McCarl (eds), *Rural lands, agriculture and climate beyond 2015: Usage and management responses*. Kluwer Academic Publishers, Dordrecht, in press.

Glantz, M. (1977), *Desertification*, Westview Press, Boulder.

Glantz, M. (ed) (1987), *Drought and famine in Africa*, Cambridge University Press, Cambridge.

Glantz, M.H. (1994), *Drought follows the plow: Cultivating marginal areas*, Cambridge University Press, Cambridge, New York.

Glazovsky, N.F. (1995), 'The Aral sea basin', in J.X. Kasperson, R.E. Kapserson and B.L. Turner II (eds), Regions at risk: Comparisons of Threatened Environments, UNU Studies on Critical Regions, United Nations University Press, Tokyo, New York, Paris, pp. 92-139.

Grainger, A. (1990), *The threatening desert*, Earthscan, London.

Graetz, R.D. (1991), 'Desertification: A tale of two feedbacks', in H.A. Mooney, E. Medina and D.W. Schindler (eds), *Ecosystem experiments*, Wiley, Chichester, pp. 59-87.

Gray, L.C. (1999), 'Is land being degraded? A multi-scale investigation of landscape change in southwestern Burkina Faso', *Land Degradation and Development*, Vol. 10, pp. 329-43.

Hiernaux, P., Turner, M.D. (2002), 'The influence of farmer and pastoralist management practices on desertification processes in the Sahel', in J.F. Reynolds and D.M. Stafford Smith (eds), *Global desertification: Do humans cause deserts?*, Dahlem Workshop Report No. 88, Dahlem University Press, Berlin, pp. 135-48.

Hootsmans, R., Bouwman, A.F., Leemans, R., Kreileman, E. (2001), *Modeling land degradation in IMAGE 2*, RIVM Report No. 481508009, National Institute of Public Health and the Environment, Bilthoven.

Houghton, R.A., Hackler, J.L., Lawrence, K.T. (1999), 'The US carbon budget: Contribution from land-use change', *Science*, Vol. 285, pp. 574-78.

Howorth, C., O'Keefe, D.J. (1999), 'Farmers do it better: Local management of change in southern Burkina Faso', *Land Degradation and Development*, Vol. 10, pp. 93-109.

Ibrahim, F.N. (1978), 'Anthropogenic causes of desertification in Western Sudan', *Geographical Journal*, Vol. 2, pp. 243-54.

Ibrahim, F.N. (1984), *Ecological imbalance in the Republic of Sudan: With reference to desertification in Darfur*, Bayreuther Geowissenschaftliche Arbeiten No. 6, Druckhaus Bayreuth, Bayreuth.

Jiang, H. (2002), 'Culture, ecology, and nature's changing balance: Sandification on Mu Us Sandy Land, Inner Mongolia, China', in J.F. Reynolds and D.M. Stafford Smith (eds), *Global desertification: Do humans cause deserts?*, Dahlem Workshop Report No. 88, Dahlem University Press, Berlin, pp. 181-96.

Jiang, H., Zhang, P., Zheng, D., Wang, F. (1995): 'The Ordos plateau of China', in J.X. Kasperson, R.E. Kapserson and B.L. Turner II (eds), *Regions at risk: Comparisons of Threatened Environments*, UNU Studies on Critical Regions, United Nations University Press, Tokyo, New York, Paris, pp. 420-59.

Johnson, D.L. (1977), 'The human dimensions of desertification', *Economic Geography*, Vol. 53, pp. 317-18.

Kadomura, H., Miyazaki, T., Fujimori, M., Sauer, U., Kawaguchi, S. (1997), *Data book of desertification/land degradation*, CGER Report DO13, Center for Global Environmental Research, National Institute for Environmental Studies, Tokyo.

Kasperson, R.E. (1993), 'Critical environmental regions and the dynamics of change', in H.-G. Bohle, T.E. Downing, J.O. Field and F.N. Ibrahim (eds), Coping with vulnerability and criticality: Case studies on food-insecure people and places, Freiburg Studies in Development Geography No. 1, Breitenbach, Saarbrücken, Ft. Lauderdale, p. 115-26.

Kasperson, J.X., Kasperson, R.E., Turner, B.L. II (eds) (1995), *Regions at risk: Comparisons of threatened environments*, United Nations University Press, Tokyo, New York, Paris.

Kasperson, R.E., Kasperson, J.X., Turner, B.L. II (1999), 'Risk and criticality: Trajectories of regional environmental degradation', *Ambio*, Vol. 28, pp. 562-8.

Kassas, M. (1995), 'Desertification: A general review', *Journal of Arid Environments*, Vol. 30, pp. 115-28.

Kates, R.W., Haarmann, V. (1992), 'Where the poor live: Are the assumptions correct?', *Environment*, Vol. 34, pp. 4-28.

Keita, N. (1998), 'Combating the creeping sands in northern Mali', *The Courier*, Vol. 172, p. 79.

Kharin, N. (2003), *Vegetation degradation in Central Asia under the impact of human activities*, Kluwer Academic Publishers, Dordrecht.

Lal, R. (1988), 'Soil degradation and the future of agriculture in Sub-Saharan Africa', *Journal of Soil and Water Conservation*, Vol. 43, pp. 444-51.

Lal, R. (2002), 'Carbon sequestration in dryland ecosystems of West Asia and North Africa', *Land Degradation and Development*, Vol. 13, pp. 45-59.

Lambin, E.F., Ehrlich, D. (1997), 'Land-cover changes in sub-Saharan Africa, 1982-1991: Application of a change index based on remotely sensed surface temperature and vegetation indices at a continental scale', *Remote Sensing of the Environment*, Vol. 61, pp. 181-200.

Lambin, E.F., Geist, H.J. (2003), 'Regional differences in tropical deforestation', *Environment*, Vol. 45, pp. 22-36.

Lambin, E.F., Geist, H.J., Lepers, E. (2003), 'Dynamics of land-use and land-cover change in tropical regions', *Annual Review of Environment and Resources*, Vol. 28, pp. 205-41.

Lambin, E.F., Baulies, X., Bockstael, N., Fischer, G., Krug, T., Leemans, R., Moran, E.F., Rindfuss, R.R., Sato, Y., Skole, D., Turner, B.L. II, Vogel, C. (1999), *Land-use and land-cover change (LUCC): Implementation strategy*, IGBP Report No. 48, IHDP Report No. 10, International Geosphere-Biosphere Programme, Stockholm.

Lambin, E.F., Turner, B.L. II, Geist, H.J., Agbola, S.B., Angelsen, A., Bruce, J.W., Coomes, O., Dirzo, R., Fischer, G., Folke, C., George, P.S., Homewood, K., Imbernon, J., Leemans, R., Li, X., Moran, E.F., Mortimore, M., Ramakrishnan, P.S., Richards, J.F., Skånes, H., Steffen, W., Stone, G.D., Svedin, U., Veldkamp, T.A., Vogel, C., Xu, J. (2001), 'The causes of land-use and land-cover change: Moving beyond the myths', *Global Environmental Change*, Vol. 11, pp. 261-69.

Lambin, E.F., Chasek, P.S., Downing, T.E., Kerven, C., Kleidon, A., Leemans, R., Lüdeke, M., Prince, S.D., Xue, Y. (2002), 'The interplay between international and local processes affecting desertification', in J.F. Reynolds and D.M. Stafford Smith (eds), *Global desertification: Do humans cause deserts?*, Dahlem Workshop Report No. 88, Dahlem University Press, Berlin, pp. 387-401.

Lamprey, H.F. (1975), *Report on the desert encroachment reconnaissance in northern Sudan 21 Oct. to 10 Nov. 1975*, UNESCO/UNEP (mimeo).

Lamprey, H.F. (1988), 'Report on the desert encroachment reconnaissance in northern Sudan 21 Oct. to 10 Nov. 1975', *Desertification Control Bulletin*, Vol. 17, pp. 1-7.

Lappé, F., Collins, M.J., Rosset, P. (1998), *World Hunger: Twelve myths*, Earthscan, London.

Lavauden, L. (1927), 'Les forêts du Sahara', *Revue des Eaux et Forêts*, Vol. 7, pp. 329-41.

Leach, M., Mearns, R. (eds) (1996), *The lie of the land: Challenging received wisdom on the African environment*, James Currey, Heinemann, Oxford, Portsmouth.

Leemans, R., Kleidon, A. (2002), 'Regional and global assessment of the dimensions of desertification', in J.F. Reynolds and D.M. Stafford Smith (eds), *Global desertification: Do humans cause deserts?*, Dahlem Workshop Report No. 88, Dahlem University Press, Berlin, pp. 215-31.

Le Houérou, H.N. (1991), 'La Méditerranée en l'an 2050: Impacts respectifs d'une éventuelle évolution climatique et de la démographie sur la végétation, les écosystèmes et l'utilisation des terres', *La Météorologie VII serie*, Vol. 36, pp. 4-37.

Le Houérou, H.N. (1993), 'Changements climatiques et désertisation', *Secheresse*, Vol. 4, pp. 95-111.

Le Houérou, H.N. (1995), *Climate change, drought, and desertification*, Report to Intergovernmental Panel on Climate Change, Final version, IPCC, Washington, D.C.

Le Houérou, H.N. (1996), 'Climate change, drought, and desertification', *Journal of Arid Environments*, Vol. 34, pp. 133-85.

Le Houérou, H.N. (2002), 'Man-made deserts: Desertization processes and threats', *Arid Land Research and Management*, Vol. 16, pp. 1-36.

Leonard, H.J. (1989), 'Environment and the poor: Development strategies for a common agenda', *US Third World Policy Perspective*, Vol. 11, pp. 3-45.

Lepers, E., Lambin, E.F., Janetos, A.C., DeFries, R., Achard, F., Ramankutty, N., Scholes, R.J. (2003), *Areas of rapid land-cover change of the world*, Millennium Ecosystem Assessment Report, MEA, Penang.

Lohnert, B., Geist, H. (1999), 'Endangered ecosystems and coping strategies: Towards a conceptualization of environmental change in the developing world', in B. Lohnert and H. Geist (eds), Coping with changing environments: Social dimensions of endangered ecosystems in the developing world, Ashgate, Aldershot, Brookfield, pp. 1-53.

Lüdeke, M.K.B., Moldenhauer, O., Petschel-Held, G. (1999), 'Rural poverty driven soil degradation under climate change: The sensitivity of the disposition towards the Sahel syndrome with respect to climate', *Environmental Modeling and Assessment*, Vol. 4, pp. 315-26.

Lupo, F., Reginster, I., Lambin, E.F. (2001), 'Monitoring land-cover changes in West Africa with SPOT vegetation: Impact of natural disasters in 1998-1999', *International Journal of Remote Sensing*, Vol. 22, pp. 2633-9.

Mabbutt, J.A. (1984), 'A new global assessment of the status and trends of desertification', *Environmental Conservation*, Vol. 11, pp. 103-13.

Mabbutt, J.A. (1986), 'Desertification indicators', *Climatic Change*, Vol. 9, pp. 113-22.

Mabbutt, J.A., Floret, C. (eds) (1980), *Case studies on desertification*, Vol. 18, United Nations Educational, Scientific and Cultural Organization, Paris.

Mainguet, M. (1999), *Aridity: Droughts and human development*. Springer Verlag, Berlin.

Mainguet, M., Chemin, M.C. (1991), 'Wind degradation on the sandy soils of the Sahel of Mali and Niger and its part in desertification', *Acta Mechanica*, Vol. 2, pp. 113-30.

Mannion, A.M. (2002), Dynamic world: Land-cover and land-use change, Arnold, London, Oxford University Press, New York.

Margaris, N.S., Koutsido, E., Giourga, C. (1996), 'Changes in traditional Mediterranean land-use systems', in J.C. Brandt and J.B. Thornes (eds), Mediterranean desertification and land use, John Wiley and Sons, Chichester, 29-42.

Matarazzo, B., Nijkamp, P. (1997), 'Meta-analysis for comparative environmental case studies: Methodological issues', *International Journal of Social Economics*, Vol. 24, pp. 799-811.

McGuire, A.D., Sitch, S., Clein, J.S., Dargaville, R., Esser, G., Foley, J., Heimann, M., Joos, F., Kaplan, J., Kicklighter, D.W., Meier, R.A., Melillo, J.M., Moore, B., Prentice, I.C., Ramankutty, N., Reichenau, T., Schloss, A., Tian, H., Williams, L.J., Wittenberg, U. (2001), 'Carbon balance of the terrestrial bioshpere in the twentieth century: Analyses of $CO_2$, climate and land use effects with four process-based ecosystem models', *Global Biogeochemical Cycles*, Vol. 15, pp. 183-206.

Mensching, H., Ibrahim, F.N. (1976), 'The problem of desertification in and around arid lands', *Applied Science and Development*, Vol. 10, Institute for Scientific Cooperation, Tübingen, pp. 1-43.

Middleton, N.J., Thomas, D.S.G. (eds) (1997), *World atlas of desertification*, Arnold, United Nations Environmental Programme, London, New York.

Mooney, H., Cropper, A., Reid, W. (eds) (2003), *Ecosystems and human well-being: A framework for assessment*, Island Press, Washington, D.C.

Mortimore, M. (1993), 'Population growth and land degradation', *GeoJournal*, Vol. 31, pp. 15-21.

Mortimore, M. (1998), *Roots in the African dust: Sustaining the drylands*, Cambridge University Press, Cambridge.

Mortimore, M.J., Adams, W.M. (2001), 'Farmer adaptation, change and "crisis" in the Sahel', *Global Environmental Change*, Vol. 11, pp. 49-57.

Nachtergaele, F. (2002), 'Land degradation assessment in drylands (LADA project', *LUCC Newsletter*, Vol. 8, p. 15.

Niamir-Fuller, M. (ed) (1999), *Managing mobility in African rangelands: The legitimization of transhumance*, Intermediate Technology Publications, London.

Nicholson, N.E. (2002), 'What are the key components of climate as a driver of desertification?', in J.F. Reynolds and D.M. Stafford Smith (eds), *Global desertification: Do humans cause deserts?* Dahlem Workshop Report No. 88, Dahlem University Press, Berlin, pp. 41-57.

Nielsen, T.L., Zöbisch, M.A. (2001), 'Multi-factorial causes of land-use change: Land-use dynamics in the agropastoral village of Im Mial, northwestern Syria', *Land Degradation and Development*, Vol. 12, pp. 143-61.

Niemeijer, D., Mazzucato, V. (2002), 'Soil degradation in the West African Sahel: How serious is it?', *Environment*, Vol. 44, pp. 20-31.

Norusis, M.J. (2000), SPSS 10.0 guide to data analysis, Prentice Hall.
Oba, G., Stenseth, N.C., Lusigi, W.J. (2000), 'New perspectives on sustainable grazing management in arid zones of sub-Saharan Africa', *BioScience*, Vol. 50, pp. 35-51.
Oba, G., Weladji, R.B., Lusigi, W.J., Stenseth, N.C. (2003), 'Scale-dependent effects of grazing on rangeland degradation in northern Kenya: A test of equilibrium and non-equilibrium hypotheses', *Land Degradation and Development*, Vol. 14, pp. 83-94.
Ojima, D.S., Galvin, K.A., Turner, B.L. II (1994), 'The global impact of land-use change', *BioScience*, Vol. 44, pp. 300-304.
Oldeman, L.R., Hakkeling, R.T.A., Sombroek, W.G. (1990), *World map of the status of human-induced soil degradation: An explanatory*, International Soil Reference and Information Center, Wageningen.
Olson, G. W. (1981), 'Archaeology: Lessons on future soil use', *Journal of Soil and Water Conservation*, Vol. 36, pp. 261-4.
Ostrom, E. (1990), *Governing the commons: The evolution of institutions for collective action*, Cambridge University Press, Cambridge.
Otterman, J. (1974), 'Baring high-albedo soils by overgrazing: A hypothesised desertification mechanism', *Science*, Vol. 86, pp. 531-3.
Oyowe, A. (1998), 'Lessons from the Sahel', *The Courier*, Vol. 172, pp. 72-3.
Pacala, S.W., Hurtt, G.C., Baker, D., Peylin, P., Houghton, R.A., Birdsey, R.A., Heath, L., Sundquist, E.T., Stallard, R.F., Ciais, P., Moorcroft, P., Caspersen, J.P., Shevliakova, E., Moore, B., Kohlmaier, G., Holland, E., Gloor, M., Harmon, M.E., Fan, S.M., Sarmiento, J.L., Goodale, C.L., Schimel, D., Field, C.B. (2001), 'Consistent land- and atmosphere-based US carbon sink estimates', *Science*, Vol. 292, pp. 2316-20.
Parmesan, C., Yohe, G. (2003), 'A globally coherent fingerprint of climate change impacts across natural systems', *Nature*, Vol. 421, pp. 37-42.
Pereira, J.M.C., Pereira, B.S., Barbosa, P., Stroppiana, D., Vasconcelos, M.J.P., Grégoire, J.M. (1999), 'Satellite monitoring of fire in the EXPRESSO study area during the 1996 dry season experiment: Active fires, burnt area, and atmospheric emissions', *Journal of Geophysical Research-Atmospheres*, Vol. 104, pp. 30701-12.
Petschel-Held, G. (2004), 'The syndromes approach to place-based assessment', in W. Steffen, A. Sanderson, P.D. Tyson, J. Jäger, P.A. Matson, B. Moore III, F. Oldfield, K. Richardson, H.J. Schellnhuber, B.L. Turner II and R.J. Wasson (eds), *Global change and the earth system: A planet under pressure*, The IGBP Book Series. Springer: Berlin, Heidelberg, New York, pp. 92-3.
Petschel-Held, G., Lüdeke, M.K.B., Reusswig, F. (1999), 'Actors, structures and environments: A comparative and transdisciplinary view on regional case studies of global environmental change', in B. Lohnert and H. Geist (eds), *Coping with changing environments: Social dimensions of endangered ecosystems in the developing world*, Ashgate, Aldershot, pp. 255-92.
Phillips, J.D. (1993), 'Biophysical feedbacks and the risks of desertification', *Annals of the Association of American Geographers*, Vol. 83, pp. 630-40.
Pieri, C., Dumanksi, J., Hamblin, A., Young, A. (1995), *Land Quality Indicators*, World Bank Discussion Paper No. 315, World Bank, Washington, D.C.
Plisnier, P.D., Serneels, S., Lambin, E.F. (2000), 'Impact of ENSO on East African ecosystems: A multivariate analysis based on climate and remote sensing data', *Global Ecology and Biogeography*, Vol. 9, pp. 481-97.
Polanyi, K. (1944), *The great transformation: The political and economic origins of our times*, Beacon Press, Boston.
Prince, S.D. (2002), 'Spatial and temporal scales for detection of desertification', in J.F. Reynolds and D.M. Stafford Smith (eds), *Global desertification: Do humans cause deserts?*, Dahlem Workshop Report No. 88, Dahlem University Press, Berlin, pp. 23-40.

Prince, S.D., De Colstoun, E.B., Kravitz, L.L. (1998), 'Evidence from rain-use efficiencies does not indicate extensive Sahelian desertification', *Global Change Biology*, Vol. 4, pp. 359-74.

Puigdefábregas, J. (1995), 'Desertification: Stress beyond resilience, exploring a unifying process structure', *Ambio*, Vol. 24, pp. 311-13.

Puigdefábregas, J. (1998), 'Ecological impacts of global change on drylands and their implications for desertification', *Land Degradation and Development*, Vol. 9, pp. 393-406.

Ragin, C.C. (1989), *The comparative method: Moving beyond qualitative and quantitative strategies*, University of California Press, Berkeley.

Ragin, C.C. (2003), *Making comparative analysis count*, COMPASSS Working Paper No. 3 (www.compasss.org).

Raynaut, C. (1997), *Societies and nature in the Sahel: Rethinking environmental degradation*, Routledge, London.

Redman, C.L. (1999), *Human impact on ancient environments*, University of Arizona Press, Tucson.

Reining, P. (1978), *Handbook on desertification indicators*, American Association for the Advancement of Science, Washington, D.C.

Reynolds, J.F. (2001), 'Desertification', in S. Levin (ed), *Encyclopaedia of Biodiversity*, Vol. 2, Academic Press, San Diego, pp. 61-78.

Reynolds, J.F., Stafford Smith, D.M. (2002), 'Do humans cause deserts?', in J.F. Reynolds, D.M. Stafford Smith (eds), *Global desertification: Do humans cause deserts?* Dahlem Workshop Report No. 88, Dahlem University Press, Berlin, pp. 1-21.

Reynolds, J.F., Stafford Smith, D.M., Lambin, E.F. (2003), 'ARIDnet: Seeking novel approaches to desertification and land degradation', *IGBP Global Change Newsletter*, Vol. 54, pp. 5-9.

Rindfuss, R.R., Walsh, S.J., Mishra, V., Fox, J., Dolcemascolo, G.P. (2003), 'Linking household and remotely sensed data: Methodological and practical problems', in J. Fox, R.R. Rindfuss, S.J. Walsh and V. Mishra (eds), *People and the environment: Approaches for linking household and community surveys to remote sensing and GIS*, Kluwer Academic Publishers, Boston, Dordrecht, London, pp. 1-19.

Robbins, P.F., Abel, N., Jiang, H., Mortimore, M., Mulligan, M., Okin, G.S., Stafford Smith, D.M., Turner, B.L.II (2002), 'Desertification at the community scale: Sustaining dynamic human-environment systems', in J.F. Reynolds, D.M. Stafford Smith (eds), *Global desertification: Do humans cause deserts?* Dahlem Workshop Report No. 88, Dahlem University Press, Berlin, pp. 325-55.

Rocheleau, D., Benjamin, P., Diang'a, A. (1995): 'The Ukambani Region of Kenya', in J.X. Kasperson, R.E. Kapserson and B.L. Turner II (eds), *Regions at risk: Comparisons of Threatened Environments*, UNU Studies on Critical Regions, United Nations University Press, Tokyo, New York, Paris, pp. 186-254.

Root, T.L., Price, J.T., Hall, K.R., Schneider, S.H., Rosenzweig, C., Pounds, J.A. (2003), 'Fingerprints of global warming on wild animals and plants', *Nature*, Vol. 421, pp. 57-60.

Rosenfeld, D., Rudich, Y., Lahav, R. (2001), 'Desert dust suppressing precipitation: A possible desertification loop', *Proceedings of the National Academy of Sciences of the United States of America*, Vol. 98, pp. 5975-80.

Runnels, C. (1995), 'Environmental degradation in ancient Greece', *Scientific American*, Vol. 272, pp. 72-5.

Sagan, C., Toon, O.B., Pollack, J.B. (1979), 'Anthropogenic albedo changes and the earth's climate', *Science*, Vol. 206, pp. 1363-8.

Sala, O.E., Chapin, F.S., Armesto, J.J., Berlow, E., Bloomfield, J., Dirzo, R., Huber-Sanwald, E., Huenneke, L.F., Jackson, R.B., Kinzig, A., Leemans, R., Lodge, D.M., Mooney, H.A., Oesterheld, M., Poff, N.L., Sykes, M.T., Walker, B.H., Walker, M.,

Wall, D.H. (2000), 'Biodiversity: Global biodiversity scenarios for the year 2100', *Science*, Vol. 287, pp. 1770-74.
Schlesinger, W.H., Gramenopoulos, N. (1996), 'Archival photographs show no climate-induced changes in woody vegetation in the Sudan, 1943-1994', *Global Change Biology*, Vol. 2, pp. 137-41.
Schlesinger, W.H., Reynolds, J.F., Cunnigham, G.L., Huenneke, L.F., Jarrell, W.M., Virginia, R.A., Whitford, W.G. (1990), 'Biological feedbacks in global desertification', *Science*, Vol. 247, pp. 1043-8.
Schulz, E. (1994), 'Changing use of the Sahara desert', in N. Roberts (ed), *The changing global environment*, Blackwell, Oxford, Cambridge, pp. 371-96.
Sharma, K.D. (1998), 'The hydrological indicators of desertification', *Journal of Arid Environments*, Vol. 39, pp. 121-32.
Shaw Thacher, P. (1979), 'Desertification: The greatest single environmental threat', *Desertification Control Bulletin*, Vol. 2, pp. 7-9.
Shi, H., Shao, M. (2000), 'Soil and water losses from the Loess Plateau in China', *Journal of Arid Environments*, Vol. 45, pp. 9-20.
Smith, S.E. (1986), 'Drought and water management: The Egyptian response', *Journal of Soil and Water Conservation*, Vol. 41, pp. 297-300.
Sneath, D. (1998), 'State policy and pasture degradation in inner Asia', *Science*, Vol. 281, pp. 1147-8.
Solbrig, O. (1993), Ecological constraints to savanna land use, in M.D. Young and O.T. Solbrig (eds), *The world's savannas: Economic driving forces, ecological constraints and policy options for sustainable land use*, Man and the Biosphere Series 12, Parthenon Press, Paris, pp. 21-48.
Soulé, M.E. (1991), 'Conservation: Tactics for a constant crisis', *Science*, Vol. 253, pp. 744-50.
Stafford Smith, D.M., Reynolds, J.F. (2002), 'The Dahlem desertification paradigm: A new approach to an old problem', in J.F. Reynolds, D.M. Stafford Smith DM (eds), *Global desertification: Do humans cause deserts?* Dahlem Workshop Report No. 88, Dahlem University Press, Berlin, pp. 403-24.
Stamp, L. (1940), 'The southern margin of the Sahara: Comments on some recent studies on the question of desiccation in West Africa', *The Geographical Review*, Vol. 30, pp. 297-300.
Stebbing, E.P. (1935), 'The encroaching Sahara: The threat to the West African colonies', *The Geographical Journal*, Vol. 85, pp. 506-24.
Steffen, W., Sanderson, A., Tyson, P.D., Jäger, J., Matson, P.A., Moore III, B., Oldfield, F., Richardson, K., Schellnhuber, H.J., Turner, B.L. II, Wasson, R.J. (2004), *Global change and the earth system: A planet under pressure*, The IGBP Book Series. Springer: Berlin, Heidelberg, New York.
Stern, P.C., Young, O.R., Druckman, D. (1992), *Global environmental change: Understanding the human dimensions*, National Academy Press, Washington, D.C.
Stocking, M. (1985), 'Soil conservation policy in colonial Africa', in D. Helms and S.L. Flader (eds), *The history of soil and water conservation*, University of California Press, Berkeley, pp. 46-59.
Stocking, M. (1996), 'Soil erosion: Breaking new ground', in M. Leach and R. Mearns (eds), *The lie of the land: Challenging received wisdom on the African environment*, Heinemann, James Currey, Portsmouth, Oxford, pp. 140-155.
Suliman, M.M. (1988), 'Dynamics of range plants and desertification monitoring in the Sudan', *Desertification Control Bulletin*, Vol. 16, pp. 27-31.

Swift, J. (1996), 'Narratives, winners and losers', in M. Leach and R. Mearns (eds), *The lie of the land: Challenging received wisdom on the African environment*, Heinemann, James Currey, Portsmouth, Oxford, pp. 73-90.
Szabolcs, I. (1991), 'Effects of predicted climatic changes on European soils, with particular regard to salinization', in M.M. Boer and R.S. De Groot (eds), *Landscape ecological impact of climatic change*, IOS Press, Amsterdam, pp. 177-93.
Taylor, C.M., Lambin, E.F., Stephenne, N., Harding, R.J., Essery, R.L.H. (2002), 'The influence of land use change on climate in the Sahel', *Journal of Climate*, Vol. 15, pp. 3615-29.
Thébaud, B., Toulmin, C. (1994), *Causes and socio-economic consequences of desertification in Africa: The pastoral issue*, International Panel of Experts on Desertification, Secretariat of the Intergovernmental Negotiating Committee for a Convention to Combat Desertification, INCD-UN, Geneva.
Thomas, D.S.G. (1997), 'Science and the desertification debate', *Journal of Arid Environments*, Vol. 37, pp. 599-608.
Thomas, D.S.G., Middleton, N.J. (1994), *Desertification: Exploding the myth*, John Wiley and Sons, Chichester.
Tiffen, M., Mortimore, M., Gichuki, F. (1994), *More people, less erosion: Environmental recovery in Kenya*, Wiley, New York.
Tomich, T.P., Cattaneo, A., Chater, S., Geist, H.J., Gockowski, J., Lambin, E.F., Lewis, J., Palm, C., Stolle, F., Valentim, J., Noordwijk, M. van and S.A. Vosti (2004), 'Balancing agricultural development and environmental objectives. Assessing tradeoffs in the humid tropics', in C.A. Palm, P.A. Sanchez, S.A. Vosti and P.J. Ericksen (eds), *Slash and Burn: The Search for Alternatives*. Columbia University Press, New York, in press.
Tomich, T.P., Noordwijk, M. van, Vosti, S.A., Witcover, J. (1998), 'Agricultural development with rainforest conservation: Methods for seeking best bet alternatives to slash-and-burn, with applications to Brazil and Indonesia', *Agricultural Economics*, Vol. 19, pp. 159-74.
Toulmin, C. (1998), 'The convention to combat desertification: A code of good practice', *The Courier*, Vol. 172, pp. 66-8.
Trimble, S.W., Crosson, P. (2000), 'Land use: US soil erosion rates – Myth and reality', *Science*, Vol. 289, pp. 248-50.
Tucker, C.J., Dregne, H.E., Newcomb, W.W. (1991), 'Expansion and contraction of the Sahara desert from 1980 to 1990', *Science*, Vol. 253, pp. 299-301.
Tucker, C.J., Nicholson, S.E. (1999), 'Variations in the size of the Sahara desert from 1980 to 1997', *Ambio*, Vol. 28, pp. 587-91.
Tucker, C.J., Newcomb, W.W., Dregne, H.E. (1994), 'AVHRR data sets for determination of desert spatial extent', *International Journal of Remote Sensing*, Vol. 15, pp. 3547-65.
Turner, B.L.II, Hyden, G., Kates, R. (eds) (1993), *Population growth and agricultural change in Africa*, University of Florida Press, Gainesville.
Turner, B.L. II, Skole, D., Sanderson, S., Fischer, G., Fresco, L., Leemans, R. (1995), *Land-use and land-cover change science/research plan*, IGBP Report No. 35, HDP Report No. 7, International Geosphere-Biosphere Programme, Human Dimensions of Global Environmental Change Programme, Stockholm, Geneva.
Turner, B.L. II, Kasperson, R.E., Matson, P.A., McCarthy, J.J, Corell, R.W., Christensen, L., Eckley, N., Kasperson, J.X., Luers, A., Martello, M.L., Polsky, C., Pulsipher, A., Schiller, A. (2003), 'A framework for sustainability analysis in sustainability science', *Proceedings of the National Academy of Sciences of the United States of America*, Vol. 100, pp. 8074-9.

Turner, M.D. (1999), 'Merging local and regional analyses of land-use change: The case of livestock in the Sahel', *Annuals of the Association of American Geographers*, Vol. 89, pp. 191-219.

Turner, M. (2003), 'Environmental science and social causation in the analysis of Sahelian pastoralism', in K.S. Zimmerer and T.J. Bassett (eds), Political ecology: An integrative approach to geography and environment-development studies, The Guilford Press, New York, London, pp. 159-78.

United Nations (1994), *Earth Summit. Convention on Desertification. United Nations Conference on Environment and Development, Rio de Janeiro, Brazil, 3-14 June 1992*, Report DPI/SD/1576. United Nations, New York.

United Nations Environmental Programme (1977), *Draft plan of action to combat desertification*, UNCOD Background Document A/CONF, 74/L36, UNEP, Nairobi.

United Nations Environmental Programme (1994), *United Nations Convention to Combat Desertification*, UNEP, Nairobi.

Vitousek, P.M., Mooney, H.A., Lubchenco, J. Melillo, J.M. (1997), 'Human domination of earth's ecosystems', *Science*, Vol. 277, pp. 494-9.

Vogel, C.H. (1995), 'People and drought in South Africa: Reaction and mitigation', in A. Binns (ed), People and environment in Africa, Wiley, London, pp. 249-56.

Vogel, C.H., Smith, J. (2002), 'Building social resilience in arid ecosystems', in J.F. Reynolds, D.M. Stafford Smith DM (eds), *Global desertification: Do humans cause deserts?* Dahlem Workshop Report No. 88, Dahlem University Press, Berlin, pp. 149-66.

Walker, B.H. (1993), 'Rangeland ecology: Understanding and managing change', *Ambio*, Vol. 22, pp. 80-87.

Warren, A. (1996), 'Desertification', in W.M. Adams, A.S. Goudie and A.R. Orme (eds), *The physical geography of Africa*, Oxford University Press, Oxford, pp. 342-55.

Warren, A. (2002), 'Land degradation is contextual', *Land Degradation and Development*, Vol. 13, pp. 449-59.

Warren, A., Agnew, C. (1988), *An assessment of desertification and land degradation in arid and semi-arid areas*, International Institute for Environment and Development Paper No. 2, University College London, London.

Watts, M.J. (1985), 'Social theory and environmental degradation', in Y. Gradus (ed), *Desert development: Man and technology in sparse lands*, Reidel, Dordrecht, pp. 14-32.

Watts, M. (1987), 'Drought, environment and food security: Some reflections on peasants, pastoralists and commoditization in dryland West Africa', in H.H. Glantz (ed), *Drought and hunger in Africa: Denying famine a future*, Cambridge University Press, Cambridge, pp. 171-211.

Watts, M. (2001), 'Desertification', in R.J. Johnston, D. Gregory, G. Pratt and M. Watts (eds), The dictionary of human geography, Blackwell Publishers, 4$^{th}$ edition, Oxford, Malden, pp. 165-6.

Watts, M.J., Bohle, H.-G. (1993), 'The space of vulnerability: The causal structure of hunger and famine', *Progress in Human Geography*, Vol. 17, pp. 43-67.

World Bank (1998), *World development report 1999*, The World Bank, Washington, D.C.

Young, A. (1999), 'Is there really spare land? A critique of estimates of available cultivable land in developing countries', *Environment, Development and Sustainability*, Vol. 1, pp. 3-18.

Zeng, N., Neelin, J.D., Lau, K.M., Tucker, C.J. (1999), 'Enhancement of interdecadal climate variability in the Sahel by vegetation interaction', Science, Vol. 286, pp. 1537-40.

Zhu, Z., Wang, T. (1993), 'Trends in desertification and its rehabilitation in China', *Desertification Control Bulletin*, Vol. 22, pp. 27-30.

## Case Studies

Aagesen, D. (2000), 'Crisis and conservation at the end of the world: Sheep ranching in Argentine Patagonia', *Environmental Conservation*, Vol. 27, pp. 208-15.

Ayoub, A.T. (1998), 'Extent, severity and causative factors of land degradation in the Sudan', *Journal of Arid Environments*, Vol. 38, pp. 397-409.

Basso, F., Bove, E., Dumontet, S., Ferrara, A., Pisante, M., Quaranta, G., Taberner, M. (2000), 'Evaluating environmental sensitivity at the basin scale through the use of geographic information systems and remotely sensed data: an example covering the Agri basin (Southern Italy)', *Catena*, Vol. 40, pp. 19-35.

Bastin, G.N., Pickup, G., Pearce, G. (1995), 'Utility of AVHRR data for land degradation assessment: A case study', *International Journal of Remote Sensing*, Vol. 16, pp. 651-72.

Benjaminsen, T.A. (1993), 'Fuelwood and desertification: Sahel orthodoxies discussed on the basis of field data from the Gourma Region in Mali', *Geoforum*, Vol. 24, pp. 397-409.

Brown, G., Schoknecht, N. (2001), 'Off-road vehicles and vegetation patterning in a degraded desert ecosystem in Kuwait', *Journal of Arid Environments*, Vol. 49, pp. 413-27.

Brown, J.H., Valone, T.J., Curtin, C.G. (1997), 'Reorganization of an arid ecosystem in response to recent climate change', *Proceedings of the National Academy of Sciences of the United States*, Vol. 94, pp. 9729-33.

Dube, O.P., Pickup, G. (2001), 'Effects of rainfall variability and communal and semi-commercial grazing on land cover in southern African rangelands', *Climate Research*, Vol. 17, pp. 195-208.

Feng, Q., Endo, K.N., Cheng, G.D. (2001), 'Towards sustainable development of the environmentally degraded arid rivers of China: A case study from Tarim River', *Environmental Geology*, Vol. 41, pp. 229-38.

Fredrickson, E., Havstad, K.M., Estell, R., Hyder, P. (1998), 'Perspectives on desertification: South-western United States', *Journal of Arid Environments*, Vol. 39, pp. 191-207.

Gauquelin, T., Bertaudière, V., Montes, N., Badri, W., Asmode, J.-F. (1999), 'Endangered stands of thuriferous juniper in the western Mediterranean basin: Ecological status, conservation and management, *Biodiversity and Conservation*, Vol. 8, pp. 1479-98.

Genxu, W., Guodong, C. (1999), 'Water resource development and its influence on the environment in arid areas of China: The case of the Hei River basin', *Journal of Arid Environments*, Vol. 43, pp. 121-31.

Gonzalez, P. (2001), 'Desertification and a shift of forest species in the West African Sahel', *Climate Research*, Vol. 17, pp. 217-28.

Helldén, U. (1991), 'Desertification: Time for an assessment', *Ambio*, Vol. 20, pp. 372-83.

Hill, J., Hostert, P., Tsiourlis, G., Kasapidis, P., Udelhoven, T., Diemer, C. (1998), 'Monitoring 20 years of increased grazing impact on the Greek island of Crete with earth observation satellites', *Journal of Arid Environments*, Vol. 39, pp. 165-78.

Ho, P. (2001), 'Rangeland degradation in north China revisited? A preliminary statistical analysis to validate non-equilibrium range ecology', *Journal of Development Studies*, Vol. 37, pp. 99-133.

Holzner, W., Kriechbaum, M. (2001), 'Pastures in south and central Tibet (China) II: Probable causes of pasture degradation', *Bodenkultur*, Vol. 52, pp. 37-44.

Keya, G.A. (1998), 'Herbaceous layer production and utilization by herbivores under different ecological conditions in an arid savanna of Kenya', *Agriculture, Ecosystems and Environment*, Vol. 69, pp. 55-67.

Khresat, S.A., Rawajfih, Z., Mohammad, M. (1998), 'Land degradation in north-western Jordan: Causes and processes', *Journal of Arid Environments*, Vol. 39, pp. 623-9.

Kosmas, C., Gerontidis, S., Marathianou, M. (2000), 'The effect of land use change on soils and vegetation cover over various lithological formations on Lesvos (Greece)', *Catena*, Vol. 40, pp. 51-68.

Lin, N.F., Tang, J. (2002), 'Geological environment and causes for desertification in arid-semiarid regions in China', *Environmental Geology*, Vol. 41, pp. 806-15.

Lin, N.F., Tang, J., Han, F.X. (2001), 'Eco-environmental problems and effective utilization of water resources in the Kashi Plain, western Terim Basin, China', *Hydrogeology Journal*, Vol. 9, pp. 202-7.

Liu, Z.M., Zhao, W.Z. (2001), Shifting-sand control in central Tibet, *Ambio*, Vol. 30, pp. 376-80.

Ludwig, J.A., Muldavin, E., Blanche, K.R. (2000), 'Vegetation change and surface erosion in desert grasslands of Otero Mesa, Southern New Mexico: 1982 to 1995', *The American Midland Naturalist*, Vol. 144, pp. 273-85.

Manzano, M.G., Návar, J., Pando-Moreno, M., Martinez, A. (2000), 'Overgrazing and desertification in northern Mexico: Highlights on northeastern region', *Annals of Arid Zone*, Vol. 39, pp. 285-304.

McAuliffe, J.R., Sundt, P.C., Valiente-Banuet, A., Casas, A., Viveros, J.L. (2001), 'Pre-columbian soil erosion, persistent ecological changes, and collapse of a subsistence agricultural economy in the semi-arid Tehuacan Valley, Mexico's 'cradle of maize', *Journal of Arid Environments*, Vol. 47, pp. 47-75.

Mortimore, M., Harris, F.M.A., Turner, B. (1999), 'Implications of land use change for the production of plant biomass in densely populated Sahelo-Sudanian shrub-grasslands in north-east Nigeria', *Global Ecology and Biogeography Letters*, Vol. 8, pp. 243-56.

Mouat, D., Lancaster, J., Wade, T., Wickham, J., Fox, C., Kepner, W., Ball, T. (1997), 'Desertification evaluated using an integrated environmental assessment model', *Environmental Monitoring and Assessment*, Vol. 48, pp. 139-56.

Mwalyosi, R.B.B. (1992), 'Land-use changes and resource degradation in south-west Masailand, Tanzania', *Environmental Conservation*, Vol. 19, pp. 145-52.

Okin, G.S., Murray, B., Schlesinger, W.H. (2001), 'Degradation of sandy arid shrubland environments: Observations, process modelling, and management implications', *Journal of Arid Environments*, Vol. 47, pp. 123-44.

Olsson, L. (1993), 'On the causes of famine: Drought, desertification and market failure in the Sudan', *Ambio*, Vol. 22, pp. 395-403.

Palmer, A.R., Van Rooyen, A.F. (1998), 'Detecting vegetation change in the southern Kalahari using Landsat TM data', *Journal of Arid Environments*, Vol. 39, pp. 143-53.

Pickup, G. (1998), 'Desertification and climate change: The Australian perspective', *Climate Research*, Vol. 11, pp. 51-63.

Ram, K.A., Tsunekawa, A., Sahad, D.K., Miyazaki, T. (1999), 'Subdivision and fragmentation of land holdings and their implication in desertification in the Thar Desert, India', *Journal of Arid Environments*, Vol. 41, pp. 463-77.

Rango, A., Chopping, M., Ritchie, J., Havstad, K., Kustas, W., Schmugge, T. (2000), 'Morphological characteristics of shrub coppice dunes in desert grasslands of southern New Mexico derived from scanning LIDAR', *Remote Sensing of Environment*, Vol. 74, pp. 26-44.

Rasmussen, K., Fog, B., Madsen, J.E. (2001), 'Desertification in reverse? Observations from northern Burkina Faso', *Global Environmental Change*, Vol. 11, pp. 271-82.

Rozanov, B.G. (1991), 'Once again on desertification', *Soviet Soil Science*, Vol. 23, pp. 22-31.

Ringrose, S., Matheson, W. (1992), 'The use of Landsat MSS imagery to determine the aerial extent of woody vegetation cover change in the west-central Sahel', *Global Ecology and Biogeography Letters*, Vol. 2, pp. 16-25.

Ringrose, S., Vanderpost, C., Matheson, W. (1996), 'The use of integrated remotely sensed and GIS data to determine causes of vegetation cover change in southern Botswana.', *Applied Geography*, Vol. 16, pp. 225-42.

Runnström, M.C. (2000), 'Is northern China winning the battle against desertification? Satellite remote sensing as a tool to study biomass trends on the Ordos Plateau in semiarid China', *Ambio*, Vol. 29, pp. 468-76.

Saiko, T.A., Zonn, I.S. (2000), 'Irrigation expansion and dynamics of desertification in the Circum-Aral region of Central Asia', *Applied Geography*, Vol. 20, pp. 349-67.

Seixas, J. (2000), 'Assessing heterogeneity from remote sensing images: The case of desertification in southern Portugal', *International Journal of Remote Sensing*, Vol. 21, pp. 2645-63.

Sheehy, D.P. (1992), 'A perspective on desertification of grazingland ecosystems in north China', *Ambio*, Vol. 21, pp. 303-7.

Tsunekawa, A., Kar, A., Yanai, J., Tanaka, U., Miyakazi, T. (1997), 'Influence of continuous cultivation on the soil properties affecting crop productivity in the Thar Desert, India', *Journal of Arid Environments*, Vol. 36, pp. 367-84.

Turner, M.D. (1999a), 'Labor process and the environment: The effects of labor availability and compensation on the quality of herding in the Sahel', *Human Ecology*, Vol. 27, pp. 267-96.

Turner, M.D. (1999b), 'Conflict, environmental Change, and social institutions in dryland Africa: Limitations of the community resource management approach', *Society and Natural Resources*, Vol. 12, pp. 643-57.

Valle, H.F. del, Elissalde, N.O., Gagliardini, D.A., Milovich, J. (1998), 'Status of desertification in the Patagonian Region: Assessment and mapping from satellite imagery', *Arid Soil Research and Rehabilitation*, Vol. 12, pp. 95-122.

Vandekerckhove, L., Poesen, J., Wijdenes, D.O., Nachtergaele, J., Kosmas, C., Roxo, M.J., Figueiredo, T. de (2000), 'Thresholds for gully initiation and sedimentation in Mediterranean Europe', *Earth Surface Processes and Landforms*, Vol. 25, pp. 1201-20.

Venema, H.D., Schiller, E.J., Adamowski, K., Thizy, J.M. (1997), 'A water resources planning response to climate change in the Senegal River Basin', *Journal of Environmental Management*, Vol. 49, pp. 125-55.

Wang, G., Qian, J., Cheng, G., Lai, Y. (2001), 'Eco-environmental degradation and causal analysis in the source region of the Yellow River', *Environmental Geology*, Vol. 40, pp. 884-90.

Weiss, E., Marsh, S.E., Pfirman, E.S. (2001), 'Application of NOAA-AVHRR NDVI time-series data to assess changes in Saudi Arabia's rangelands', *International Journal of Remote Sensing*, Vol. 22, pp. 1005-27.

Wijdenes, D.J.O., Poesen, J., Vandekerckhove, L., Ghesquiere, M. (2000), 'Spatial distribution of gully head activity and sediment supply along an ephemeral channel in a Mediterranean environment', *Catena*, Vol. 39, pp. 147-67.

Yang, X. (2001), 'The oases along the Keriya River in the Taklamakan Desert, China, and their evolution since the end of the last glaciation', *Environmental Geology*, Vol. 41, pp. 314-20.

Zhou, W.J., Dodson, J., Head, M.J., Li, B.S., Hou, Y.J., Lu, X.F., Donahue, D.J., Jull, A.J.T. (2002), 'Environmental variability within the Chinese desert-loess transition zone over the last 20,000 years', *The Holocene*, Vol. 12, pp. 107-12.

# Index

Aagesen, D. 36, 91-3, 205, 209-10
Abel, N. 11
ability, *see* capability
abrupt change, *see* rate
Achard, F. 7, 9, 15, 217, 221, 228
Adamowski, K. 17, 35, 64, 68, 73-4, 132, 204, 208
Adams, C.R. 8, 9, 23
Adams, W.M. 11
Adger, W.N. 18
adaptation, adaptive capacity, *see* vulnerability
African drylands 4, 9-20, 22, 34-40, 42, 47-8, 58, 63-74, 95-9, 100-107, 109-16, 120-21, 125-9, 136-44, 162, 167-9, 176-85, 198-9, 202-14, 217-18, 227; *see also* cause; desert; indicator; initial condition; progression; Sahelian drought and famine
Algeria 35, 73
Atlas Region 35, 73, 132
Botswana 35, 65, 68, 71, 132, 144, 153, 159
Burkina Faso 35, 68, 128, 131, 144, 147, 152, 158, 215
East Africa 35, 63-7, 95, 116, 120, 126, 168, 203, 210
Kalahari 35, 65-8, 71, 95, 126, 132, 144, 153, 159
Kenya 13-4, 35, 66-7, 218
Mali 35, 71-3, 126-8, 143, 152, 158, 162
Masailand 35, 65-6, 72
Mauritania 35
Morocco 35, 73, 132
Niger 35, 67, 70, 126-8, 143, 152, 158, 162
Nigeria 35, 68, 70, 126, 144-6, 159, 212, 215
North Africa 10, 35
Okavango Delta 35, 68, 71-2, 126, 144
Sahara 4, 9, 12, 16-9, 22, 64, 216-7
Sahel 7, 9, 11-18, 20-22, 34-5, 64, 68-73, 95, 103, 106, 120-21, 125, 128, 144-7, 152-3, 158-9, 162, 202-3, 206, 211-14, 217, 222
Senegal 14-5, 19, 35, 64, 68, 74, 128, 132, 144, 147, 162, 210
South Africa 14, 35, 71
southern Africa 35, 63-9, 71-2, 95, 143, 163
Sudan (Republic) 15, 18-9, 22, 35, 63-4, 67-9, 71, 127, 144, 160-62, 216
Sudan-Sahel 18, 20, 34-5, 64, 68-70, 95, 120-21, 125, 144-47, 158-9, 168, 202-3, 206, 211-14
Tanzania 35, 72, 65-6, 126, 146, 153, 160
Turkana Steppe 35, 66-7, 218
West Africa 14-9, 34-5, 63-6, 69-71, 74, 95, 121, 125, 128, 144, 162, 168, 202-3, 206, 211-14
Agbola, B. 1, 12, 25, 214
Agnew, C. 17
agricultural suitability, *see* land quality; (process) rate
agro-industrialization, *see* underlying cause (commercialization, industrialization)
albedo, *see* multiple impact
amplifying feedback, *see* system property
analytical framework, *see* research design
ancient desertification 4-7, 180; *see also* contemporary desertification; historical desertification; scales
Angelsen, A. 1, 12, 25, 214
animal species, *see* multiple impact (biodiversity)
Annan, Kofi. 23
antiquity, *see* ancient desertification; collapse (of ancient societies)
Archer, E.R.M. 211, 212

aridization, *see* proximate cause (increased aridity); underlying cause (desiccation)
Armesto, J.J. 10
Ash, A.J. 211, 212
Asian drylands 4-5, 9, 15, 20, 34, 37-40, 42, 47-48, 53-54, 58, 61, 63, 95-9, 100-104, 106-12, 114-16, 121, 129-130, 133-134, 136-45, 149, 152, 163, 166-168, 175-85, 198-199, 204, 206, 216-218; *see also* cause; collapse of (ancient societies); indicator; initial condition; progression
   Arabian Peninsula 34, 55, 135, 151, 153, 198
   Aral Sea Basin 13, 35, 54, 59-60, 62, 130, 154-155, 157, 161, 164, 166-167, 206, 216
   Caspian Sea Region (and Plain) 35, 57, 60, 62, 155 ,164
   Central Asian desert and steppe region 9, 34, 61, 97, 144, 175, 198, 207, 212, 216-218
   China 3, 5-7, 13, 15, 34, 35, 45, 54, 55, 56, 57, 60, 62, 116, 120-121, 126-127, 124, 144-152, 154-166, 202, 203, 217-218
   East Mediterranean (steppe zone) 4-5, 15, 34-35, 135, 203
   India 5, 34-35, 58, 135, 138, 160-161
   Jordan 35, 135
   Kazakstan 34, 63, 130, 135, 151-152
   Kuwait 35, 135
   Russia 34, 56-57, 120, 126, 135, 150, 163
   Saudi Arabia (Saudi Desert) 35, 58, 114, 135, 146, 151, 153
   Turkmenistan 34, 63, 130, 135, 151-152
   Uzbekistan 14, 63, 130, 135, 151-152
Asmode, J.F. 35, 36, 73, 76, 78, 79, 132, 204, 205, 209
assessment, *see* LADA
attenuating feedback, *see* system property
Aubreville, A. 17
Australian drylands 13, 15, 36-40, 42, 47-49, 79-82, 96-8, 100-101, 104, 107, 109, 111-12, 115, 136-7, 139, 141-3, 170, 176-85, 206, 209, 210, 213, 218; *see also* cause; indicator; initial condition; pathways
   Northern Territory 36, 80, 198
   New South Wales 36, 79-80, 160, 198
   Queensland 36, 80, 160, 198
   South 36, 80-81, 198
   West 36, 80, 198
Ayoub, A.T. 35, 67, 127, 144, 204, 208

Badri, W. 35, 36, 73, 76, 78, 79, 132, 204, 205, 209
Baker, D. 11
Ball, T. 36, 83, 84, 86, 145, 147, 205, 209
Barbier, E.B. 12
Barbosa, P. 16
Bassett, T.J. 11, 18, 21
Basso, F. 36, 76, 77, 127, 209
Bastin, G.N. 36, 80, 81, 82, 205
Baulies, X. 25
Behnke, R. 24, 218
Behrenfeld, M.J. 16
Benjamin, P. 13
Benjaminsen, T.A. 11, 18, 35, 64, 67, 70, 71, 73, 143, 204, 208
Berlow, E. 10
Bertaudière, V. 35, 36, 73, 76, 78, 79, 132, 204, 205, 209
Bilsborrow, R.E. 211
Birdsey, R.A. 11
Biswas, A.K. 22
Blaikie, P. 2, 3, 211, 215
Blanche, K.R. 34, 36, 82, 83, 84, 146, 147, 205
Bloomfield, J. 10
Bockstael, N. 25
Bohle, H.-G. 3, 7, 21
Boserup, E. 211
Bouwman, A.F. 15
Bove, E. 36, 76, 77, 127, 209
Bras, R.L. 10
Brasseur, G. xvii
Breckle, S.W. 17, 201
Briceño, S. 8, 10
Brookfield, H. 2, 3, 211, 215
Brooks, E. 13
Brovkin, V. 10

Brown, G. 35, 55, 59, 60, 82, 83, 84, 204, 205
Brown, J.H. 36
Brown, K. 18
Brown, L.R. 17, 18, 214, 218
Bruce, J.W. 1, 12, 25, 214
Bunney, S. 4

capability; *see also* indicator; (process) rate; syndrome
  of land 2-3, 9-10, 21, 23, 129, 132, 158-162
  of people 3, 13, 21, 161-162
carbon, *see* multiple impact
carrying capacity 3, 18, 20-21, 56, 89, 159-162, 172; *see also* misconception
Casas, A. 4, 9, 36, 90, 91, 126, 130, 209
Caspersen, J.P. 11
Cattaneo, A. 224, 225
causal pattern, *see* cause; mode
cause 1, 4-8, 11, 13-22, 24-31, 34, 40-42, 46, 49-51, 95-116, 201, 213, 215, 222, 227; *see also* proximate cause; multiplicity; underlying cause
  framing of 40-42
  irreducible complexity 20-22, 24
  multiple cause 11-13, 21, 25, 216, 222
  single factor causation 20-21, 103, 117, 119, 124, 201, 213, 222
  synergy, *see* mode
CCD, *see* UNCCD
Chapin, F.S. 10
Charney, J. 10
Chasek, P.S. 8, 14
Chater, S. 224, 225
Chemin, M.C. 17, 218
Cheng, G.D. 35, 45, 54, 55, 56, 58, 59, 60, 61, 62, 138, 145, 146, 147, 203, 204, 207
Chopping, M. 36, 83, 84, 146, 205
Chou, N. 22
Christensen, L. 2
Ciasis, P. 11
Claussen, M. 10
Cleaver, K. 9, 17, 214
Clein, J.S. 10
climate, *see* cause; initial condition; multiple impact

climatic variability, *see* cause; multiple impact
collapse; *see also* importance (of desertification)
  of ancient societies 4-7, 54
  of local economies 7-10, 80, 90, 116, 161
Collins, M.J. 18
Colstoun, E.B. 16
concept, *see* definition
contemporary desertification 7-11, 24, 62, 82, 87, 166-167, 181, 222; *see also* ancient desertification; historical desertification
control points, *see* system property
controversy 2, 16-22
conversion (of dryland) 4, 11-12, 15, 21-3, 57-8, 152, 171; *see also* modification
coping, *see* vulnerability
Coomes, O. 1, 12, 25, 214
Corell, R.W. 2, 8
coupled process, *see* system property
criticality, *see* vulnerability
crop loss, *see* multiple impact
Cropper, A. 2, 10, 15, 26, 211, 221
cropping, *see* proximate cause (crop production)
Crosson, P. 10, 22
Cunningham, G.L. 11, 128, 216
Curtin, C.G. 36, 82-4, 204-5

Dahlem Desertification Paradigm 26-7, 223-7
dampening feedback, *see* system property
Dargaville, R. 10
Darkoh, M.B.K. 8
data (analysis, bias, selection, statistics), *see* research design
DDP, *see* Dahlem Desertification Paradigm
De Camara, C.C. 16
definition
  of degradation 3
  of desertification 1-3, 201
  of drylands 2, 47-8
  of land use 38-42, 178
  of syndrome 14, 42-4
deforestation, *see* proximate cause

degree (of desertification) 9-10, 15, 45, 71, 83, 164, 187-97, 213
DeFries, R. 7, 9, 15, 217, 221, 228
desert (semi-, sub-) 2, 4-8, 16-22, 31, 34-6, 40, 45, 63-5, 82-7, 91-3, 126-8, 130-5, 138, 144-50, 152-67, 171-6, 180-1, 198-9, 208, 212, 216-8
  advancing ('marching') 17, 22, 87-9, 170-72; *see also* misconception
  'browning of the Sahara' 4; *see also* Africa (Sahara)
  encroachment 17-9, 22, 69, 126-8, 130, 143, 152; *see also* misconception; multiple impact (sandification)
  'man-made' 17-20; *see also* misconception
Diamond, J. 5
Diemer, C. 36, 75, 77-8, 126, 205
Dinag'a, A. 13
Diouf, A. 23, 215, 227
direct cause, *see* proximate cause
Dirzo, R. 1, 10, 12, 25, 214
disaster, *see* hazard
discrepancy, *see* controversy
disturbance, *see* perturbation
diversity, *see* multiple impact (biodiversity)
Dodson, J. 5, 35, 54-7, 127, 204, 207, 211
Dolcemascolo, G.P. 25, 201
Donahue, D.J. 5, 35, 54-7, 127, 204, 207, 211
Döös, B.R. 223
Dowlatabadi, H. 211-12
Downing, T.E. 3, 7, 13-4, 21
Dregne, H.E. 2, 7- 8, 14-6, 21-2, 30, 201, 217
driving forces, *see* underlying cause
drought, *see* proximate cause; Sahelian drought and famine; underlying cause; vulnerability
Druckman, D. 30
Dube, O.B. 35, 65-6, 68, 72, 126, 144, 203-4, 208
Dumanski, J. 36, 40
Dumontet, S. 36, 76-7, 127, 209
dust bowl, *see* syndrome; United States (Midwest)
Dwyer, E. 16

Eckholm, E. 17-8, 214, 218
Eckley, N. 2
ecological dimensions 1, 2, 24-6, 44-6, 49, 51, 56, 59, 70, 73, 108, 114, 143, 163-7, 170, 176-7, 189-93, 201, 212-8, 221-5; *see also* cause; coupled process; indicator; (process) rate; syndrome
ecosystem goods and services, *see* multiple impact
Ehrlich, D. 16
Elissalde, N.O. 36, 92-3, 205
Ellis, J. 21
El Niño Southern Oscillation, *see* underlying cause (climatic factor)
Eltahir, E.A.B. 10
Emel, J. 13
endangerment, *see* vulnerability
Endo, K.N. 35, 45, 54, 56, 58-62, 138, 146-7, 203-4, 207
environmental history, *see* initial condition
Esaias, E. 16
Esser, G. 10
Essery, R.L.H. 11
Estell, R. 17, 36, 83-7, 127, 146, 205
Eswaran, H. 8-9, 23
Euphrates, *see* collapse (of ancient societies)
European Commission 22-3
European drylands 5, 15, 17, 35-42, 47-8, 74-84, 91-116, 119, 136-42, 170-85, 198-9, 2056, 209-12, 217-8; *see also* cause; collapse (of ancient societies); indicator; initial condition; progression
  Greece 5, 36, 75, 77, 126, 131, 175
  Italy 36, 77, 116, 127
  Mediterranean Region (Basin) 5, 15, 34-6, 58, 60, 73-8, 110, 114, 121, 131, 135, 173-5, 206, 211-13, 217-8; *see also* Africa; Asia
  Portugal 35-6, 75, 77, 128, 144, 146
  Spain 13, 36, 75, 77, 127, 152-3
export (orientation), *see* underlying cause (commercialization)
exposure, *see* vulnerability
extent (of desertification) 13-6, 22-7, 45-7, 54, 65, 76, 83, 87, 146-53, 158-9, 162-4, 172-4, 202-3, 206, 217

extraction, *see* proximate cause

Fairhead, J. 17-8
Falkowski, P.G. 16
Fan, S.M. 11
fast change, *see* rate
feedback, *see* system property
Feldman, G.C. 16
Feng, Q. 35, 45, 54, 56, 58-9, 60-2, 138, 146-7, 203-4, 207
Fernandez, R.J. 211-12
Ferrara, A. 36, 76-7, 209
Fertile Crescent, *see* collapse (of ancient societies)
Field, C.B. 11, 16
Field, J.O. 3, 7
Figueiredo, T. de 36, 75-8, 205, 209
Fischer, G. 1, 12, 21, 25, 214
fire; *see also* proximate cause
 and climate 16, 100
 and deforestation or forest degradation 78, 87-8, 173-4
 and vegetation mosaic 78-9
 anthropogenic (land use) 16, 74-5, 78, 87-8, 168, 172
 bush fires (wildfires) 64, 79-80, 140, 142, 168
 changes in fire regime 40, 80-1, 99-100, 170, 177
 frequency and impact 78, 80, 99-100, 128, 175
 suppression 10, 80, 85
firewood collection, *see* proximate cause
flexibility, *see* vulnerability
fluctuation, *see* pathway (reversible)
Ferrara, A. 36, 76, 127
Floret, C. 14, 23, 30
Fog, B. 16, 35, 64-5, 68-70, 131, 144, 147, 204, 208, 211
Foley, J. 10
Folke, C. 1, 12, 25, 214
Forse, B. 1
Fox, C. 36, 83-4, 86, 145, 147, 205, 209
Fox, J. 25, 201
Fredrickson, E. 17, 36, 83-7, 127, 146, 205
Fresco, L. 22, 25
Frouin, R. 16
framework, *see* research design
fuelwood collection, *see* proximate cause

Fujimori, M. 19
fundamental cause, *see* underlying cause

Gagliardini, D.A. 36, 92-3, 205
Galvin, K.A. 21, 25
Ganopolski, A. 10
Gauquelin, T. 35-6, 73, 76, 78-9, 132, 204-5, 209
Geist, H.J. 1, 3, 10, 12-5, 21-2, 24-6, 29, 42-4, 46, 201, 211, 213-4, 216, 221-5, 227-8
Genxu, W. 17, 35, 59-60, 62, 130, 146-7, 204, 207, 210
George, P.S. 1, 12, 25, 214
Gerontidis, S. 36, 75-8, 131, 205, 209-11
Ghesquiere, M. 36, 75-7, 127, 205, 209
Gichuki, F. 11
Giourga, C. 202
Glantz, M.H. 8, 13, 20
GLASOD, *see* Global Assessment of Human-induced Soil Degradation
Glazovsky, N.F. 13, 17
Global Assessment of Human-induced Soil Degradation 15, 23
Gloor, M. 11
Gockowski, J. 224-5
Goodale, C.L. 11
Gonzalez, P. 15, 17, 35, 64-5, 68, 73, 128, 144, 147, 204, 208
gradual change, *see* rate
Graetz, R.D. 2
Grainger, A. 22
Gramenopoulos, N. 15, 217
grass(land), *see* vegetation
Gray, L.C. 11
grazing, *see* proximate cause (livestock production)
Great Plains, *see* United States
Grégoire, J.M. 16
Guodong, C. 17, 35, 59-60, 62, 130, 146-7, 204, 207, 210

Haarmann, V. 12, 211
Hackler, J.L. 10
Hakkeling, R.T.A. 15
Hall, K.R. 29
Hamblin, J. 36
Han, F.X. 35, 58-62, 203-4, 207
Harding, R.J. 11
Harmon, M.E. 1

Harris, F.M.A. 35, 64-5, 68-70, 126, 144, 146, 204, 208, 211-12
harvest loss, *see* multiple impact
Havstad, K.M. 17, 36, 83-7, 127, 146, 205
hazard, *see* vulnerability
Head, M.J. 5, 35, 54-7, 127, 204, 207, 211
Heath, L. 11
Heimann, M. 10
Helldén, U. 1, 15, 19, 64-5, 71, 201, 208
heterogeneity, see multiplicity
Hiernaux, P.H.Y. 11, 211-12
Hill, J. 36, 75, 77-8, 126, 205
historical desertification 5-7, 13, 170-3, 180; *see also* ancient and contemporary desertification
Ho, P. 35, 55-7, 61, 204, 207, 218
Holland, E. 11
Holzner, W. 35, 54-6, 128, 132, 204
Homewood, K. 1, 12, 25, 214
Hootsmans, R. 15
Hostert, P. 36, 75, 77-8, 126, 205
hot spots (of desertification) 7, 9, 15, 27, 207, 221
Hou, Y.J. 5, 35, 54-7, 127, 204, 207, 211
Houghton, R.A. 10-1
Howorth, C. 11
Huber-Sanwald, E. 10
Huenneke, L.F. 10-1, 128, 216
human dimensions 1-31, 36, 40, 48, 54, 64, 67, 73, 77, 83-6, 90, 103, 106, 128, 132, 162, 166, 187-8, 193-6, 202, 207, 213-8, 222, 224; *see also* cause; coupled process; indicator; multiple impact; (process) rate; syndrome
human welfare, *see* human dimensions
human well-being, *see* human dimensions
Hurtt, G.C. 10
Hyden, G. 211
Hyder, P. 17, 36, 83-7, 127, 146, 205
hydrology, *see* water cycle

Ibrahim, F.N. 3, 7, 15, 17
ICASALS, *see* The·International Center for Arid and Semiarid Land Studies
IGBP, *see* International Geosphere-Biosphere Programme

IHDP, *see* International Human Dimensions Programme on Global Environmental Change
Imbernon, J. 1, 12, 25, 214
impact, *see* multiple impact
importance (of desertification) 4-10, 15-6, 20, 78, 167, 210, 224
impoverishment, *see* underlying cause
independent, *see* mode
indicator (of desertification) 23-4, 44-5, 133, 187, 201, 215-7, 223-8
    broad clusters 187-9
    ecological 189-93
    meteorological 196-7
    regional degradation types 198-9
    socio-economic 193-6
indirect cause, *see* underlying cause
initial condition 24-6, 29, 31, 34, 40, 42, 46, 176-9, 213, 225-7; *see also* climate, land use, land tenure, soil, vegetation
    African drylands, 63-74, 176
    Asian drylands 53-63, 176
    Australian drylands 79-82, 176
    European drylands 74-9, 176
    Latin American drylands 87-93, 176
    North American drylands 82-7, 176
intensification, *see* land use transition; underlying cause (economic factor, technological factor)
interaction, *see* mode; scale
intermediate factor 21, 26, 34, 40, 93-5, 124-5, 163, 174, 184, 213, 225-7
International Geosphere-Biosphere Programme xvii, 11, 222
International Human Dimensions Programme on Global Environmental Change xvii, 11, 222
intervention 14, 24, 27-8, 43-4, 134, 137, 140, 164, 223-5; *see also* trade-off; sustainability; syndrome
irrigated croplands, *see* cropping
irrigation, *see* proximate cause

Jackson, R.B. 10
Jäger, J. 4, 10, 24, 221
Janetos, A.C. 7, 9, 15, 217, 221, 228
Jarrell, W.M. 11, 128, 216
Jiang, H. 5, 11, 13, 212
Joos, F. 10

# Index

Jull, A.J.T. 5, 35, 54-7, 127, 204, 211

Kadomura, H. 19
Kaplan, J. 10
Kar, A. 36
Kasapidis, P. 36, 75, 77-8, 126, 205
Kasperson, J.X. 2, 13, 215-6, 228
Kasperson, R.E. 2, 13, 215-6, 228
Kassas, M. 2, 8, 217
Kates, R.W. 12, 211
Kawaguchi, S. 19
Keita, N. 17
Kepner, W. 36, 83-4, 86, 145, 147, 205, 209
Kerven, C. 14, 24, 218
Keya, G.A. 35, 65-8, 204, 210, 216, 218
key cause, *see* cause; misconception
key indicator, *see* indicator; misconception
Kharin, N. 211
Khresat, S.A. 35, 58-9, 60, 62, 204, 207
Kicklighter, D.W. 10
Kinzig, A. 10
Kleidon, A. 14-5
Kohlmaier, G. 11
Kolber, D.D. 16
Kosmas, C. 36, 75-8, 131, 205, 209-11
Koutsidou, E. 202
Kravitz, L.L. 16
Kreileman, E. 15
Kriechbaum, M. 35, 54-6, 128, 132, 204
Krug, T. 25
Kustas, W. 36, 83, 146, 205

LADA, *see* Land Degradation Assessment in Drylands
Lahav, R. 11
Lai, Y. 35, 54-5, 145-7, 204
Lal, R. 10
Lambin, E.F. xvii, 1, 3, 7, 9-16, 21-6, 29, 42-4, 46, 201, 211, 213-7, 221-5, 227-8
Lamprey, H.F. 14-5, 17-8, 22, 216
Lancaster, J. 36, 83-4, 86, 145, 147, 205, 209
Land Degradation Assessment in Drylands 14, 27, 225
land quality, *see* capability (of land)
land use, *see* proximate cause
Land-Use/Cover Change project xvii, 11

land use history, *see* initial condition
land use transition 178, 202-10, 223, 226-8
land surface change, *see* multiple impact
land tenure 3, 19, 43, 67-8, 103, 108-9, 120, 124, 137-40, 143, 163, 179, 183, 211, 226; *see* also underlying cause (policy and institutional factor)
Lappé, F. 18
Latin American drylands 9, 15, 36-42, 47-8, 53, 87, 95-104, 107-12, 115-6, 136-43, 176-85, 198-9, 205-6, 209-10; *see also* cause; collapse (of ancient societies); indicator; initial condition; progression
   Argentina 14, 36-7, 87, 91-3, 97, 103, 108, 125, 160, 163, 169-73, 198, 213, 217-8
   Mexico 4, 8, 14, 36, 84, 87-91, 120, 126-7, 146-9, 153, 159-60, 163, 169, 171-3, 198, 206, 213, 218
Lau, K.M. 11, 211
Lavauden, L. 17
Lawrence, K.T. 10
Leach, M. 11, 17-8
Leemans, R. 1, 10, 12, 14-5, 22, 25, 214
Le Houérou, H.N. 17, 30, 201, 213
Leonard, H.J. 211
Lepers, E. 1, 7, 9-10, 13-5, 22, 25-6, 42-4, 201, 211, 216-7, 221-2, 227-8
Lewis, J. 224-5
Li, B.S. 5, 35, 54-7, 127, 204, 207, 211
Li, X. 1, 12, 214
Lin, N.F. 5, 34-5, 54-62, 144, 203-4, 207
Liu, Z.M. 35, 54-5, 58, 126, 204, 207
livestock (production), *see* proximate cause
Lodge, D.M. 10
Lohnert, B. 3
Loos, S.Q. 16
Lu, X.F. 5, 35, 54-7, 127, 204, 207, 211
Lubchenco, J. 10
LUCC, *see* Land-Use/Cover Change project
Ludwig, J.A. 34, 36, 82-4, 146-7, 205
Lüdeke, M. 3, 13-4, 21, 42, 216, 228
Luers, A. 2
Lupo, F. 16
Lusigi, W.J. 21

Mabbutt, J.A. 14, 23, 30
McAuliffe, J.R. 4, 8, 36, 90-1, 126, 130, 209
McCarthy, J.J. 2
McClaine, C.R. 16
McGuire, A.D. 10
Madsen, J.E. 16, 35, 64-5, 68-70, 131, 144, 147, 204, 208, 211
Mainguet, M. 8, 17, 218
Mannion, A.M. 20, 214
Manzano, M.G. 17, 36, 88-9, 127, 203, 205, 209
Marathianou, M. 36, 75-8, 131, 205, 209-11
Margaris, N.S. 202
Marsh, S.E. 35, 55, 60, 146, 204, 207
Martello, M.L. 2
Martinez, A. 17, 36, 88-9, 127, 203, 205, 209
Masakhalia, Y.F.O. 22
Matheson, W. 17, 35, 64-9, 73, 128, 204, 208
Matson, P.A. 2, 4, 10, 24, 221
Mattarazzo, B. 29, 46, 51
Mazzucato, V. 10
MEA, *see* Millennium Ecosystem Assessment
Mearns, R. 11, 18
mediating factor, *see* intermediate factor
Mediterranean Region (Basin), *see* African drylands (Algeria, Morocco); Asian drylands (Jordan); European drylands (Italy, Greece, Portugal, Spain)
Meier, R.A. 10
Melillo, J.M. 10
Mensching, H. 17
Mesopotamia, *see* collapse (of ancient societies)
meta-analysis, *see* research design
meteorological dimensions 1-3, 20, 44, 187, 196-7, 215-7, 228; *see also* cause; coupled process; indicator; multiple impact; (process) rate; syndrome
methodology, *see* research design
Middleton, N.J. 2, 8, 15, 213
migration, *see* underlying cause (demographic factor)
Millennium Ecosystem Assessment 15

Milovich, J. 36, 92-3, 205
misconception 2, 11, 16, 18-21, 23
Mishra, V. 25, 201
mitigating feedback, *see* system property
Miyazaki, T. 19, 35-6, 58, 138, 204, 207
mobility, *see* flexibility; migration
mode (of factor operation and interaction) 13, 106, 116
  chain-logical conncetion 117-22
  concomitant occurrence 54, 73, 83, 89, 106, 117-9, 168, 172
  synergistic manner 27, 73, 97, 106, 117-9, 126-8, 165-8
modelling 15-6, 22, 221
modification (of drylands) 12, 15-6, 21-3, 52, 162, 171-2, 215, 224, 227; *see also* conversion
Mohammad, M. 35, 58-60, 62, 204, 207
Moldenhauer, O. 14
monitoring 14, 23, 45, 187, 215-6
Montes, N. 35-6, 73, 76, 78-9, 132, 204-5, 209
Mooney, H.A. 2, 10, 15, 26, 211, 221
Moore, B. III 4, 10-11, 24, 221
Moorecroft, P. 11
Moran, E.F. 1, 12, 25, 214
Mortimore, M.J. 1, 11-12, 20, 35, 64-5, 68-70, 126, 144, 146, 204, 208, 211-12, 214
Mouat, D. 36, 83-4, 86, 145, 147, 205, 209
Muldavin, E. 34, 36, 82-4, 146-7, 205
Mulligan, M. 11
multiple impact
  albedo 3, 10, 124, 128, 168-9, 224
  alkalinization 93
  biodiversity 10-12, 15, 20, 37, 40, 44, 88-91, 54-5, 60, 65-9, 73, 78-9, 82, 86, 128, 147, 158-9, 168-9, 171, 174, 177, 189-90, 193-4, 215, 222-5
  carbon 10, 27
  climate (rainfall) 1-7, 10-11, 16-21, 24-6, 29, 40, 44-5, 54, 58-9, 63-7, 72-83, 87-90, 99, 106, 119-21, 124-32, 138, 143-5, 150, 156, 163-76, 188, 197, 201, 208, 211-12, 217-9, 222-4, 227

economic livelihood (harvest, yield) 9, 23, 27, 63, 69-74, 78, 87-91, 138, 159-61, 173, 188-9, 193-5, 199, 224
encroachment (grass, scrub, bush) 44, 64, 83, 87-9, 172, 188, 190, 198-9
(human) health 9, 44, 90, 142, 162, 196
salinization 4-5, 26, 46, 59, 62, 74, 93, 120, 130, 143, 154-7, 161, 166-7, 189, 192, 213, 217-8, 228
sand encroachment 17-9, 69, 126-8, 130, 143, 152
soil crusting (compaction) 3, 56, 89, 93, 188, 190
soil degradation 9-15, 22-3, 72-4, 77-8, 123, 170, 173, 190, 199, 214-5, 221
soil erosion (wind, water, sheet, gully) 2-9, 13, 18, 22-3, 43-5, 55-7, 63, 69, 72, 76-7, 82-3, 89, 93, 99, 127-8, 131-2, 140, 143, 150, 152, 166, 172-5, 187-91, 198-9, 208, 214
vegetation degradation 15, 20-21, 54, 60, 91-3, 164-7, 170-1, 174
water degradation 10, 13, 44-5, 59, 82, 85, 143, 154-7, 161, 164-6, 173, 177, 187-9, 192; *see also* salinization
multiplicity 11, 24-5, 201, 215-6, 223, 227; *see also* (multiple) cause; multiple impact; scales
Murray, B. 36, 82-4, 86, 128, 130, 209
Mwalyosi, R.B.B. 17, 35, 65-6, 72-3, 126, 146, 204, 208, 210
myth 1, 11, 17, 214; *see also* misconception; simplification

Nachtergaele, F. 14, 23, 27
Nachtergaele, J. 36, 75-8, 205, 209
narrative 18, 20-1, 24-5, 29-31, 34, 40, 53, 63, 74, 79, 82, 87, 174, 201, 210, 222, 227
natural (environmental) change, natural variability, *see* underlying cause (climatic factor)
Návar, J. 17, 36, 88-9, 127, 203, 205, 209

Neelin, J.D. 11, 211
Newcomb, W.W. 16, 217
Niamir-Fuller, M. 11, 13, 202-3, 212
Nicholson, S.E. 16, 26, 40, 211, 217
Nielsen, T.L. 11, 52
Niemeijer, D. 10
Nijkamp, P. 29, 46, 51
Noordwijk, M. van 224-5
nomadism, *see* livestock (production); misconception
non-equilibrium trajectory, *see* pathway
North American drylands; *see* United States
Norusis, M.J. 34

Oba, G. 21
Odego-Ogwal, L.A. 22
Oesterheld, M. 10
Ojima, D.S. 25
O'Keefe, D.J. 11
Okin, G.S. 11, 36, 82-6, 128, 130, 209
Oldeman, L.R. 15
Oldfield, F. 4, 10, 24, 221
Olson, G.W. 4
Olsson, L. 35, 71, 144, 204
oscillation, *see* pathway (reversible); underlying cause (climatic variability)
Ostrom, E. 211
Otterman, J. 10
overcultivation, *see* proximate cause
overgrazing, *see* proximate cause
overpopulation, *see* misconception; underlying cause (demographic factor)
Oyowe, A. 7

Pacala, S.W. 10
Palm, C. 22, 223-5, 228
Palmer, A.R. 35, 66, 71, 204, 208
Palnagyo, E.P. 22
Pando-Moreno, M. 17, 36, 88-9, 127, 203, 205, 209
pan formation, *see* soil crusting
Parmesan, C. 29
past work, *see* previous studies
pastoralism, *see* livestock (production)
pastoral suitability, *see* land quality; (process) rate
pasture, *see* livestock

pathway (of desertification) 29-31, 34, 40-42, 49, 128, 163-75, 213-4, 216-9, 222-7
  Africa 167-9
  Australia 170
  Central Asia 163-7
  conceptualization 11, 24-7, 46
  irreversible 11, 17, 20, 22-4, 61, 77, 91, 93, 128, 130-32, 164, 166-7, 174, 201, 207, 218
  Mediterranean 173-5
  Mexico (north) 171-3
  non-equilibrium 10, 16, 21, 24, 52, 218-9
  Patagonia 171-3
  reversible 6, 16, 21-3, 58, 63-5, 73, 76, 80, 99, 121, 127, 161, 165-9, 173, 195, 202-3, 206, 211-18, 227
  US Southwest 171
pattern, *see* causal pattern
Pearce, G. 36, 80-2, 205
Pereira, B.S. 16
Pereira, J.M.C. 16
perturbation, *see* vulnerability (endangerment, hazard, stressor)
Petschel-Held, G. 14, 42, 216, 228
Peylin, P. 11
Pfirman, E.S. 35, 55, 60, 146, 204, 207
Phillips, J.D. 22
Pickup, G. 16-7, 21, 35-6, 65-6, 68, 72, 79, 80-2, 126, 144, 203-5, 208-9, 213
Pieri, C. 36, 40
Pisante, M. 36, 76, 77, 127, 209
plant species, *see* multiple impact (biodiversity)
Plato 5
Plisnier, P.D. 16
Poesen, J. 35-6, 75-8, 127, 205, 209
Poff, N.L. 10
Polanyi, K. 7
policy, *see* intervention; syndrome; underlying cause
Pollack, J.B. 10
Pollack, N.H. 16
Polsky, C. 2
Pounds, J.A. 29
poverty, *see* underlying cause
pre-historic, *see* ancient
Prentice, I.C. 10

previous studies 10-16
Price, J.T. 29
primary cause, *see* misconception; cause
Prince, S.D. 14, 16, 23
progression 25-7, 218; *see also* extent; pathways; rate; system property
progressive change, *see* rate
proximate cause 40-42, 95-102, 182, 213, 226-7; *see also* cause
  broad clusters 95-7, 182
  agricultural activities 96-8, 182
    crop production ('overcultivation') 7-10, 20, 37-40, 44, 48-9, 53, 58, 69-75, 81, 98, 125, 164, 167-75, 195, 202-3, 206-11, 226
    livestock production ('overgrazing') 1, 5-13, 18-21, 24-7, 36-40, 53-8, 62-4, 67-78, 80-93, 97-8, 108-10, 114-21, 125-7, 132, 138, 146, 150, 163-7, 170-5, 194, 202-6, 210-13, 218, 226
  increased aridity 26, 40, 95-100, 103, 106, 117-19, 122, 168, 182, 188, 196-9, 226; *see also* fire
  infrastructure extension 27, 36-40, 56, 95-99, 101-3, 117-8, 121-2, 126-7, 137, 165-9, 182, 226
  wood extraction (and related extractional activities)
    deforestation 1, 5, 19-26, 29, 45, 64-8, 73, 128, 180, 182, 213-5, 221-2, 227
    collection of plant/animal products 37, 67, 73-4, 95-6, 102, 129, 162, 167-9, 174, 180, 182, 196, 213
    digging (for peat, turf, herbs) 67, 95-6, 102, 118, 121-2, 132, 180, 182
Puigdefábregas, J. 1, 2, 7, 13, 21, 213
Pulsipher, A. 2

Qian, J. 35, 54-5, 145-7, 204
Quaranta, G. 36, 76-7, 127, 209

Ragin, C.C. 30-31, 51
rainfed croplands, *see* cropping
rain use efficiency, *see* indicator
Ram, K.A. 35, 58, 138, 204, 207

## Index

Ramakrishnan, P.S. 1, 12, 214
Ramankutty, N. 7, 9-10, 15, 217, 221, 228
ranching, see livestock
Randerson, J.T. 16
rangelands, see livestock
Rango, A. 36, 83-4, 146, 205
rapid change, see rate
Rasmussen, K. 16, 35, 64-5, 68-70, 131, 144, 147, 204, 208, 211
rate (of desertification) 13-6, 22-3, 29-31, 40-2, 46, 49, 57, 59-60, 65-7, 85, 133, 201, 216-8, 221, 224, 228
  abrupt change 11, 24, 131, 152, 222
  fast (rapid, progressive) change 4, 7-9, 11, 15, 18-9, 24, 26-7, 42-6, 63, 71-2, 80-1, 85-6, 91, 108, 114, 131, 134-5, 138, 140, 143-52, 153-61, 164, 167-9, 174, 206, 212, 217-8, 221, 226-7
  process rate 134-5, 143-64, 174
    decline in agricultural suitability 161-2
    decline in pastoral suitability 158-60
    meteorological change 143-5
    vegetation change 145-50
    water degradation 154-8
    wind and water erosion 150-3
  slow (gradual) change 11, 24-7, 42-6, 81, 90, 133-40, 143-6, 149-61, 173-4, 201, 212, 216-7, 226-8
Raven, P.H. 10
Rawajfih, Z. 35, 58-9, 60, 62, 204, 207
Raynaut, C. 11
Redman, C.L. 4
Reginster, I. 16
Reichenau, T. 10
Reid, W. 2, 10, 15, 26, 211, 221
Reining, P. 23
relevance, see importance
research design (meta-analysis) 11, 25-52
  data analysis 34-46
  data bias 46-52
  data selection 31-3
  data statistics 32-46
    frequency of occurrence of land uses 36-40, 49
    location of case studies 33-6
    methodology 29-31
resilience, see vulnerability

response 3, 215 ; see also system property (feedback)
  ecological (biological) 22, 24-6, 44-6, 58, 66, 81-2, 128, 131, 150, 175, 222
  social 13, 24-6, 46, 110, 116, 172, 222, 226
Reusswig, F. 14, 42, 216, 228
Reynolds, J.F. 1-3, 11, 14, 20-26, 34, 42, 44, 47-8, 52, 128, 201-2, 211-17, 223, 225, 227-8
Richards, J.F. 1, 12, 214
Richardson, K. 4, 10, 24, 221
Rindfuss, R.R. 25, 201
Ringrose, S. 17, 35, 64-9, 73, 128, 132, 204, 208
Ritchie, J. 36, 83-4, 146, 205
Robbins, P.F. 11
Rocheleau, D. 13
root cause, see underlying cause
Root, T.L. 29
Rooyen van, A.F. 35, 66, 71, 204, 208
Rosenfeld, D. 11
Rosenzweig, C. 29
Rosset, P. 18
Roxo, M.J. 36, 75-8, 205, 209
Rozanov, B.G. 2, 35, 56-7, 59-62, 126, 203-4, 207, 214, 217
Rudich, Y. 11
Runnels, C. 4
Runnström, M.C. 35, 54-8, 60, 127, 131, 203-4, 207

Sagan, C. 10
Sahad, D.K. 35, 58, 138, 204, 207
Sahelian drought/famine 7, 17, 214, 222
Saiko, T.A. 17, 35, 54, 59-62, 128, 130, 145, 204, 207
Sala, O.E. 10
salinity, see multiple impact
Sanderson, A. 4, 10, 24
Sanderson, S. 21, 25, 221
sandification, see multiple impact (sand encroachment)
Sarmiento, J.L. 11
Sato, Y. 25
Sauer, U. 19
scales (of desertification)

space (spatial, hierarchical) 1, 15, 22, 24-7, 122-4, 201, 211, 216-8
time (temporal) 1, 3-16, 19-21, 24-7, 30-31, 34, 42, 45-6, 54-6, 64, 71-2, 80, 84, 88-93, 130, 163-5, 168, 170, 201-11, 216-7, 221-8; *see also* ancient, contemporary and historical desertification
Schellnhuber, H.J. 4, 10, 24, 221
Schiller, A. 2
Schiller, E.J. 17, 35, 64, 68, 73-4, 132, 204, 208
Schimel, D. 11
Schlesinger, W.H. 11, 15, 36, 82-4, 86, 128, 130, 209, 216-7
Schloss, A. 10
Schmugge, T. 36, 83, 146, 205
Schneider, S.H. 29
Schoknecht, N. 35, 55, 59, 60
Scholes, R.J. 7, 9, 15, 217, 221, 228
Schreiber, G. 9, 17, 214
Schulz, E. 16, 217
Scoones, I. 24, 218
Seixas, J. 36, 75-8, 128, 144, 146, 205, 209, 211, 216
sensitivity, *see* vulnerability
Serneels, S. 16
severity (of desertification) 14-6, 22, 27, 45, 145, 164
Shao, M. 3
shaping factor, *see* intermediate factor
Sharma, K.D. 23
Shaw Thacher, P. 8
Sheehy, D.P. 5, 35, 54-8, 60, 126, 144, 204, 207
Shevliakova, E. 11
Shi, H. 3
shrub, *see* vegetation
simplification 3, 11, 16-9, 132, 213
Sitch, S. 10
Skånes, H. 1, 12, 214
Skole, D. 21, 25
slow change, *see* rate
Smith, J. 2-3, 9, 13, 17, 21, 212, 215
Smith, S.E. 22
Sneath, D. 21
socio-economic dimensions, *see* cause; human dimensions
socio-political factor, *see* underlying cause

soil compaction, crusting, erosion, degradation, surface sealing, *see* multiple impact
Solbrig, O.T. 21
Sombroek, W.G. 15
Soulé, M.E. 1
South Africa 14
species, *see* multiple impact (biodiversity)
speed, *see* rate
Stafford Smith, D.M. 1, 3, 11, 14, 20-23, 25-6, 34, 42, 47-8, 52, 201-2, 213-4, 215, 217, 223, 225, 227-8
Stallard, R.F. 11
Stamp, L. 17
Stebbing, E.P. 17
Steffen, W. 1, 4, 10, 12, 24, 214, 221
Stenseth, N.C. 21
Stephenne, N. 11
Stern, P.C. 30
Stocking, M. 17
Stolle, F. 224-5
Stone, G.D. 1, 12, 214
Stone, P.H. 10
stressor, *see* vulnerability
Stroppiana, D. 16
Suliman, M.M. 17, 22
Sumerian civilization, *see* collapse of ancient societies
Sundquist, E.T. 11
Sundt, P.C. 4, 8, 36, 90-91, 126, 130, 209
sustainable development and land use, *see* sustainability
sustainability 2, 8, 15, 27, 80, 83, 89, 128, 148, 172, 206, 214, 217, 222-4, 228
Svarstadt, H. 18
Svedin, U. 1, 12, 214
switch and choke points, *see* control points
Swift, J. 1, 3, 17
Sykes, M.T. 10
synchronous, *see* mode
syndrome 133-43, 184; *see also* cause; definition; rate
changes in social organization, resource access, beliefs, values and attitudes 3, 26, 40-4, 57, 68-75, 91,

110, 116, 125-6, 139-43, 189, 212, 215
  changing opportunities created by markets 3, 7, 12, 43-4, 68, 80, 84-5, 92, 103, 108, 110, 114, 121, 130, 134, 137-41, 164, 171-2, 202, 211, 226
  'dust bowl' 7, 9, 14, 17
  increased vulnerability, *see* vulnerability
  loss of adaptive capacity, *see* vulnerability
  (outside) policy intervention 24, 28, 43-4, 134, 137, 140-1, 164
  resource scarcity causing a gradual pressure of production on resources 43-4, 134-8, 141, 171, 211, 226
  typology 43-4, 184
synergy, *see* causal factor synergy; mode
synthesis (of case studies), *see* research design
system property
  control points 24, 46, 95, 129, 131-2, 165, 169-70, 185, 218
  coupled process 2-5, 10-11, 21-7, 40-4, 54, 131-3, 165, 198, 215-8, 222-3
  feedbacks 2, 24-7, 40-42, 46, 95, 116, 150, 165-7, 201, 213, 227
    amplifying (or attenuating) 11-13, 24-6, 125-8, 168-9, 174, 185, 222
    mitigating (or dampening) 24-6, 125-7, 168, 170-71, 185, 222
  thresholds 10, 19, 24-6, 46, 95, 106, 128-32, 167-71, 175, 185, 208, 218, 226
Szabolcs, I. 17
switch and choke points, *see* system property (control points)

Taberner, M. 36, 76-7, 127, 209
Tanaka, U. 36
Tang, J. 5, 34-35, 54-62, 144, 203-4, 207
Tanzania
Taylor, C.M. 11
Thébaud, B. 17
The International Center for Arid and Semiarid Land Studies 23
Thizy, J.M. 17, 35, 64, 68, 73-4, 132, 204, 208

Thomas, D.S.G. 1-2, 8, 15, 201, 213
threshold, *see* system property
Tian, H. 10
Tiffen, M. 11
Tigris, *see* collapse of ancient societies
Tomich, T. 22, 223-5, 228
Toon, O.B. 10
Toulmin, C. 2, 17
trade-off 12, 29, 223-8
trajectory, *see* pathway
transhumance, *see* livestock
Trimble, S.W. 10, 22
Tsiourlis, G. 36, 75, 77-8, 126, 205
Tsunekawa, A. 35-6, 58, 138, 204, 207
Tucker, C.J. 11, 16, 211, 217
Turner, B. 35, 64-5, 68-70, 126, 144, 146, 204, 208, 211-12
Turner, B.L. II 1-2, 4, 10-13, 21, 24-5, 211, 214-6, 221, 228
Turner, M.D. 3, 11, 35, 64, 68-70, 126, 204, 208, 211-12
Tyson, P.D. 4, 10, 24, 221

Udelhoven, T. 36, 75, 77-8, 126, 205
ultimate cause, *see* underlying cause
UN, *see* United Nations
UNCCD, *see* United Nations Convention to Combat Desertification
UNCOD, *see* United Nations Conference on Desertification
underlying cause 40-42, 50-51, 102-16, 183, 226-7; *see also* cause; proximate cause
  broad clusters 102-5, 183
  climatic factor 2, 4-6, 11, 14, 19-20, 23-6, 50-1, 58-9, 63-4, 74, 87, 91, 100-105, 117-20, 123, 150, 163-5, 173-6, 183, 213, 224
  climate change 1, 10, 29, 76, 99, 124-6, 128, 132, 144-5, 167, 222
  desiccation 17-8, 21, 128-30, 197
  drought 1, 3, 7-18, 21-3, 26, 40, 54, 57, 65-8, 71-3, 79-85, 89-90, 93, 99-100, 106, 116, 122, 125-8, 131-2, 140-45, 161, 167-72, 188, 212-5, 222
  variability (fluctuation, oscillation, variation) 1, 12-3, 16, 19-21, 24, 40, 66, 70, 76, 79-82, 99-100, 126-31, 170, 196, 218, 221

cultural (socio-political) factor 104-5, 114-6, 123, 139
public attitudes, values and beliefs 10, 40-4, 56, 63, 67-9, 82-6, 92, 110, 114-6, 124-6, 137, 140-42, 164, 167, 171-2, 178, 183-4, 188-9, 193-5, 214, 222-3
individual and household behaviour 3, 9, 12-4, 17-9, 21, 24, 43, 67-70, 73, 86, 114-6, 121, 139-40, 143, 167, 174, 212
demographic factor 104-5, 110, 112-4, 123, 137, 183
age structure (imbalance of) 113-4, 125, 129, 174
density 55-7, 67-8, 71, 113-4, 121-3, 138, 164, 174, 183
growth ('overpopulation', 'pressure') 12, 17-21, 27, 85, 112-4, 121, 136-8, 184
life cycle change 21, 43, 69-70, 112-3, 121, 136, 139, 167, 227
migration 9, 12, 19-20, 40, 44, 56, 61-2, 72, 84, 110, 112-4, 121-2, 130, 138, 141, 164, 174, 183, 189, 194-5
natural increase (fertility) 20, 112-5, 122-3, 125, 136-8, 217, 228
size (increase of) 6, 20, 54, 57, 61-2, 71-3, 85-6, 90, 110, 112-4, 121-3, 138, 164, 174, 207, 226
economic factor 104-5, 110-11, 123, 137, 183
commercialization 12, 37, 61, 68-71, 79-80, 90, 111-12, 121, 137-8, 166-9, 172, 178, 184, 202, 206
demand 13, 61-3, 67, 73-4, 92, 110-11, 121, 124, 130, 165-7, 172, 207, 212, 223, 226
industrialization 62, 110-11, 124, 137-8, 143, 164, 178
intensification 12, 21, 37-8, 44, 53, 56, 62-4, 67-75, 80, 83, 86-93, 98, 108, 110-11, 120, 124, 138, 141, 164-8, 172, 178, 182-4, 201-13, 216, 223-7; *see also* land use transition
prices 7, 11, 43-4, 80, 85, 92, 110-11, 127, 137, 140-41, 169-70

market (failure) 3, 7, 12, 43-4, 68, 80, 84-5, 92, 103, 108-11, 114, 121, 130, 134, 137-41, 164, 168, 171-2, 183-4, 202, 211, 226
poverty 2, 4, 8-9, 12, 17-20, 34, 43, 81, 87-91, 111, 124-5, 130, 139-42, 169-74, 183-4, 189, 211, 214, 225-7
surplus extraction 20, 43, 136, 172, 184
urbanization 3, 9, 12, 62-3, 69, 74, 82, 85-8, 110-14, 121, 124, 127, 139, 141-3, 164, 169, 172, 178, 181-3, 189, 194, 213-4
policy and institutional factor 20, 104-5, 108-10, 123, 137, 179, 183
agricultural development policies 12, 18, 43, 57, 61, 67, 71-5, 79-80, 99, 108-9, 117-19, 124, 127, 137, 140, 164, 168, 202, 211, 224-6
European colonization 17, 79-85, 103, 108-9, 112, 141, 170, 206
'free grazing' 84, 88, 109-12
lack of effective rule 109, 116, 120, 124
land zoning 68, 108-9, 120, 141, 178, 202
private management 62, 68-70, 74, 88, 108-9, 112, 116, 120-21, 135, 143, 172-3, 179, 189, 214-5
sedentarization 37, 56, 63, 67-9, 97-8, 109-10, 167, 178, 203-6, 214
state management 20, 56, 74, 88, 108-9, 116, 120, 124, 163, 179
subsidies 55, 74-5, 109-11, 121, 124, 137, 140, 173-4, 183
succession law (traditional) 55-6, 67, 103, 108-9, 120-21, 124, 137-8, 183
technological factor 104-8, 123, 137, 183
introduction of new (agro) technologies 5, 12, 17, 61, 67-8, 85, 92, 103, 106-9, 120, 138, 141, 164, 166-72, 174-5, 178, 182-5, 195, 207-8, 211-12, 225-6; *see also* underlying cause (economic factor, intensification)

deficiency of technological applications 3, 63, 12, 107-8, 120, 130, 164-5
UNEP, *see* United Nations Environmental Programme
United Nations 8
United Nations Conference on Desertification 2, 8, 14, 20
United Nations Convention to Combat Desertification 2, 8, 18-9, 23, 214
United Nations Environmental Programme 2, 4, 8, 14, 18, 22, 34, 47, 202, 213-4, 227
United States 15, 22, 38-9, 96-101, 104, 107-12, 115, 136-42, 176-85; *see also* initial condition
  Llano Estacado 13
  Midwest 7, 17
  Southwest (Great Basin Region)
    Chihuahua Desert 36, 83-7, 146-50, 153, 177-85, 198; *see also* Latin American drylands (Mexico)
    Colorado (Plateau) 36, 83, 145-7, 177-85
    Mojave Desert 36, 83, 86, 128-30, 152-3, 177-85, 198
    Sonora (Desert) 36, 87-9, 121, 146, 149, 159, 177-85

Valentim, J. 224, 225
Valiente-Banuet, A. 4, 8-9, 36, 90-91, 126, 130, 209
Valle, H.F. del 36, 92-3, 205
Valone, T.J. 36, 82-4, 204-5
Vandekerckhove, L. 35-6, 75-8, 127, 205, 209
Vanderpost, C. 35, 66-8, 73, 132, 204, 208
Vasconcelos, M.J.P. 16
vegetation *see* initial condition; multiple impact
Veldkamp, T.A. 1, 12, 214
Venema, H.D. 17, 35, 64, 68, 73-4, 132, 204, 208
Veste, M. 17, 201
Virginia, R.A. 11, 128, 216
Vitousek, P.M. 10
Viveros, J.L. 4, 9, 36, 90-91, 126, 130, 209

Vogel, C.H. xvii, 1-3, 9, 12-3, 17, 21, 25, 211-12, 214-5
Vosti, S.A. 224-5
vulnerability 2, 3, 7, 13-4, 139-40; *see also* sustainability
  adaptive capacity (adaptation) 3, 9, 13, 21, 25, 37, 43-4, 55, 67-9, 78, 88, 121, 128, 139-42, 167-9, 172, 184, 201-2, 210-15
  coping 9, 73, 79-80, 169-70, 184
  criticality 13-4, 26, 77, 87, 129, 131-2, 173-5, 203, 211, 215
  endangerment 3, 9, 90
  exposure 2, 26; *see also* initial condition
  flexibility (mobility) 13, 63, 67-9, 108, 118-20, 126, 129, 168-9, 202-3, 212-4
  hazard 2, 12-6, 106, 140-42, 167, 184, 193; *see also* drought
  resilience 2-3, 16, 66, 68, 74, 77-9, 130-1, 164, 168-70, 174-6, 212, 218, 225-7
  sensitivity 2, 43, 135-6, 216
  stressor 2-3, 74, 83, 128, 132, 169, 175

Wade, T. 36, 83-4, 86, 145, 147, 205, 209
Walker, B.H. 10, 24, 211-12, 218
Walker, M. 10
Wall, D.H. 10
Walsh, S.J. 25, 201
Wang, F. 13
Wang, G. 35, 54-5, 145-7, 204
Wang, T. 15
Warren, A. 1, 17, 21, 201
water (cycle), *see* climate, initial condition; multiple impact
Watts, M.J. 3, 7-8, 11, 17, 20-22, 201, 211, 213
Wasson, R.J. 4, 10, 24, 221
Weiss, E. 35, 55, 60, 146, 204, 207
Weladji, R.B. 21
wetland cropping, *see* cropping
Whitford, W.G. 11, 128, 216
Wickham, J. 36, 83-4, 86, 145, 147, 205, 209
Wiegand, T. 211-12
Wijdenes, D.O. 36, 75-8, 127, 205, 209
Williams, L.J. 10

Witcover, J. 224
Wittenberg, U. 10
wood extraction (harvesting), *see* proximate cause
World Bank 17-8
Wucherer, W. 17, 201

Xu, J. 1, 12, 214
Xue, J. 14

Yanai, J. 36
Yang, X. 35, 54, 56, 58-62, 130, 204, 207
yield loss, *see* multiple impact
Yohe, G. 29

Young, A. 36, 223
Young, O.R. 30

Zeng, N. 11, 211
Zhang, P. 13
Zhao, W.Z. 35, 54-5, 58, 126, 204, 207
Zheng, D. 13
Zhou, W.J. 5, 35, 54-7, 127, 204, 207, 211
Zhu, Z. 15
Zöbisch, M.A. 12, 52
Zonn, I.S. 17, 35, 54, 59-62, 128, 130, 145, 204, 207
Zuéli, K.B. 11, 18, 21